Pioneers in Arts, Humanities, Science, Engineering, Practice

Volume 32

Series Editor

Hans Günter Brauch, Peace Research and European Security Studies (AFES-PRESS), Mosbach, Germany

http://www.afes-press-books.de/html/PAHSEP.htm
http://afes-press-books.de/html/PAHSEP_Robinson.htm

Henrietta Wilson · Mary Kaldor ·
Josh Kaldor-Robinson · Peter Pringle ·
Alexander Ghionis · Richard Guthrie
Editors

Julian Perry Robinson: Pioneer in the Prohibition of Chemical and Biological Warfare

Springer

Editors
Henrietta Wilson
King's College London
Bristol and London, UK

Mary Kaldor
London School of Economics and Political Science (LSE)
Brighton and London, UK

Josh Kaldor-Robinson
Brighton, UK

Peter Pringle
New York, USA

Alexander Ghionis
Harvard Sussex Program, Science Policy Research Unit (SPRU)
University of Sussex
Falmer, Brighton, UK

Richard Guthrie
CBW Events
Bradford on Avon, UK

UNIDIR's Weapons of Mass Destruction Program
Brighton, UK

Acknowledgement:
The photo on the book cover was taken by © Richard Guthrie, who gave permission to use it here.

ISSN 2509-5579 ISSN 2509-5587 (electronic)
Pioneers in Arts, Humanities, Science, Engineering, Practice
ISBN 978-3-032-04063-3 ISBN 978-3-032-04064-0 (eBook)
https://doi.org/10.1007/978-3-032-04064-0

Language and copy editing: Henrietta Wilson, King's College London

Final review and layout: Dr. Hans Günter Brauch, HGB Foundation for Peace and Ecology in the Anthropocene (HGBF)

This Springer imprint is published by the registered company Springer Nature Switzerland AG
The registered company address is: Gewerbestrasse 11, 6330 Cham, Switzerland

Dedicated to the memory of
Julian Perry Robinson:
life partner, father,
friend, colleague and inspiration

Acknowledgements

We would like to thank everyone who sent tributes to Julian, including those whose words we have been able to use in Chap. 3, as well as the people that supported our work selecting papers to include in Chaps. 7 and 8. In particular, we are grateful to Brian Balmer, Hans Günter Brauch, Francesco Calogero, Malcolm Dando, Daniel Feakes, Alastair Hay, Kai Ilchmann, Patrick Lamb, Filippa Lentzos, Roy MacLeod, Susan Martin, Bob Mathews, Caitríona McLeish, Bob Mikulak, Jordi Molas-Gallart, George Papandreou, James Revill, Oliver Robinson, Nicholas Sims, Katie Smallwood, Ralf Trapp, Rim Turkmani, John Walker, William Walker, Simon Whitby and Jean Pascal Zanders. Julian co-wrote some of the papers in Chap. 7; as detailed in the text, we are indebted to his co-authors Malcolm Dando and Matthew Meselson for giving us permission to reprint these works, and to Graham Pearson. We also appreciate the organizers of and participants in the Julian Perry Robinson Annual Seminars that took place in November 2022, November 2023 and November 2024, giving us opportunities to discuss the book.

We give deep thanks to Professor Matthew Meselson for his warm and insightful chapter detailing his work and friendship with Julian, and reflecting on the significance of their endeavours to prevent the misuse of chemistry and biology—we are extremely grateful for this and for his wider support for our work.

A big thank you, too, for the generosity of the funders enabling our work, including the Open Society, Carnegie Endowment for International Peace, the Tolkien Trust, Seth Carus and the many individuals that have made donations to the Julian Perry Robinson Fund.

We are also indebted to the people that helped prepare the manuscript for the book, particularly Edward Bullock, Milo Bullock and Kai Ilchmann for their painstaking work preparing usable formats of Julian's papers, Milo Bullock for helping with the photographs in Part II and Lydia Wilson for preparing a transcript of Julian's memorial service.

The team at Springer Nature has helped in numerous ways, especially Series Editor Dr. Hans Günter Brauch, Dr. Christian Witschel, Editorial Director with

Springer Nature in Heidelberg (Germany), and Production Editor Ms. Jayanthi Krishnamoorthi and Production Supervisor in India Ms. Mano Priya S; we appreciate their invaluable advice and support.

Bristol and London, UK — Henrietta Wilson
Brighton and London, UK — Mary Kaldor
Brighton, UK — Josh Kaldor-Robinson
New York, USA — Peter Pringle
Brighton, UK — Alexander Ghionis
Bradford on Avon, UK — Richard Guthrie
January 2025

Introduction

Mary Kaldor

Julian Perry Robinson, my life partner for over half a century, died from COVID on April 22, 2020. It was right at the beginning of the pandemic before vaccinations and before treatments and we were not allowed to visit him in hospital. It is unbearable to think of him suffering alone even though I know he was barely conscious. The COVID pandemic showed the world what an attack with biological weapons might be like – something Julian had devoted his life to trying to prevent.

This book has been put together by members of his family, former students and colleagues and friends. It is both about his work and scholarship and about his life and the kind of person he was. His influence stretched far beyond his writings. He touched so many peoples' lives. He was an inspiration to many people, as our son Josh describes in his eulogy in Chap. 2. He was tireless in his support for students and colleagues: he helped people to use and navigate his archives, the Sussex Harvard Information Bank (SHIB); he was painstaking in commenting and correcting papers and manuscripts (everyone remembers the misplaced comma!); and he was always active in engaging in discussions with fellow scientists as well as government and intergovernmental officials, journalists and civic activists.

The book combines a collection of his writings and a list of his publications, with chapters about his life and work. In this short introduction, I will try to portray Julian as a person. Other chapters will cover his biography, his research and his considerable contribution to making the world a better place.

Julian was, of course, very brilliant and also very good at sports. I have a photograph of him as a young boy surrounded by the trophies he had won for both his academic and athletic achievements.[1] Jonathan Dimbleby, the broadcaster, who attended the same school, was in awe of him. I have always wondered whether it was because he won so many prizes and scholarships as a boy that he was so disdainful of honours and awards in later life.

[1] This photo is reproduced in Part II, Chap. 4.

In addition, he was very good-looking. "Impossibly handsome" my friend Emma said. My grandmother thought he was too handsome; she thought it was a great mistake to marry a good-looking man. Luckily, she turned out to be wrong, at least in Julian's case.

He was very empathetic. He was very sensitive to other people and seemed to know instinctively what they needed – a quality he managed to hand down to his sons. He always listened attentively to other people, especially if they seemed uneasy at social events, for example. He was a soothing presence. Children and animals loved him. I remember once he was bitten by a dog when we were in France. What upset him was not the pain of the bite nor the rabies injection he had to have but the thought that any dog would want to bite him.

He had a wide range of interests, which he pursued with scientific precision. He was an amazing cook, and our kitchen is overflowing with recipe books. We have files of recipes, which he annotated to adjust the amounts or cooking times after trying them out. At Christmas he always prepared a timetable listing what had to be done when – a timetable which we still use and adapt every year.

He loved butterflies and our butterfly books have notes at the back with the times and dates that we observed particular species each year. As a student he was into vintage cars and when I met him, he owned a beautiful Bugatti. Later we bought a Morgan, which was eventually stolen owing to our failure to mend the garage lock. He also loved music and poetry. He set up elaborate speakers in our kitchen. At breakfast, we would listen to Vivaldi's Four Seasons and Handel's Messiah on Sunday; these were replaced later in the day by Leonard Cohen and Stevie Wonder. He always played Django Reinhart and Stefan Grapelli when cooking and, in recent years, he became fond of the piano, so we listened to Mozart's Sonatas and Chopin's Nocturnes in the evenings in our house in France.

He had a knack for everyday life. He loved games, especially cards and mahjong; he didn't like parlour games. We have a beautiful mahjong set, which his grandfather brought from China, and it is such a pleasure to play. He spent a lot of time finding the right book for everyone, or before streaming, the right CD or record for Christmas and birthdays. He always had an objective for walks and expeditions, which made them much more fun, and he always made delicious and well-organized picnics. Once soon after we met, he decided that we should drive to Scotland (in the Morgan) to see Highland cattle in the rain (they have very long and woolly fur) but unfortunately it was sunny!

Despite all this, Julian was very shy. He agonized about social life, at least in anticipation. He hated the telephone and always made me ring up about appointments or repairs or whatever needed to be asked. He refused to ask people the way when we got lost – this was before the days of GPS, which, even when it became available, he never actually trusted. He never liked giving public lectures or speeches, and spent a long time preparing them when he did agree to give them, even though he could be inspirational. He resisted talking to the media, except as a source.

Above all, Julian was a very good person. He had a strong sense of morality, which I think he inherited from his mother. At an early stage of our life together, he decided that his life mission was to prevent the militarization of biology and chemistry and that the means to achieve this was research that made a difference. He was never interested in an academic career as such, even though he became a distinguished professor. He saw the university as a place where he could conduct the research he needed to do. And as the chapters that follow show, he put a lot of effort into communicating that research to fellow scientists, often through the Pugwash movement, to journalists and activists as well as people in government and international institutions. Through these activities, he created what might be called an epistemic community of people concerned about the issues of chemical and biological weapons.

This is why his family and members of that community are determined to keep his mission going. This book is part of that project. The first chapter by his closest colleague and friend Matthew Meselson, who co-founded the Harvard Sussex Program, introduces their work together. The next chapter by Josh Kaldor-Robinson (our son) traces his life and ideas and also includes the eulogy Josh gave at Julian's funeral. The following chapter by Peter Pringle, another close friend, summarizes the tributes and commentaries from a range of colleagues and friends that we received after he died. Chapter 4 is a collection of photographs.

These four chapters are followed by a collection of Julian's writings – we have tried to cover a variety of pieces, from academic articles to blogs and op-eds, as well as evidence before Congressional committees, and we based our choice on nominations from members of the epistemic community; this is explained in Chap. 5. Chapter 6 by Peter Pringle is about Julian's very special style of writing. Chapters 7 and 8 contain Julian's work. The writings in Chap. 7 are about chemical and biological warfare, and in Chap. 8 we have also included some of his work on broader issues, such as trends in conventional munitions and environmental modification weapons. Finally, in Chap. 9 we include a bibliography of his publications relating to chemical and biological warfare.

Julian's work was so comprehensive that a single volume cannot capture the entirety of his life's work – rather, we have tried here to assemble a snapshot of who he was, what he did, and why his work still matters. We hope that the book serves as an introduction and signpost to all his writings, and that it inspires readers by providing examples of what independent specialists can do – in terms of the quality of outputs they can derive from careful acquisition and analysis of open sources, and the ways that such expertise can be used to make a difference to national and international policy-making on arms control, disarmament and non-proliferation.

As well as producing this book, we are engaged in a project to digitalize Julian's archives so that they are available to researchers in the future and we hope to be able to keep it as a living information base that will be updated as part of an ongoing research programme. We organize an annual Julian Perry Robinson Seminar on or around his birthday, November 11, that brings together the epistemic community. And we are trying to find ways to sustain and grow the overall research programme.

The mission is more important than ever. Chemical weapons have been used in Syria and Ukraine. The Conventions that prohibit biological and chemical warfare that Julian did so much to promote are currently being undermined. And, partly as a consequence of the COVID pandemic, there has been an acceleration of the life sciences. This is why research on the life sciences, peace and health is so important. We hope this book will encourage others to participate in Julian's mission.

Julian and Mary in their flat in Brighton. From © Mary Kaldor's private photo collection

Finally, a word about how much I owe Julian. I met him when I was 22 and he was 26, so we really shaped each other's lives. He taught me how to work – I got so bored because he worked so much of the time that I did the same! He was incredibly supportive – morally, psychologically, intellectually and in practical terms. He was a participating husband and father long before it became the norm, doing more than his share of cooking and childcare. He enabled me to work, have children, be an activist, and to spend lovely summers together in France (working most of the time!). He was, I can say, the underpinning of my life.

Prof. Mary Kaldor, CBE
Professor Emeritus of Global Governance
London School of Economics and Political Science (LSE)
Brighton and London, UK

Contents

Acronyms and Abbreviations

AAAS	American Association for the Advancement of Science
ACDA	[US] Arms Control Disarmament Agency
ACDRU	[UK Foreign and Commonwealth Office] Arms Control and Disarmament Research Unit
ADIU	Armament Disarmament Information Unit
ADTAM	Artillery-Delivered Target-Activated Munitions
AGM	Annual General Meeting
ARCAT	Advanced RCA Technology
BCSIA	Belfer Center for Science and International Affairs
BW	Biological Weapons
BWC	Biological Weapons Convention
CBW Conventions	Chemical and Biological Weapons Conventions
CBW	Chemical and Biological Warfare (and, in some of Julian's papers reproduced in Chaps. 7 and 8, chemical/biological weapons; chemical and biological weapons)
CD	Conference on Disarmament (and one reference to Compact Disc!)
CIA	[US] Central Intelligence Agency
CW	Chemical Weapons
CWC	Chemical Weapons Convention
DEDM	Downward-Ejection Dispenser Munition
DIA	[US] Defense Intelligence Agency
EnMod Treaty	Convention on the Prohibition of Military or Any Other Hostile Use of Environmental Modification Techniques
ESRC	Economic and Social Research Council
EU	European Union
FAE	Fuel-Air Explosives
FCO	UK Foreign and Commonwealth Office
FGA	Fourth Generation Agent
GC/MS	A technique combining gas chromatography and mass spectrometry

GO	Government Organization
GPS	Global Positioning System
HE	High Explosive
HSP	Harvard Sussex Program on CBW Arms and Arms Limitation
IAEA	International Atomic Energy Agency
IC	Intelligence Community
ICM	Improved Conventional Munition
ICRC	International Committee of the Red Cross
IR	International Relations
JIM	OPCW-UN Joint Investigative Mechanism
KGB	Komitet Gosudarstvennoy Bezopasnosti (USSR Committee for State Security)
LSD	Lysergic Acid Diethylamide
N, C and B	Nuclear, Chemical and Biological
NATO	North Atlantic Treaty Organization
NGO	Non-Governmental Organization
NIE	National Intelligence Estimate
NIO	National Intelligence Officer
NLW	Non-Lethal Warfare
OPCW	Organisation for the Prohibition of Chemical Weapons
PIB	Polyisobutene
R&D	Research and Development
RCA	Riot Control Agent
REDM	Rearward-Ejection Dispenser Munition
S&T	Science and Technology
SALT	Strategic Arms Limitation Talks
SHIB	Sussex Harvard Information Bank
SIPRI	Stockholm International Peace Research Institute
SPRU	Science Policy Research Unit
STLC	Supertoxic Lethal Chemicals
TEA	Triethylaluminium
TIC	Toxic Industrial Chemical
TPA	Thickened Pyrophoric Agents
UK	United Kingdom
UN	United Nations
UNMOVIC	United Nations Monitoring, Verification and Inspection Commission
UNSCOM	United Nations Special Commission
US	United States of America
USA	United States of America
USAMIIA	U.S. Army Medical Intelligence and Information Agency
USSR	Union of Soviet Socialist Republics (Soviet Union)

WHO	World Health Organization
WMD	Weapons of Mass Destruction
WTO	World Trade Organization

List of Tables

Part I
Life and Work

Chapter 1
Reflections from a Friend and Colleague

Matthew Meselson

For half a century, until his death in 2020, Julian was the world's preeminent historian of chemical and biological weapons and warfare. His deep knowledge of the subject and his unfailing objectivity gained the respect and trust of government officials and others concerned with chemical and biological warfare (CBW) matters throughout the world. His writings and personal interactions contributed to the accomplishment of the international conventions prohibiting such weapons, the 1972 Biological Weapons Convention (BWC) and the 1993 Chemical Weapons Convention (CWC).

After reading chemistry at Oxford and learning international patent law in a London patent office, Julian's first major contributions to CBW scholarship were as a researcher at the Stockholm International Peace Research Institute (SIPRI) during 1968–1971. While at SIPRI, he was lead author of the first two volumes of the six-volume study *The Problem of Chemical and Biological Warfare*, still the most comprehensive and reliable work on the subject for the period it covers.[1] Three more books written by Julian and published by SIPRI are *Chemical Weapons: Destruction*

[1] SIPRI. *The Problem of Chemical and Biological Warfare* (6 volumes). Stockholm: Almqvist & Wiksell. 1971–1975. Published in digital format April 2000. Available from: https://www.sipri.org/publications/2000/problem-chemical-and-biological-warfare [Accessed 28 February 2024]. Editors' note. The first two volumes of SIPRI's *The Problem of Chemical and Biological Warfare* are listed in Chap. 9 as items 42 and 52. The fifth volume is item 43.

Professor Matthew Meselson, Thomas Dudley Cabot Professor, Harvard University.

M. Meselson (✉)
Harvard University, Cambridge, MA, USA

H. Wilson et al. (eds.), *Julian Perry Robinson: Pioneer in the Prohibition of Chemical and Biological Warfare*, Pioneers in Arts, Humanities, Science, Engineering, Practice 32, https://doi.org/10.1007/978-3-032-04064-0_1

and Conversion (1980);[2] *Chemical and Biological Warfare Developments: 1985*;[3] and *Chemical Warfare Arms Control: A Framework for Considering Policy Alternatives* (1985).[4] It was while at SIPRI that Julian met his lifelong partner, Mary Kaldor, now Professor Emeritus of Global Governance at the London School of Economics and Political Science.

In 1971, upon joining the Science Policy Research Unit (SPRU) at the UK's University of Sussex, Julian began to create what was to become the foremost academic centre for research, teaching, and outreach regarding CBW history and policy. His former students now occupy positions in academia, government, and international organizations. And his meticulously organized and eventually immense archive of public and official documents, the Sussex Harvard Information Bank (SHIB), became and is now the world's foremost archival resource for scholarly research in CBW matters.[5] A satellite component of SHIB, focused on herbicide warfare in Vietnam, the Yellow Rain allegations, the 1979 Sverdlovsk anthrax outbreak, and international criminalization of CBW weapons is kept at Harvard.[6]

Julian and I met in November 1969 in London at a conference on CBW organized by the Women's International League for Peace and Freedom.[7] I had come to meet and exchange views with British scholars and government officials interested in CBW matters, a topic on which I had been an advisor to the US Arms Control and Disarmament Agency in Washington. Believing that by coordinating our efforts we could contribute more effectively to CBW scholarship and the prevention of CBW, Julian and I agreed to stay in close touch. There followed forty years of collaboration and best friendship.

---

[2] Jozef Goldblat, Karlheinz Lohs, Julian Perry Robinson, Arne Kjellstrand, Rajesh Kumar, Johan Lundin and Theodor Nemec. *Chemical Weapons: Destruction and Conversion*. SIPRI and Taylor & Francis. 1980. Details at https://www.sipri.org/publications/1980/chemical-weapons-destruction-and-conversion [Accessed 28 February 2024]. Editors' note. This is listed in Chap. 9 as item 96.

[3] Julian Perry Robinson. *Chemical and Biological Warfare Developments: 1985*. SIPRI Chemical & Biological Warfare Studies 6. Oxford University Press, 1986. Details at https://www.sipri.org/publications/1986/chemical-and-biological-warfare-developments-1985 [Accessed 28 February 2024]. Editors' note. This is listed in Chap. 9 as item 205.

[4] Julian Perry Robinson. *Chemical Warfare Arms Control: A Framework for Considering Polity Alternatives*. SIPRI Chemical & Biological Warfare Studies 2. Taylor & Frances. 1985. Details at https://www.sipri.org/publications/1985/chemical-warfare-arms-control-framework-considering-policy-alternatives [Accessed 5 March 2024]. Editors' note. This is listed in Chap. 9 in item 181.

[5] HSP. SHIB [online]. Available at http://hsp.sussex.ac.uk/new/what-we-do/shib [Accessed 4 March 2024].

[6] Harvard University. Meselson CBW Archive [online]. Available at https://meselsonarchive.hsites.harvard.edu/ [Accessed 4 March 2024].

[7] Brian Balmer. ' "Science was digging its own grave": the Women's International League for Peace and Freedom and the campaign against chemical and biological warfare'. *The Nonproliferation Review*. 27 (4–6). 4 March 2021. Available at https://www.tandfonline.com/doi/abs/10.1080/10736700.2020.1838703.

Starting in the 1970s we attended and worked closely together at the many biological weapons (BW) and chemical weapons (CW) working group meetings organized initially under the auspices of SIPRI and soon taken over by the Pugwash Conferences on Science and World Affairs.[8]

These meetings went through several manifestations: the Pugwash BW Study Group [1964–69]; the Pugwash CW Study Group [1974–1992]; and the Pugwash Study Group on Implementation of the CBW Conventions [1993–2010]. The Study Groups were timed to immediately precede sessions of international negotiating bodies – the Eighteen Nation Disarmament Committee (1961–1969), the Conference of the Committee on Disarmament (1969–1978), the Committee on Disarmament (1979–1984) and the Conference on Disarmament (1984–present). This arrangement facilitated small informal meetings between us and individual delegates to the negotiations, at which Julian's ability to identify manageable topics for discussion helped to clarify matters at issue in the negotiations.

An early joint publication was a 1980 article in *Scientific American* describing chemical warfare agents, chemical weapons, and the immense threat to civilian populations if nerve agent warfare were to occur in central Europe or similarly populated areas. We warned that "Mankind has entered a period of rapid and accelerating understanding of the fundamental biochemical and cellular processes of life... As this knowledge expands, so too will the range of possible applications for good and ill".[9] Another collaborative effort at the time was a review of US government plans to produce a new generation of chemical weapons, the so-called binaries. Testifying jointly before a committee of the US Senate, we argued that the binary programme was technically flawed and an obstruction to negotiations for verified chemical disarmament. The binary programme was eventually abandoned.

Starting in 1981, we undertook an extensive examination of evidence for the allegation by the Reagan Administration that the Soviet Union was conducting chemical warfare against Hmong villagers in Laos and Khmer Rouge soldiers in Cambodia. According to interrogations by US investigators of Hmong refugees interned in Thailand, a poisonous substance was sprayed from aircraft, leaving yellowish spots on leaves and rocks and causing diverse symptoms and sometimes death. The alleged

[8] J. P. Perry Robinson. The Impact of Pugwash on the Debates over Chemical and Biological Weapons'. *Annals of the New York Academy of Sciences*. 866 (1). 2006. pp. 224–252. Available at https://doi.org/10.1111/j.1749-6632.1998.tb09155.x. This is listed in Chap. 9 in item 336

[9] Matthew Meselson and Julian Perry Robinson. 'Chemical Warfare and Chemical Disarmament'. *Scientific American*. 242 (4). April 1980. Available at: https://projects.iq.harvard.edu/files/meselsonarchive/files/scientificamerican0480cbwd-1.pdf [Accessed 4 March 2024]. Editors' note. This is listed in Chap. 9 in item 102.

agent, the so-called "Yellow Rain", was said by US officials at a 1982 State Department briefing to be a mycotoxin-containing liquid that also contained pollen. This was the first public association of the alleged agent with pollen.[10]

Samples of the yellowish material turned in by Hmong refugees to Western investigators were found upon light and electron microscopic examination to be the pollen-rich faeces of wild Southeast Asian honeybees.[11] Yale entomologist Thomas Seeley first suggested that these were the result of mass honeybee "cleansing flights", and Seeley and I subsequently experienced the phenomenon during a 1983 field investigation in Thailand. Also, the reports of mycotoxins from two US university laboratories could not be confirmed by US and UK government laboratories. Further interviews of Hmong refugees by US investigative teams in Thailand consistently reported failure to substantiate earlier claims of chemical warfare, as did interrogation of captured and defecting North Vietnamese chemical officers by a CIA case officer. [12,13,14,15]

In 1988, we initiated the quarterly publication, *The Chemical Weapons Convention Bulletin*, later renamed *The CBW Conventions Bulletin*, eventually comprising 87 issues, each running to some 50 pages.[16] Initially, the *Bulletin* was published by the Federation of American Scientists Fund in Washington. Then, in 1990, we created HSP, the Harvard Sussex Program on CBW Arms and Arms Limitation. This provided a distinct entity for purposes of fund-raising, an in-house publisher for the *Bulletin* and a scholarly presence within our respective universities.

[10] U.S. Department of State. 'On-the-Record Briefing on Chemical Warfare in Afghanistan, Laos, and Kampuchea, Washington, DC, Monday, November 29, 1982, 2:30 p.m.' [official transcript].

[11] Joan W Nowicke and Matthew Meselson. 'Yellow rain – a palynological analysis'. *Nature*. 309. 1984. pp. 205–206. https://doi.org/10.1038/309205a0.

[12] Julian Robinson, Jeanne Guillemin, and Matthew Meselson. 'Yellow Rain: The Story Collapses'. *Foreign Policy*. 68. Fall 1987. Available at: https://projects.iq.harvard.edu/files/meselsonarchive/files/1987_yr_story_collapse_for_pol.pdf [Accessed 6 March 2024], listed in Chap. 9 in item 219.

[13] Julian Robinson, Jeanne Guillemin, and Matthew Meselson. 'Yellow Rain in Southeast Asia: The Story Collapses'. Chapter 10 in Susan Wright (ed). *Preventing a Biological Arms Race*. Cambridge MA: The MIT Press. 1990. Available at: https://projects.iq.harvard.edu/files/meselsonarchive/files/1990_yellow_rain_in_southeast_asia.pdf [Accessed 6 March 2024], listed in Chap. 9 in item 219.

[14] Matthew Meselson and Julian Perry Robinson. 'The Yellow Rain Affair: Lessons from a Discredited Allegation'. Chapter 4 in Anne L Clunan, Peter R Lavoy and Susan B Martin (eds). *Terrorism, War, or Disease? Unraveling the use of Biological Weapons*. Stanford Security Studies: Stanford University Press. 2008. Available at: https://projects.iq.harvard.edu/files/meselsonarchive/files/2008_twd_meselson_robinson_yr_0.pdf [Accessed 6 March 2024]. Editors' note. This is reproduced in Chap. 7 of this book (paper 7.7), listed in Chap. 9 in item 454.

[15] Merle L. Pribbenow. ' "Yellow Rain": Lessons from an Earlier WMD Controversy'. *International Journal of Intelligence and CounterIntelligence*. 19 (4). 2006. pp. 737–745. Available at: https://www.tandfonline.com/ https://doi.org/10.1080/08850600600656525.

[16] HSP. *CBW Conventions Bulletin* [online]. http://hsp.sussex.ac.uk/new/resources/cbw-conventions-bulletin [Accessed 13 October 2025].

Among its regular features, the *Bulletin* included: an editorial by a guest or by ourselves; a 20–25 page "News Chronology" of recent events relevant to CBW weapons and CBW arms control written by Julian; "Report From Geneva" – a detailed report from the negotiation in Geneva or, beginning in 1993, "Progress in The Hague" – usually written by an HSP researcher selected and supported by HSP but seconded to the Organisation for the Prohibition of Chemical Weapons (OPCW) for a period of a year or more. Our idea was to provide readers of the *Bulletin* with reliable, timely information about the implementation of the CWC, while at the same time providing the OPCW with an additional staff position supported by HSP.

We would often work together at Julian's office at SPRU, or at Harvard, or at my family's summer house at nearby Woods Hole on Cape Cod. There, during the summer of 1986, starting with an outline I had made at Harvard the previous January, Julian and I prepared the first full draft of an international convention that would make the development, production, possession and use of biological and chemical weapons crimes under international law, patterned after existing treaties such as those that criminalize airline hijacking, piracy, and torture. We subsequently refined the text of the HSP Draft Convention on the basis of suggestions from two meetings of legal scholars that we organized at Harvard in January 1997[17] and at the Lauterpacht Center for International Law at Cambridge University.[18]

The present version of the Draft Convention was published in the December 1998 issue of *The CBW Conventions Bulletin* and in a special issue of the *Bulletin* in 2011,[19] together with a chronology and bibliography. The Draft Convention has been translated into the several official languages of the UN,[20] and an extensive legal commentary on the Draft Convention was prepared with the aid of several generations

[17] 'Criminalization of Biological and Chemical Weapons: A Project of the Harvard Sussex Program and the Center for Science and International Affairs Harvard University' [meeting paper]. Available at: https://projects.iq.harvard.edu/sites/projects.iq.harvard.edu/files/meselsonarchive/files/1997_convention_aaas.pdf [Accessed 6 March 2024].

[18] 'Workshop on International Criminalization of Biological and Chemical Weapons' [meeting paper]. Available at: https://projects.iq.harvard.edu/sites/projects.iq.harvard.edu/files/meselsonarchive/files/1998_lauterpacht_report.pdf [Accessed 6 March 2024].

[19] 'Draft Convention on the Prevention and Punishment of the Crime of Developing, Producing, Acquiring, Stockpiling, Retaining, Transferring or Using Biological or Chemical Weapons'. *CBW Conventions Bulletin Special Issue.* pp. 6-9. February 2011. Available at http://hsp.sussex.ac.uk/new/_uploads/bulletin/CBWCB_Special_Issue.pdf [Accessed 13 October 2025]. Editors' note. The 1998 version of this paper is reproduced as '7.10 A Draft Convention to Prohibit Biological and Chemical Weapons under International Criminal Law' in Chap. 7 of this book.

[20] 'HSP Draft Convention' [online]. Available at: http://hsp.sussex.ac.uk/new/about/draft-convention [Accessed 13 March 2024].

of Harvard law students.[21] Although there is no current initiative to hold individuals criminally liable for activities prohibited by the CBW Conventions, the HSP Draft Convention may in future provide an impetus for such an enhancement of the regime against CBW weapons and warfare, thereby supporting the existing Chemical and Biological Weapons Conventions' prohibitions on states.

The CWC entered into force on April 29, 1997. It prohibits the development, production, acquisition, stockpiling, retention, and use of chemical weapons. There remained dispute regarding the legality of the use in war of so-called riot-control agents such as CS, massively used in military operations by the US in the Vietnam War. In editorials in the *Bulletin* and in other writings we argued against any interpretation of the CWC that would exempt such chemicals from use as weapons of war, believing that any such exemption seriously risks unravelling of the CWC, that the most durable prohibition is "no gas".[22]

In September 2012 Robert Neild, founding director of SIPRI, wrote to Julian and me proposing that he and Martin Rees, the UK's Astronomer Royal and former President of the Royal Society, nominate the two of us for the Nobel Peace Prize, asking us to provide details of our work. Julian wrote back that

> I would not want you to go ahead but thank you most warmly for suggesting it. Matt I can't speak for, but I shall send him a copy of this message in case he thinks I am being over hasty. Writing all those things about myself would be an awful distraction from the project that he and I are currently pursuing – trying to nail down the Yellow Rain once and for all.

In response, I wrote to Julian,

> The detailed and novel thinking that went into the BWC and the CWC is applicable at least in part to other weapons, including nuclear weapons...The idea of a Peace Prize for CBW disarmament and the underlying regimes should therefore be strongly supported...Another way, if acceptable to the Norwegian committee, would be to give the prize to the BWC and CWC regimes themselves...

Accordingly, the Organisation for the Prohibition of Chemical Weapons (OPCW) was nominated for and was awarded the 2013 Nobel Peace Prize for their "extensive work to eliminate chemical weapons".

---

[21] 'Commentary to the Draft Convention on the Prevention and Punishment of the Crime of Developing, Producing, Acquiring, Stockpiling, Retaining, Transferring or Using Biological or Chemical Weapons'. Available at: https://projects.iq.harvard.edu/sites/projects.iq.harvard.edu/files/meselsonarchive/files/legal_commentary_hsp_site.pdf [Accessed 6 March 2024].

[22] Editors' note. See 7.12 ' "Non-Lethal" Weapons, the CWC and the BWC' in this book.

Matthew Meselson and Julian Perry Robinson outside The Swan Inn in Falmer, Sussex. From © Matthew Meselson's collection, included with his kind permission.

Chapter 2
Intellectual Contributions and Working Methods

Josh Kaldor-Robinson

> "One factor which has served to constrain CBW activities over the last 50 years is that for a great many people, whether military or non-military, CB weapons represent a peculiarly loathsome and despicable method of fighting".
> *The Problem of Chemical and Biological Warfare*, SIPRI, volume 5, p.18.

My father, Julian Perry Robinson (born Jerusalem on November 11, 1941) made the prevention of the misuse of science, particularly in relation to chemical and biological warfare (CBW), his vocation. His interest in CBW started while studying chemistry at Oxford, where, instead of carrying out a more traditional laboratory-based thesis, he chose to produce a dissertation entitled *Some Aspects of Chemical Warfare: An Investigation into the Influence of the Second World War on the Development of Chemistry, With Particular Reference to the War Gases.*[1] After graduating and with a new young family to support, rather than pursuing post-graduate work, he undertook legal training and started working as a patent lawyer.

After two (according to family legend, fairly miserable) years at a London Patent Office (although he did manage to identify patents taken out by the UK government for nerve agents), Dad was invited to a conference convened by the British group Scientists for Peace, where he met John H. Humphrey[2] who made a great impression on him. Humphrey went on to become his mentor, recommending him to Robert

[1] 'Some aspects of chemical warfare: An investigation into the influence of the Second World War on the development of chemistry, with particular reference to the war gases'. March 1964. Dissertation, Oxford University, 264 p. This is listed in Chap. 9 as item 1.

[2] Professor of Biology at Cambridge and founder of the British Society for Immunology.

Josh Kaldor-Robinson, Julian and Mary's son, and researcher on identities and representations of diaspora communities on the Internet. Brighton, United Kingdom.

J. Kaldor-Robinson (✉)
Brighton, United Kingdom

H. Wilson et al. (eds.), *Julian Perry Robinson: Pioneer in the Prohibition of Chemical and Biological Warfare*, Pioneers in Arts, Humanities, Science, Engineering, Practice 32, https://doi.org/10.1007/978-3-032-04064-0_2

Neild, then head of the Stockholm International Peace Research Institute (SIPRI) and looking for someone to run the Institute's CBW programme. Dad left his job in London and moved to Sweden to take up the position.

While Dad found his work as a patent lawyer unsatisfying, the combination of chemistry and law would prove to be invaluable for the rest of a career based around the development of legal constraints against CBW, through both international treaties and national legislation. The interplay of science policy, technology and regulation was a major theme in his writing, in over 500 papers and monographs written over 50 years (see Chap. 9 for a bibliography of his CBW-related writings).

During Dad's three years at SIPRI, he became active within the Pugwash Conferences on Science and World Affairs, notably founding the Pugwash CBW working group. He also started working with the World Health Organization, and met Matthew Meselson,[3] with whom he later established the Harvard Sussex Program (HSP). It was at SIPRI that he met my mother, Mary Kaldor, who would be his partner for the rest of his life. In 1971 they moved back to the UK to take up jobs at the Science Policy Research Unit (SPRU) at the University of Sussex, the institution where Dad would spend the rest of his career.

At the heart of my father's work were the related notions of "taboo" and "assimilation". The use of poison in warfare had long been seen as beyond the pale and subject to prohibitions from the *Mahabharata* onwards. Following the First World War and the horrific experiences of the use of chemical weapons, as encapsulated in Wilfred Owen's poem *Dulce et Decorum Est*, the taboo was seemingly strengthened, leading to the Protocol for the Prohibition of the Use in War of Asphyxiating Poisonous or other Gases, and of Bacteriological Methods of Warfare (more commonly referred to as the Geneva Protocol) in 1925.[4] The idea of upholding and strengthening the taboo against the use of CBW underpinned Dad's whole approach, especially his work concerning disarmament.[5]

Linked to the idea of taboo was the idea of assimilation that he introduced in the influential six-volume work *The Problem of Chemical and Biological Warfare* (SIPRI 1971), which is still a standard reference source for scholars and policy practitioners.[6] He argued that, contrary to conventional wisdom, the reason that chemical weapons remained unused during the Second World War, despite their extensive employment in the First, was not to deter others' use but because military

[3] Thomas Dudley Cabot Professor of the Natural Sciences at Harvard University, and author of Chap. 1 of this book.

[4] UNODA. 1925 Geneva Protocol [online]. Available at https://disarmament.unoda.org/wmd/bio/1925-geneva-protocol/#:~:text=The%201925%20Geneva%20Protocol%20prohibits,force%20on%208%20February%201928. [Accessed 5 April 2024].

[5] Several papers in Chap. 7 include Julian's account of the CBW taboo, for example, '7.1 Near-Term Development of the Governance Regime for Biological and Chemical Weapons'.

[6] SIPRI. *The Problem of Chemical and Biological Warfare* (6 volumes). Stockholm: Almqvist & Wiksell. 1971–1975. Published in digital format April 2000. Available at: https://www.sipri.org/publications/2000/problem-chemical-and-biological-warfare [Accessed 28 February 2024]. Chapter 9 of this book lists three of the six volumes: volume 1 is item 42, volume 5 is item 43 and volume 3 is item 52.

institutions had become structurally predisposed against them. In the process of adapting to such other revolutionary technologies as the tank and the combat aircraft, armed forces became unable to accommodate chemical weapons adequately within their doctrine or structure. Specialist institutions established during the First World War to develop chemical-warfare technology continued to do so, but they had become sidelined, with their technology remaining largely unintegrated. The assimilation of chemical weapons was therefore constrained and incomplete.[7]

Dad developed these ideas at SPRU in his work on the "weapons succession process" – a concept that used notions of bureaucratic politics and organizational behaviour taken from decision-making theory to explain military-technological change in terms of the influence of existing technologies on competing successor technologies as they proceeded through their life-cycle phases of research, development, production, integration and obsolescence. A revolutionary technology that achieved the phase of integration was, by definition, becoming assimilated. The theory enabled researchers to describe constraints on assimilation in structural and therefore policy-sensitive terms, for example, with reference to the conflict between institutional interests prominent at successive phases in the progress of that technology, or assimilation promoters. Dad thus created a novel theoretical foundation for policy-prescription.

He applied the concept in numerous studies of particular policy issues, including his work on binary chemical munitions in the mid-1970s and again in the mid-1980s. Binary munitions were based on the concept of combining two less toxic chemicals to produce nerve agents and as such they were easier to handle and transport than unitary chemical weapons. In some eyes, this made them closer to conventional munitions than to nerve gas and so their development could be interpreted as a technological attempt to ease constraints on the assimilation of chemical weapons. Dad's work on the 1979 outbreak of anthrax in Sverdlovsk[8] and on the "Yellow Rain" biological warfare[9] reports from Southeast Asia also relied on the concept of the "weapons succession process". This work was carried out in collaboration with Matthew Meselson and Jeanne Guillemin;[10] it addressed the vulnerability of scientific inquiry to misrepresentation by affected interests, in these two cases mainly within the intelligence community but also interests that were influential within decision-making processes on Western chemical and biological rearmament. Publications in

[7] For an example of Julian's writing on assimilation, see '7.3 Chemical arms control and the assimilation of chemical weapons' in Chap. 7 of this book.

[8] See '7.4 "The Soviet Union and the Biological Weapons Convention" and a Guide to Sources on the Sverdlovsk Incident: Discussion Of The Paper By Philip Towle' in Chap. 7 of this book.

[9] The alleged active materials in the 'Yellow Rain' allegations were mycotoxins which fall into the overlap between chemical and biological weapons (being poisonous substances produced by living things). The 'Yellow Rain' accusations have at various times been couched in terms of allegations of biological and/or chemical warfare.

[10] Jeanne Guillemin (March 6, 1943 – November 15, 2019) was a Professor of Sociology and Senior Fellow in the Security Studies Program at Massachusetts Institute of Technology. She and Matthew Meselson married in 1986.

Science (1983),[11] *Nature* (1985)[12] and *Foreign Policy* (1987)[13] were largely responsible for the now generally accepted view that the Yellow Rain was bee-faeces, not communist germ warfare. And their Sverdlovsk inquiries, which, like their Yellow Rain work, included field investigations by teams of scientists headed by Matthew Meselson as well as Jeanne Guillemin's research, eventually led to a conclusive demonstration that the anthrax outbreak did indeed originate in a Soviet military biological laboratory.[14]

After the end of the Cold War, one of Dad's major concerns was that new utilities would be developed for CBW, which would both undermine the taboo against their usage and enable the assimilation of chemical and biological weapons into modern militaries. He investigated this in his work on the 1988 Halabja chemical attack (in which Saddam Hussein used chemical weapons against the Kurdish population in Iraq),[15] a major European Union project on bio-terrorism, an Economic and Social Research Council (ESRC) project with the London School of Economics and Political Science (LSE) exploring security technologies,[16] and his scrupulous collection of data on the use of chemical weapons in Syria.[17]

My father worked all the time. He was a collector and a hoarder – as a child he collected stamps and was a car-spotter. He once said that the reason he chose his field was that he could read everything that had been written on the subject; of course, this is no longer true, especially after the advent of the Internet, but he tried. He was painstaking about verifying information. He never accepted oral sources unless they were backed up by written sources and he cross-checked all written sources. On principle, he only used information from the public domain and thus he could be said to have been a pioneer of open source investigations, demonstrating how expertise derived from open sources could be used to access and inform national and

---

[11] Peter Ashton, Matthew Meselson, J. P. Perry Robinson and Thomas Seeley. 'Origin of Yellow Rain' in letters columns, *Science*. vol 222. 28 October 1983. pp. 366 & 368. Available at https://www.science.org/doi/epdf/10.1126/science.222.4622.366 [Accessed 30 January 2025]. This paper is listed in Chap. 9 as item 167.

[12] Peter Ashton, Matthew Meselson, Joan Nowicke, J. P. Perry Robinson and John Seeley. 'Chemical weapons evidence unconvincing' in 'Scientific Correspondence' columns, *Nature*. vol 315. 23 May 1985, p. 284. Available at: https://www.nature.com/articles/315284a0.pdf [Accessed 30 January 2025]. This is listed in Chap. 9 as item 185.

[13] Julian Robinson, Jeanne Guillemin, and Matthew Meselson. 'Yellow Rain: The Story Collapses'. *Foreign Policy*. 68. Fall 1987. Available at: https://projects.iq.harvard.edu/files/meselsonarchive/files/1987_yr_story_collapse_for_pol.pdf [Accessed 6 March 2024]. This is listed in Chap. 9 as item 219.

[14] See also the examples of Julian's writing on Sverdlovsk and Yellow Rain in Chap. 7 of this book (papers 7.4 and 7.7).

[15] See for example Julian's work listed in item 488 in Chap. 9 of this book.

[16] See Chap. 9 item numbers 480, 485, 486, 489.

[17] For an example of Julian's analysis of chemical weapons use in Syria, see '7.18 Presentation: "Does Arms Control Have a Future?" ' in Chap. 7 of this book. Chapter 9 lists this as item 493. Other examples are listed in Chap. 9 in items 478, 479, 481, 482, 483, 484, 490, 492, 494, 496, 497.

international policy-making.[18] He never wanted security clearance. A good example of his method of working was the way he uncovered information about the development of new Soviet agents, the Novichoks, by studying Soviet scientific journals. He referred to Novichoks within months of the first English-language use of the term,[19] and identified them in the early 2000s. He applied the same meticulous approach to supervising student theses as he used in his own research.

He kept everything, even though everyday administration like letters about income tax were relegated unopened to a drawer in his study. His study was so full of books and papers and objects that my brother Charlie[20] compared it to Merlin's cave. At the University of Sussex, he built up an incredible archive focussed on open source information relating to CBW – thought to be the largest in the world outside government – the Sussex Harvard Information Bank (SHIB). He created a complex and idiosyncratic retrieval system designed to help researchers, and especially himself, find relevant data. His students and other colleagues helped to collect, process and file information for SHIB; researchers from round the world relied on it, many of them visiting in person, where Dad warmly welcomed them. Dad's precise and meticulous arrangement of SHIB also had some less obvious beneficiaries; I was an obsessive comic reader and he taught me to file my comics on similar principles.

Dad also pioneered novel methods of making sense of the data he collected. The key method for collating information was the Rolling Chronology, which pooled data from diverse sources to glean insights and log developments related to global CBW in date order, anticipating contemporary events databases.[21] He published regular extracts from the Rolling Chronology in HSP's journal *The CBW Conventions Bulletin*. He also developed several specialist chronologies; most recently, he compiled an extraordinary chronology detailing the use of chemical weapons in Syria.

Another element of his way of working was networking. Dad was the centre of an epistemic community of people interested in CBW, which included practitioners, researchers, policy-makers, journalists and civil society activists. In the early years, his networking mostly took place through the Pugwash Conferences on Science and World Affairs, and the Pugwash CBW Study Groups detailed in Chap. 1. In later years, HSP also held regular Sussex Days, at which former students, colleagues, officials and other experts were able to discuss particular issues. Dad's network provided a continuous flow of information; by the end of his life, he had a huge email correspondence. It was also an important channel of influence. I think he

[18] Henrietta Wilson, Richard Guthrie, and Brian Balmer (2024). 'Open Source Investigations before the Age of Google: The Harvard Sussex Program'. In Henrietta Wilson, Olamide Samuel, and Dan Plesch (eds). *Open Source Investigations in the Age of Google*. Singapore: World Scientific Publishing.

[19] News Chronology 16 September 1992. *Chemical Weapons Bulletin*. 18. p. 14. Available at chrome-extension://efaidnbmnnnibpcajpcglclefindmkaj/http://hsp.sussex.ac.uk/new/_uploads/bulletin/CWCB18.PDF [Accessed 31 January 2025].

[20] Also known as Ralph Perry Robinson.

[21] See for example the Armed Conflict and Locations Events Database (ACLED), available at https://acleddata.com/ [Accessed 14 May 2024].

epitomized the salience of rigorous research in influencing policy-making. He was hugely respected both in government and civil society and this underpinned his impact on national and international policy-making, policy-shaping and advocacy groups.

He depended on soft money for his entire academic career. He never requested increases in salary or promotion as this would require him to raise more funds, although, of course, he was eventually promoted to Professor! Funders included the MacArthur Foundation, the Ford Foundation, the Carnegie Corporation, the Joseph Rowntree Charitable Trust, ESRC, and Horizon. However, the associated reliance on soft money was a source of endless anxiety. Every other Christmas was overshadowed by his next MacArthur proposal.

The big achievements of his life work were his contributions to the 1972 Biological Weapons Convention (BWC) and the 1993 Chemical Weapons Convention (CWC) – something that Matthew Meselson highlights in his contribution to this book (Chap. 1). One important aspect of Dad's influence was on both the BWC and CWC's use of a General Purpose Criterion to define the scope of their treaty's prohibitions. In the case of the CWC, the General Purpose Criterion specifies that the Convention prohibits all chemicals and activities intended for use in a chemical weapons capability, rather than outlawing particular chemicals. The great advantage of this approach is that it is future proof – it applies to new chemicals that were not anticipated at the time the Convention was negotiated.[22] More recently, as Meselson explains in Chap. 1, Dad and he were keen to propose a new treaty that would assign individual criminal responsibility for involvement in CBW manufacture, trade or use, extending the national prohibitions laid out in the BWC and CWC.[23]

Dad was also influential in other areas beyond CBW, for example, the Red Cross and the laws of war. He described and analysed the target effects of many different forms of conventional and unconventional weapons in relation to those precepts of law established in the nineteenth century that prohibit the use of weapons that cause unnecessary suffering or have indiscriminate effects or which are perfidious. In the aftermath of the Vietnam War, international humanitarian lawyers sought to develop these generic precepts into specific prohibitions of particular weapons or of how they were used. Their efforts could not succeed without a sound technical understanding such as my father and others like him tried to provide.[24] His publications here were anonymous, in the form of a succession of expert-group reports during the 1970s from the UN Secretariat, the International Committee of the Red Cross and the UN Environment Programme,[25] for which he served as rapporteur or editor as well as

---

[22] See '7.8 Some Consequences of the General Purpose Criterion Having Been Used to Define the Scope of the Chemical Weapons Convention' in Chap. 7 of this book.

[23] See '7.10 A Draft Convention to Prohibit Biological and Chemical Weapons under International Criminal Law' in Chap. 7 of this book.

[24] See '8.2 Qualitative Trends in Conventional Munitions – The Vietnam War and After' in Chap. 8 of this book.

[25] See '8.1 Possible Impacts of Weapons on Ecosystems' in Chap. 8 of this book.

contributing consultant. The 1981 UN Convention on Certain Conventional Weapons is rooted in this work and is itself the progenitor of the Land Mines Convention.

Another important focus was public health, especially in the context of the work of the World Health Organization (WHO). He was responsible for bringing together a group of consultants who compiled what is known as the Blue Book – *Public Health Response to biological and chemical weapons: report of a WHO Group of Consultants*. The first edition was published in 1970 and the second edition in 2004.[26] Of particular note is the recommendation that the best defence against biological weapons is to have effective global health systems – spanning local, national and international provisions – and, in particular, pandemic preparedness – something that has particular resonance after Covid-19, which showed us what a biological attack might look like, and tragically cost Dad his life.

We lost Dad to Covid at the beginning of the global pandemic. To finish I would like to give a rather more personal tribute; what follows is the eulogy I delivered at his funeral.

2.1 Eulogy for Julian Perry Robinson

I would like to talk for a little bit about Dad.

He never spoke much about his childhood or school days, except for often talking about the summers he and his sister Harriett spent in North Wales with his beloved Aunt Poppy, although we do know he was a keen athlete – playing for the England school boys hockey team and the Sussex junior county cricket team – and academic, winning scholarships to both his school Canford and to Merton College,[27] where he read Chemistry. After Oxford, he spent a brief period as a patent lawyer before moving to Sweden to work at the Stockholm International Peace Research Institute (SIPRI), where he began his life's work working to prevent the proliferation of chemical and biological weapons and met Mum.

There are many people much more able than me to discuss Dad's work and the impact it had. What I want to talk about is Dad as an inspiration, both personally and in the way he instilled his values in his family, friends, colleagues and wider society.

In some ways both of my brothers owe their careers to Dad. It was building model aeroplanes with Dad that started Charlie's love of model making and in turn his architectural career and Olly certainly wouldn't be an epidemiologist if he hadn't wanted to be a scientist like his dad.[28] Mum credits Dad's work ethic as inspiring

[26] World Health Organization. Health Aspects of CBW: report of a WHO Group of Consultants. 1970. Updated and reissued as Public health response to biological and chemical weapons. 2004. Available at https://www.who.int/publications/i/item/public-health-response-to-biological-and-chemical-weapons-who-guidance-(2004) [Accessed 10 April 2024]. These publications are listed in Chap. 8 as items 21 and 418.

[27] At the University of Oxford.

[28] Charles Perry Robinson (also known as Ralph), b. 19 March 1965, Oliver Robinson, b. 16 July 1979.

her own and giving her the motivation to achieve so much. It was also amazing that so many of the tributes we received from people from all over the world said that Dad had inspired them to be more like him. Another key way he inspired so many of us was in his love of cookery. He took cooking incredibly seriously and treated it almost like chemistry and there were certain recipes, particularly his fish pie, and of course the Soupe au Pistou, that he spent his life refining and trying to perfect, while never being satisfied with the final result.

As well as inspiring people to be like him, his values inspired so many of us, particularly through his love of nature and his belief in what I would term civility. The Perry Robinsons have been keen nature lovers for generations and Dad was no exception; according to Harriett, when he was young he seriously contemplated becoming a farmer. He was always much loved by any animals he encountered as well as keeping cats for most of his life, indeed Lily and Snowball, just like the rest of us, seem bereft without him, they are definitely becoming more feral without his stabilizing presence.

A country walk with him would take at least twice as long as one would expect owing to the need to stop every five minutes to examine an interesting insect or leaf. Butterflies and moths were Dad's particular passion, he was never happier than when we were in La Garde Freinet – particularly at Easter – where he could look forward to great food and the chance of seeing large numbers of butterflies. The quality of every walk he went on was measured by the number of different butterfly species we managed to see, although no walk I ever went on with him matched his legendary 27 butterfly walk in the hills around La Garde. Indeed, none of us will ever be able to see a butterfly without both remembering Dad and trying to identify the butterfly.

When I talk about Dad's belief in civility, I don't just mean treating each other in a polite and civilized manner, although Dad certainly did this – he was unfailingly polite and rarely got angry with anyone outside of his close family, and was interested, or at least gave the appearance of being interested, in everyone he met and would engage them keenly in conversation regardless of the subject and his own interest in it. But also, his civility meant that he believed that it was a moral imperative for everyone to treat each other both equally and well.

This, I think, was the motivation behind his life's work. As a scientist he couldn't bear to see the hard-won knowledge of biology and chemistry misused and turning into indiscriminate and barbaric weapons. This civility also inspired his politics, he had a deep distrust of the British establishment; despite a background and education that meant he could have easily been at the heart of it, he felt that the establishment could never be trusted to behave in a civilized manner without external pressure holding it to account.

This belief in civility as an imperative to treat each other equally and well also led to his 50 years of membership and support of the Labour Party; ten, probably deeply exasperating, years as a school governor at the difficult comprehensive Olly and I attended; far too much time on the over argumentative and slightly ridiculous Board of Directors of our building in Sussex Square; and in the last four years going on far more demonstrations than someone whose mobility had become so limited would be expected to manage.

A very fond memory for me is the day after Trump had been elected in 2016, a demonstration was called in Brighton to protest. Dad had found out, I'm not sure how, possibly from his favourite shop Seed and Sprout, or as he insisted on referring to it, leaf and stem. And he was adamant that the two of us had to go and join the demo. I then got to see Dad and two of the crustiest Brighton Anarchists you can imagine all chanting "Whose Streets? Our Streets" – definitely something I will never forget.

As well as being a source of inspiration, Dad was also happy to be inspired. The biggest inspiration in his life was of course Mum, even after 52 years, they were still utterly devoted to each other, I don't think I ever heard Dad call Mum by any other name than Sweetie Poppet, regardless of where they were, who they were talking to (indeed when they taught together it apparently caused much mirth amongst the students hearing this term of endearment from such a proper English Gentleman during seminars), or even when they were arguing. The love they shared was so palpable, and she certainly encouraged him to get over his social shyness. He would dread every social event Mum organized, whenever we had a dinner party, he would exaggerate the numbers coming and stress out about it for the whole day until the event arrived at which point he would transform into the perfect host and no one would realize his previous anxieties.

Dad was also often inspired by art and culture, he was a voracious reader of both poetry and prose, a lover of the theatre and art galleries. He also relied on a whole range of cookery writers for inspiration in his cooking, always happy to discover new writers and adjust his cooking accordingly.

Dad undoubtedly leaves the world in a better place than it would have been without him. He leaves a glorious legacy behind, his three children, five grandchildren, the transformation of the field of CBW, generations of colleagues inspired by him and of course the Chemical Weapons Convention and the Organisation for the Prohibition of Chemical Weapons. He also leaves behind a massive archive – the Sussex Harvard Information Bank – that we as a family are determined to digitize and make available for future generations of researchers.[29]

Dad, the world will never be the same again without you, but we will always love you. The way you inspired us and countless others will ensure your legacy lives on.

Goodbye Dad, we love you.

[29] This work is ongoing. See the Harvard Sussex Program website for updates: http://hsp.sussex.ac.uk/new/ [Accessed 24 July 2024].

Julian and Josh. From © Mary Kaldor's private collection.

Chapter 3
Tributes to Julian Perry Robinson

Peter Pringle

Tributes to Julian came from around the world. Obituaries appeared in *The Times, The Guardian, Nature, Arms Control Today, the Bulletin of the Atomic Scientists* and elsewhere, as detailed in the references at the end of this chapter. Personal tributes were sent by colleagues, former students, government officials, delegates to disarmament talks and peace activists.

At Julian's funeral in May 2020, when few could attend because of Covid restrictions, Julian's son, Olly,[1] talked about the many tributes, "He was recognized as a man, sometimes of few words, but who paid lots of attention to the needs of others. He was modest about his achievements and shunned the limelight and honours. Yet to so many his soft-spoken presence was wisdom incarnate. He was a perfectionist, particularly with cooking, and inspired others to attempt to reproduce dishes. Everyone enjoyed his company greatly, his understated humour, witty, a bit self-deprecating, the stories told with a twinkle in his eye. His friends remember that quizzical lift of the eyebrow – 'Do you really think that?' And he could be critical in a collegiate way that deepened not detracted from other people's understanding. Many remember being welcomed into his research group over a drink after meetings, particularly at The Swan pub near the University of Sussex campus. To many of his international friends and colleagues he epitomized the best and most endearing traits of 'Englishness.' He was seen as the quintessential English gentleman driving up to the conference center in his classic sports car, hood down".[2]

[1] Dr. Oliver Robinson is a Lecturer at the School of Public Health, Imperial College London. See https://profiles.imperial.ac.uk/o.robinson [Accessed 28 June 2024].

[2] Taken from the summary of tributes sent to Julian's family, summarized by Oliver Robinson for Julian Perry Robinson's funeral, May 2020.

Peter Pringle, Journalist, author, foreign correspondent, New York, United States of America.

P. Pringle (✉)
New York, NY, USA
e-mail: pringlep@ix.netcom.com

H. Wilson et al. (eds.), *Julian Perry Robinson: Pioneer in the Prohibition of Chemical and Biological Warfare*, Pioneers in Arts, Humanities, Science, Engineering, Practice 32, https://doi.org/10.1007/978-3-032-04064-0_3

3.1 Colleagues

***Colleagues** remembered his scholarship, his basic decency, his willingness to put others before himself, and his pervasive modesty despite his deep understanding of chemical and biological warfare (CBW) issues.*

William Walker[3] remembered, "Working alongside Julian was one of the privileges and pleasures of my two decades at SPRU [the Science Policy Research Unit at the University of Sussex]. He was the kindest, gentlest and most modest of men, a marvellously dogged, old-fashioned yet skilful researcher, an example to all on how to gain influence. He had a tremendous command of detail and became the great historian of CBW, a wise expert and friendly guide to anyone, anywhere seeking advice on curbing these dreadful weapons".[4]

Jordi Molas-Gallart[5] wrote, "I can still see him driving down the hill past the Mantell Building [the building that housed SPRU at the University of Sussex] in his old Morgan convertible, on a cold day, roof down, scarf flying in the wind. To me, used as I was to warmer climates, that was the epitome of English eccentricity. By the standards of today's dominant academic behaviour, Julian would also seem eccentric. He was concerned about what he could do to make our world better, how to influence policy for the good, completely unencumbered by considerations of his own image, the ranking of journals, citations or academic recognitions. And yet, this attitude should be the norm rather than the exception; working towards making it the norm would be a good way to celebrate Julian's memory".[6]

Richard Guthrie[7] initially planned to work with Julian at ADIU [Armament Disarmament Information Unit][8] for just six weeks and left SPRU fifteen years later. Why? "Because of Julian's personal qualities, his skills and ethos. A primary motivation behind ADIU was that the effective debate [about complex policy issues] could only take place if it was an informed debate tapping into diverse sources of understanding across different traditional disciplines of study.

---

[3] Professor William Walker, Emeritus Professor, University of St Andrews. See https://www.st-andrews.ac.uk/international-relations/people/wbw/ [Accessed 21 April 2024].

[4] Taken from William Walker's post on the Forever Missed page for Julian Perry Robinson. Available at https://www.forevermissed.com/julian-perry-robinson/about [Accessed 10 April 2024].

[5] Professor Jordi Molas Gallart, Research Professor at the Spanish Council for Scientific Research (CSIC). He is currently the director of INGENIO, a joint Institute of the Spanish Council for Scientific Research and the Polytechnic University of Valencia. See https://www.ingenio.upv.es/en/people/jordi-molas-gallart [Accessed 8 May 2024]. Jordi worked at SPRU University of Sussex, during which time he collaborated with Julian on projects for the EC Joint Research Centre and STOA-European Parliament.

[6] Taken from Jordi Molas-Gallart's comments posted on University of Sussex. "In memory of Julian Perry Robinson: Obituary" [online]. Available at: http://www.sussex.ac.uk/spru/about/history/julian-perry-robinson [Accessed 7 March 2024].

[7] Richard Guthrie, CBW Events.

[8] ADIU was a public information unit at SPRU that dealt with military technologies and arms control issues.

"When the first public allegations were made in the 1990s that new toxic compounds were being explored by government labs in the Soviet Union, Julian eventually gathered enough information to put together the general structure of these new compounds, and published these structural details in a footnote to a paper he presented to a working group [of the Pugwash Conferences on Science and World Affairs]. Julian was summoned to a meeting in Whitehall where officials expressed disbelief that there was sufficient public information for Julian to reach his conclusions, but indeed he had worked it out from multiple open sources.

"A key [aspect] of Julian's work was that you only needed to acquire a piece of information once to be able to make use of it forever. That was the core of the Sussex Harvard Information Bank (SHIB) which was and remains far more than a collection of papers, more than an archive, and even describing it as a themed repository does not do it justice... [It is] the greatest collection or store of information on CBW outside of any government, in fact, better than most governments, an unrivalled archive of reports and publications and grey literature on CBW – such that when SPRU moved into the new Freeman Centre in 2003 the building had to be constructed with a special reinforced floor to take the considerable weight of the archive.

"Julian insisted that the text of the Harvard Sussex Program (HSP) *Chemical Weapons Convention Bulletins* should be proof-read multiple times and he did not like computer spellchecks. And this was typical of his drive to produce work of the highest quality.

"With Julian's passing we have lost one of the most collaborative, intellectually generous researchers the world has seen. I was proud to call him a colleague".[9]

Brian Balmer[10] remembered, "Julian's general concern was knowing the history of CBW not as a curiosity or as an intellectual exercise, but as the central part of ongoing efforts towards CBW control and disarmament. When I was a research fellow at SPRU (1993–4), Julian steered me to the National Archives where he knew they were beginning to open up on CBW. I kept on saying I want to do something contemporary...but he knew the importance of history [and I spent the next 30 years researching the history of CBW]. The earliest archival reference I found to Julian was in 1975 when the UK government patent on VX nerve gas was found in the public library and widely reported in the press and he argued that the patent had provided a step by step set of instructions for making the VX agent. Julian wrote a short commentary of the affair in the *New Scientist*[11] pointing out that the revelation of the VX formula was entirely new".[12]

---

[9] Adapted from Richard Guthrie's speech, Julian Perry Robinson memorial service, University of Sussex, November 2022.

[10] Brian Balmer, Professor of Science Policy Studies, University College London, and HSP Sussex Researcher 1993–1994, and HSP Associate 1994–present. See https://www.ucl.ac.uk/sts/people/professor-brian-balmer [Accessed 21 April 2024].

[11] Julian's article is listed in Chap. 9 as item 73.

[12] Adapted from Brian Balmer's comments at Julian Perry Robinson's memorial service, University of Sussex, 11 November 2022.

3.2 Policy-Makers

Julian sometimes surprised government officials with his encyclopaedic knowledge and extraordinary range of contacts in the CBW world. Former officials recalled how Julian had been a trusted resource for them.

John Walker[13] formerly of the UK Foreign and Commonwealth Office (FCO), wrote, "There cannot be anyone working on CBW history, arms control and disarmament who has not benefited enormously at some point in their careers from Julian's very great expertise, knowledge and generosity of spirit. Julian was an exemplary and diligent scholar and, as such, a great example and role model for us all to aspire to. He was modest about his considerable achievements and shunned the limelight and honours – softly spoken and ever ready to listen to what others had to say. I think he was more influential than he ever knew. His impact on young academics, government officials and non-governmental organisations (NGOs) over fifty years was immense. Julian was such good company. We had many a long chat on CBW matters over the years of my time at the FCO – he came to our meetings with a list of topics written in red ink to address and we went through these one by one over a pint or two at one of Whitehall's many hostelries".[14]

Patrick Lamb,[15] formerly of the UK Foreign Office wrote, "I got to know Julian – initially as a wily academic adversary – when I became Head of the CBW Section in FCO from 1997 onwards. These were in retrospect great and optimistic days. There was nevertheless a lot of polite but serious sparring over RCAs (riot control agents and even the unmentionable FGAs (fourth generations agents). Indeed, and to my shame, I once summoned Julian to the Foreign Office for a formal dressing down on the latter subject. I wondered even then, judging by the mischievous smile that never left his face, whether he took this quite as seriously as we did and who exactly was right. I tend now to think he was! I was however forgiven and subsequently met him on numerous occasions at Wilton Park[16] and at the University of Sussex".[17]

[13] Dr. Walker OBE was Head of the Foreign and Commonwealth Office's Arms Control and Disarmament Research Unit (ACDRU) from December 2014 to May 2020 when he retired from the FCO. He joined ACDRU in March 1985. He is currently a Senior Associate Fellow at the Royal United Services Institute and the European Leadership Network and a Senior Research Fellow at the Department of Science and Technology Studies at University College, London.

[14] Taken from John Walker's post on the Forever Missed page for Julian Perry Robinson. Available at https://www.forevermissed.com/julian-perry-robinson/about [Accessed 10 April 2024].

[15] Patrick Lamb OBE worked for the Foreign Office in the Counter-Proliferation Department, and headed UK negotiations on a Verification Protocol to the BTWC between 1997 and 2001. He was later Head of the FCDO Iraq Inquiry Unit before retiring in 2017.

[16] Editors' note. For more information about Wilton Park, see https://www.wiltonpark.org.uk/ [Accessed 5 January 2025)

[17] Taken from Patrick Lamb's post on the Forever Missed page for Julian Perry Robinson. Available at https://www.forevermissed.com/julian-perry-robinson/about [Accessed 10 April 2024].

Bob Mathews[18] recalled how he loved to watch the way Julian would "so readily deconstruct complex issues during meetings; he would break them into various elements and then get discussion. So, he was a wonderful person to chair and to facilitate Pugwash and other CBW meetings [and he was] an amazingly popular person in the CBW community. Everybody liked him and trusted him. Even when they disagreed with a particular view that he was presenting, they realized that he was coming from a position of being always curious, worrying about details, looking for explanations when there were problems. It was really a very good example of academic/practitioner interaction".[19]

Bob Mikulak[20] wrote "It's been many years – several decades – since I first met Julian at a Pugwash CBW meeting. And since then we've crossed paths many times, although infrequently in the last few years, as we've pursued complementary routes toward chemical and biological weapons disarmament. It has been a great privilege to have known him. His deep commitment to the goal of eliminating the threat from chemical and biological weapons has helped inspire me and many others. On a personal basis I'll remember Julian's friendly and generous personality and impish smile. And on a professional basis I will never forget his extraordinary capacity for collecting, organizing and analysing information about chemical and biological weapons. I often learned more from his research using unclassified sources than from the highly-classified analyses to which I had access. His dedicated efforts will live on, not only in his scholarly publications and in the memories of his friends and colleagues, but also in the legacy of numerous students who have become important contributors to the cause of peace and security in their own right".[21]

3.3 Students

Several colleagues had also been Julian's doctoral students. They remembered him as one of the most self-effacing academics, a rare breed who always went out of his way to spend time with his students and include them in his activities.

Caitríona McLeish[22] remembered, "Julian did all the things you would expect a Ph.D. supervisor to do. He met with you, read your drafts, wielding his mighty red

[18] Robert J. Mathews, Honorary Associate Professor University of Melbourne Law School. Formerly Head of the Nuclear Biological and Chemical Arms Control Unit of the Australian Defence Science and Technology Organisation. See https://law.unimelb.edu.au/about/staff/robert-mathews [Accessed 21 April 2024].

[19] Adapted from Bob Mathews' comments at Julian Perry Robinson's memorial service, University of Sussex, 11 November 2021.

[20] Robert Mikulak. Expert, Biological Policy Staff, US Department of State.

[21] Taken from Bob Mikulak's post on the Forever Missed page for Julian Perry Robinson. Available at https://www.forevermissed.com/julian-perry-robinson/about [Accessed 10 April 2024].

[22] Caitríona McLeish, UK Foreign, Commonwealth and Development Office, previously co-director of the Harvard Sussex Program on Chemical and Biological Weapons, and Julian's M.Sc. and D.Phil student (1996–1997 and 1998–2002 respectively).

pen, he helped you hone your argument, encouraged you to write papers, speak at conferences, but he also brought you into his networks. Whenever visitors came to see him you were always invited to lunch, it didn't matter whether the visitor was a fellow student, a renowned scholar, an activist or a representative from government or an international organization. But if you accepted the invitation you weren't allowed to be a passive lunch mate, you were encouraged to offer your own views and to ask questions. Julian also invited us to Pugwash meetings and to London seminars with John Walker. So, you would find yourself sitting in meetings there was no way a PhD student would normally be invited to, listening to conversations that were really shaping the direction of CBW disarmament policy".[23]

Katie Smallwood[24] recalled, "Julian's knowledge was so impressive that during my more vulnerable dissertation-writing moments, I suspected he found me rather frivolous. What was I doing treating his lifetimes' work in a way he probably found both endearing and trivial? Even if [this was] the case, gentle as he was, he would never let on, save the odd expression he would let slip (somewhere between amusement and bemusement) that I was never quite sure how to read. I would wonder what he really thought of us all! Julian was a huge supporter of his students – 'Yours is a dissertation that must not hide its light under a bushel' he once wrote to me.

"People who didn't know him particularly well were somehow drawn to him and people at SPRU remember him walking through the corridors or around the building… he had this way of walking, it was sort of leisurely but purposeful, and always curious about what was around him. Those of us working at HSP will also remember Julian's quizzical look and the special metal briefcase he always carried as he made his way to and from the archive".[25]

Kai Ilchmann[26] recalled that Julian was "outstandingly generous with his time, and spent hours discussing my dissertation, meticulously exploring the etymology of curiously chosen terms. He never gave me a quick answer, as much as I wanted it. Rather, he would place the question into context, reframe it into a sensible question, address all angles and highlight the policy relevance. I was incredibly fortunate to have worked with him".[27]

As well as supervising his own students, Julian also supported other students.

---

[23] Adapted from Caitríona McLeish's comments at Julian Perry Robinson's memorial service, University of Sussex, 11 November 2022.

[24] Catherine (Katie) Smallwood, World Health Organization, formerly Harvard Sussex Program Hague Researcher 2007–08, and Julian's M.Sc. and D.Phil. student 2004–2009. See https://www.linkedin.com/in/katiesmallwood/ and http://hsp.sussex.ac.uk/new/about/hsp-associates [Accessed 21 April 2024].

[25] Taken from Katie Smallwood's post on the Forever Missed page for Julian Perry Robinson. Available at https://www.forevermissed.com/julian-perry-robinson/about [Accessed 10 April 2024].

[26] Kai Ilchmann. Harvard Sussex Program Honorary Senior Research Fellow, and Julian's D.Phil. student 2004–2010. See http://hsp.sussex.ac.uk/new/about/people [Accessed 21 April 2024].

[27] Taken from Kai Ilchmann's post on the Forever Missed page for Julian Perry Robinson. Available at https://www.forevermissed.com/julian-perry-robinson/about [Accessed 10 April 2024].

Jean Pascal Zanders[28] wrote. "The Swan Inn was my real university. Over beer (him) and wine (I) we discussed progress in my Ph.D. project. And the meetings usually ended with detailed suggestions of what I should research next. No instructions, no. Always open questions. 'Hmm, very nice, Jean Pascal.... However, have you thought of this or that aspect?' By the next time, I had followed up on his suggestions, so we found ourselves back in The Swan Inn, and with pride I summarized what progress (I thought) I had made. 'Hmm, very nice, Jean Pascal, but have you...?' Yes, I had a sense of frustration; a sense of feeling every time tantalisingly close to the answer, yet never ever being able to quite grab it. Gradually I came to realise that time and time again he was pushing the edges of my understanding of the research questions, teaching me never to be satisfied with the answers I came up with. Always question one's own findings, never relent because there may always be more to discover beyond what one thinks one knows. THAT has stayed with me ever since".[29]

Simon Whitby[30] wrote that he had "preserved Julian's External Examiner Viva copy of my Ph.D. on anti-crop biological weapons crammed with his sticky labels in the top, bottom and sides, covered with queries, questions and clarifications. Julian went through the whole thing carefully, systematically and with characteristic attention to detail. He spent ages going through the thesis in preparation for its publication as a book; and both benefited hugely from his extensive archive, from his supreme knowledge of the subject matter, and from his generosity of spirit".[31]

3.4 Subject Matter Experts

Delegates Julian encountered at international disarmament talks, and other such conferences on CBW, remembered how they often relied on his leadership and his ability to make sense of the sessions.

Alastair Hay[32] recalled Julian at a press conference just before the Iraq invasion in 2003. "We were all being a little cautious on the evidence from the UN team about Iraq WMD. I was cautious. No way was Julian cautious, he was absolutely

[28] Dr. Jean Pascal Zanders is an independent researcher/consultant on disarmament and security questions. He heads The Trench, a research initiative dedicated to the future of disarmament, and is also a non-resident Senior Research Associate at the Fondation pour la Recherche Stratégique (Paris). He first met Julian during his PhD research degree in Political Sciences, which was awarded in 1996 from the Free University of Brussels.

[29] Taken from Jean Pascal Zanders. 'JPPR – The 4 letters that shaped my career' [online]. *The Trench*, 13 May 2020. Available at https://www.the-trench.org/jppr [Accessed 8 May 2024].

[30] Taken from Dr. Simon Whitby, Associate Professor, University of Bradford. See https://www.bradford.ac.uk/staff/swhitby/ [Accessed 21 April 2024].

[31] Simon Whitby's post on the Forever Missed page for Julian Perry Robinson. Available at https://www.forevermissed.com/julian-perry-robinson/about [Accessed 10 April 2024].

[32] Professor Alastair W. M. Hay, Emeritus Professor University of Leeds. See https://medicinehealth.leeds.ac.uk/medicine/staff/408/professor-alastair-w-m-hay [Accessed 21 April 2024].

categorical, there was no evidence whatsoever that Iraq had these weapons and we now know that was the truth.... I loved his approach and his generosity. And I loved his sense of humour".[33]

Filippa Lentzos[34]: "To me, Julian was the heart and driving force of the British CBW community. He set the standard and the atmosphere. In my foreign eyes, he represented the best of the British. A deeply knowledgeable and critical thinker, an absolute gentleman, modest, encouraging, suave, always with that bemused smile".[35]

Nicholas Sims[36] described how "Julian exemplified high standards of scholarship in his own work, yet wore his learning lightly and was generous in his help and hospitality to successive generations keen to benefit from his knowledge. He has been fittingly described as the grandfather of non-governmental research into the control of CBW".[37]

Ralf Trapp[38] wrote that Julian had attended some 70 Pugwash meetings from the early 1970s onwards. He had "become a key figure in setting the agenda of the group and steering its deliberations". When the Chemical Weapons Convention (CWC) began moving from exploratory talks to addressing specific technical and legal issues, Julian became directly involved in "how the capabilities and activities of the industry could be made sufficiently transparent to ensure confidence in treaty compliance.... these exchanges were important to ensure that the industry verification system would be technically sound. In 1992, after the drafts at the CWC had been adopted, Julian took the lead in discussions on how to implement both [the Biological Weapons Convention and CWC] treaties together".[39]

Francesco Calogero[40] recalled, "During my tenure as Secretary General of Pugwash, I essentially delegated substantive decisions about anti-CBW activities

[33] Adapted from Alaistair Hay's comments at Julian Perry Robinson's memorial service, University of Sussex, 11 November 2022.

[34] Dr. Filippa Lentzos, Reader in Science and International Security. See https://www.kcl.ac.uk/people/filippa-lentzos [Accessed 21 April 2024].

[35] Taken from Filippa Lentzos' post on the Forever Missed page for Julian Perry Robinson. Available at https://www.forevermissed.com/julian-perry-robinson/about [Accessed 10 April 2024].

[36] Nicholas A. Sims is an Emeritus Reader in International Relations at the London School of Economics and Political Science, University of London. https://www.lse.ac.uk/international-relations/people [Accessed 18 April 2025]

[37] Nicholas A. Sims, 'Julian Perry Robinson, top chemical and biological arms control expert, dies of coronavirus'. *Bulletin of the Atomic Scientists.* 30 April 2020. Available at https://thebulletin.org/2020/04/julian-perry-robinson-top-chemical-and-biological-arms-control-expert-dies-of-coronavirus/#post-heading [Accessed 18 April 2025]

[38] Ralf Trapp, International Disarmament Consultant. See https://ralftrapp.net/ [Accessed 21 April 2024].

[39] Adapted from Ralph Trapp's comments at Julian Perry Robinson's memorial service, University of Sussex, 11 November 2022.

[40] Francesco Calogero, Emeritus Professor of Theoretical Physics, La Sapienza University of Rome, and participant in the European Leadership Network. Formerly Secretary General of the Pugwash Conferences of Science and World Affairs (1989–1997) and Chair of the Pugwash Council (1997–2002, and 'ex-officio' member thereafter). See https://www.europeanleadershipnetwork.org/person/francesco-calogero/ [Accessed 21 April 2024].

to Julian and Matt [Meselson] …I think this Pugwash activity was quite important for the achievement of the 1972 Biological Weapons Convention (BWC) and the 1993 Chemical Weapons Convention (CWC) ... perhaps most significantly, for the CWC by managing to involve the leadership of the American Chemical Manufacturers Association… [and] convincing them of the necessity of the CWC. The pre-eminent merit for this monumental achievement, goes, in my opinion, to Julian and Matt".[41]

Hans Günter Brauch[42] wrote "I knew Julian during my active time as a member of the Pugwash movement (1976–1992) … he has been one of my mentors who helped when I was working, writing and speaking on issues of chemical and biological weapons, warfare, arms control and disarmament, including the use of herbicides in the war in Southeast Asia. Julian has always been very open and cooperative and very careful in his judgment. He has always been a source of information, inspiration and support".[43]

Daniel Feakes[44] recalled how Julian always insisted that "the [Biological Weapons and Chemical Weapons] Conventions were negotiated based on a long-standing and widely-held belief, a taboo as he usually called it – that the use of poison and disease as weapons was beyond the pale. Julian could quote sources from ancient Greece and Rome as evidence of that historical norm.

"Julian also followed closely the discussions at the national level in the UK on the domestic implementation of the CWC, advocating for a unique national authority advisory committee of which he was a founder member. Through his work with Matthew Meselson via Pugwash both on the negotiations and subsequent implementation [of the treaties], Julian was a continuous presence over the decades and a committed supporter for the conventions. I remember celebrating the news in April 1997 that the USA had ratified the Chemical Weapons Convention. We had a bottle of champagne in Julian's office on that day".[45]

---

[41] Taken from Francesco Calogero's comments on the Forever Missed website, available at https://www.forevermissed.com/julian-perry-robinson/tributes [Accessed 10 April 2024].

[42] Hans Günter Brauch, political scientist, Professor Emeritus of the Berlin Free University and Founder of the HGB Foundation for Peace and Ecology in the Anthropocene. See https://hgb-stiftung.de/ [accessed 10 July 2024].

[43] Taken from Hans Günter Brauch, "Personal Memories on Prof. Julian P. Perry Robinson", available at http://www.afes-press-books.de/html/PDFs/2020/Obituary_Julian_%20P_Robinson.pdf [accessed 10 July 2024].

[44] Chief, BWC Implementation Support Unit, and Harvard Sussex Program Hague Researcher and Sussex Researcher 2000–09. See https://www.linkedin.com/in/daniel-feakes-1566865/ and http://hsp.sussex.ac.uk/new/about/hsp-associates [Accessed 21 April 2024].

[45] Adapted from Daniel Feakes' comments at Julian Perry Robinson's memorial service, University of Sussex, 11 November 2022.

3.5 Later Collaborations

In 2007, a celebration was held to mark Julian's formal retirement. The University of Sussex obituary notes "One of those who could not be present at his retirement but who sent letters of congratulation wrote as follows: 'I was once asked by an individual who had ploughed through quite a few papers on biological weapons issues and was well versed in the realities of international affairs a simple question: 'Who is this Julian Perry Robinson everyone writing in this area refers back to?' My response was to try and put the answer into an easily understandable context. Julian Perry Robinson is to the UK CBW community what Sir David Attenborough is to the British public when it comes to natural history. Like Attenborough, Julian had a distinctive and instantly recognizable voice, and was blessed with a gift for elegant phrasing and the ability to get messages across in a clear and cogent manner' ".[46]

But Julian never really retired, continuing to work in recent years on the use of chemical weapons in Syria and on Russian Novichoks following the Salisbury poisonings.

Rim Turkmani[47] recalled the 2018 chemical weapons attack at Ghouta, Syria. "There was complete chaos on the ground, no one knew how to handle it, what agent was used, what to do, how to protect themselves and how to take evidence. We needed somebody independent and trusted such as Julian. I got in touch with him and immediately we had two networks in action, one on the ground, the doctors and paramedics and activists in Ghouta, and Julian had his own network of international experts, and people started documenting things in a professional way during the war. Julian started an important chronicle of the events, including a very detailed archive of the chemical weapons used in Syria and the international efforts to limit this use. It spanned 40 years, from 1982 until now".[48]

George Papandreou, the former Prime Minster of Greece, added his recollections from meetings with Julian during a speech to members of the Symi symposium in 2021.

"Aristotle might have concluded he had three elements of the best of politics:

"Scientific knowledge on the one hand; a rational yet wise approach to enforcing the implementation of international law … but also emotion, pathos, a courage in expressing his indignation and sense of justice as he pushed for dialogue and international cooperation. … [He was] a rare example of selflessness, avoiding the hubris of those caught up in their narcissism or sense of genius. What a contrast to so many who today like to call themselves leaders in this world".

---

46 University of Sussex. 'In memory of Julian Perry Robinson: Obituary' [online]. Available at: http://www.sussex.ac.uk/spru/about/history/julian-perry-robinson [Accessed 7 March 2024].

47 Dr. Rim Turkmani, London School of Economics and Political Science Research Fellow and Director of Syria Conflict Research Programme (CRP). See https://www.lse.ac.uk/international-development/people/fellows/rim-turkmani [Accessed 21 April 2024].

48 Adapted from Rim Turkmani's comments at Julian Perry Robinson's memorial service, University of Sussex, 11 November 2022. See also '7.18 Presentation: "Does Arms Control Have a Future?"' for Julian's analysis of "Syrian-mode chemical warfare" ' in Chap. 7 of this book.

3.6 Obituaries

Professor Julian Perry Robinson, obituary

- John Hart. 'In Memoriam—Professor Julian Perry Robinson (1941–2020); When COVID-19 strikes close to home' Middlebury Institute of International Studies at Monterey: James Martin Center for Nonproliferation Studies [online]. (24 April 2020). Available at: https://nonproliferation.org/in-memoriam-professor-julian-perry-robinson-1941-2020/ [Accessed 7 March 2024].
- (30 April 2020): 'Obituary: SPRU remembers Professor Julian Perry Robinson, expert in chemical and biological warfare prevention'.
- University of Sussex. 'In memory of Julian Perry Robinson: Obituary' [online]. Available at: http://www.sussex.ac.uk/spru/about/history/julian-perry-robinson [Accessed 7 March 2024].
- Nicholas Sims. 'Julian Perry Robinson, top chemical and biological arms control expert, dies of coronavirus'. *Bulletin of the Atomic Scientists.* (30 April 2020). Available at: https://thebulletin.org/?s=julian+perry+robinson%2C+top+chemical [Accessed 7 March 2024].
- Richard Guthrie. 'Julian Perry Robinson obituary. Specialist in the control of chemical and biological warfare who guided experts and policymakers from east and west'. *The Guardian.* 8 May 2020). Available at: https://www.theguardian.com/world/2020/may/08/julian-perry-robinson-obituary [Accessed 7 March 2024].
- Peter Pringle. 'Professor Julian Perry Robinson obituary'. *The Times.* 19 May 2020. Available at: https://www.thetimes.co.uk/article/professor-julian-perry-robinson-obituary-5d7gtgg9j [Accessed 7 March 2024].
- Caitríona McLeish. 'Obituary: Julian Perry Robinson (1941–2020). Chemist and lawyer who shaped international weapons conventions'. *Nature* 581 (378). 21 May 2020. Available at: https://www.nature.com/articles/d41586-020-01450-1 [Accessed 7 March 2024]. Julian Perry Robinson 1941–2020 [online].
- Daniel Feakes. 'Julian Perry Robinson (1941–2020) Dedicated to Eradicating Chemical and Biological Weapons'. *Arms Control Today.* June 2020. Available at: https://www.armscontrol.org/act/2020-06/features/julian-perry-robinson-1941%E2%80%932020-martin-b-malin-1961%E2%80%932020 [Accessed 7 March 2024].

Julian with students and colleagues at The Swan pub in Falmer, Sussex in January 2003. From © Daniel Feakes' collection, included with his kind permission.

Part II
Photographs

Chapter 4
Photographs – Private Collections

Henrietta Wilson

4.1 Childhood to Early Adulthood

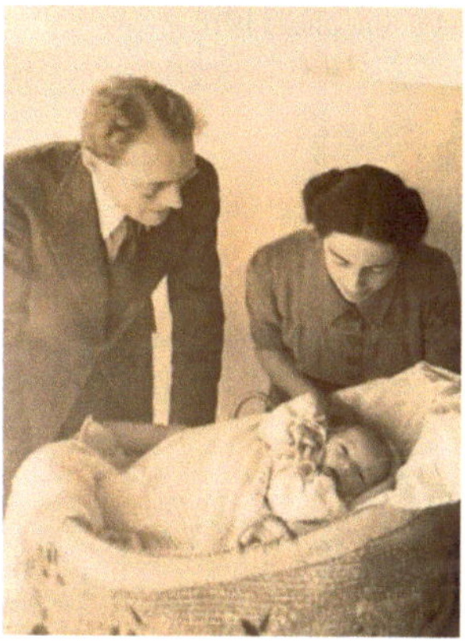

Julian Perry Robinson as a baby admired by his parents. From © Mary Kaldor's private collection.

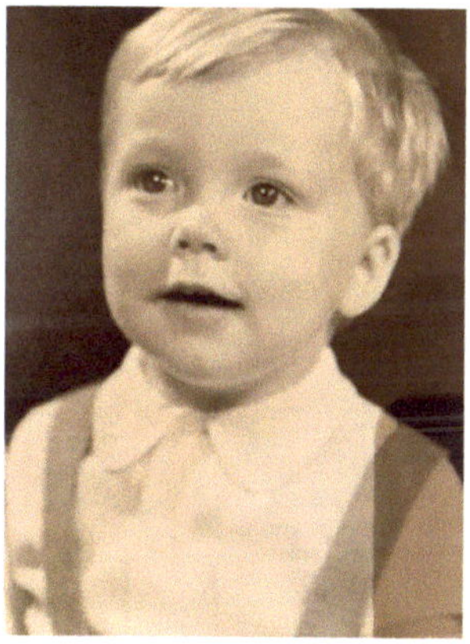

Julian as a little boy. From © Mary Kaldor's private collection.

H. Wilson (✉)
King's College London, Bristol and London, UK

H. Wilson et al. (eds.), *Julian Perry Robinson: Pioneer in the Prohibition of Chemical and Biological Warfare*, Pioneers in Arts, Humanities, Science, Engineering, Practice 32, https://doi.org/10.1007/978-3-032-04064-0_4

Julian with a rabbit. From Mary Kaldor's © private collection.

Julian playing hockey. With thanks to © Canford School for their permission to reproduce this photo here.

Julian with his prizes. From Mary Kaldor's © private collection.

Julian Perry Robinson in a lab. From © Mary Kaldor's private collection.

Julian punting in Oxford. From © Mary Kaldor's private collection.

4.2 Julian and Mary

Julian and Mary in Sweden. From © Mary Kaldor's private collection.

Julian and Mary in the Bugatti. From © Mary Kaldor's private collection.

Julian and Mary looking at the sea in Slovenia, during a break from an ISODARCO Conference. From © Mary Kaldor's private collection.

Julian and Mary on an elephant. From © Mary Kaldor's private collection.

Julian and Mary at London Zoo. From © Mary Kaldor's private collection.

© Mary Kaldor. From her private collection.

Julian and Mary at Buckingham Palace, on the occasion of Mary receiving her CBE. From © Mary Kaldor's private collection.

Julian and Mary in Brussels. From © Mary Kaldor's private collection.

4.3 Julian At Work and With Colleagues

Julian with a typewriter and Milton Leitenberg in Sweden. From © Milton Leitenberg's collection, included with his kind permission.

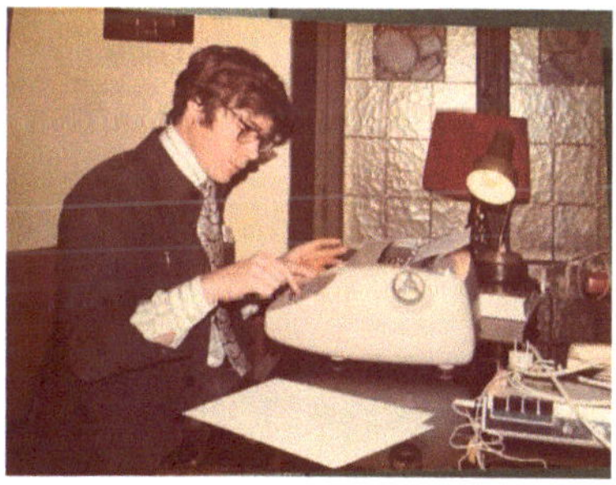

Julian with a typewriter and lamp. From © Mary Kaldor's private collection.

Milton Leitenberg, one other, Julian and Dr. M.M. Dubenin. From © Milton Leitenberg's collection, included with his kind permission.

Julian at his desk in the Mantell Building University of Sussex, approximately 1990. From © Richard Guthrie's private collection.

Julian making comments. © Richard Guthrie.

Julian with Jonathan Tucker January 2007 © Harvard Sussex Program

Julian at his retirement celebration November 2007 © Harvard Sussex Program.

Julian at dinner during a Pugwash meeting December 2009 © Richard Guthrie.

Matthew Meselson, Julian, Lenna and Martin Kaplan (right) at a Pugwash Conference © Pugwash Conferences on Science and World Affairs

Matthew Meselson and Julian Robinson at the 20th Pugwash Workshop Study Group on the Implementation of the CBW Conventions, Geneva, Switzerland. 8–9 November 2003. From © Matthew Meselson's collection, included with his kind permission.

4.4 Julian Enjoying Outdoors

Julian at the beach on Samos. From © Mary Kaldor's private collection.

Julian outside his parents' house in Wales. From © Mary Kaldor's private collection.

The Morgan. From © Mary Kaldor's private collection.

Julian walking. From © Mary Kaldor's private collection.

Julian on Symi by the sea. From © Mary Kaldor's private collection.

Julian outside the family house in France. From © Mary Kaldor's private collection.

Julian listening. From © Mary Kaldor's private collection.

Julian walking by the sea in Skiathos in Greece. From © Mary Kaldor's private collection.

Julian on Brighton Beach
© Carlos Saba.

Julian ready for a walk. From © Mary Kaldor's private collection.

4.5 Julian with Family and Friends

Julian's sister Harriett, father Brad and Julian. From © Mary Kaldor's private collection.

Julian and Mary with Adele Simmons and Emma Rothschild in the family house in France. From © Mary Kaldor's private collection.

Julian with family and friends in France. From © Mary Kaldor's private collection.

Julian in conversation with Andrew Packard. From © Mary Kaldor's private collection.

Julian, Gus and Robert Skidelsky, and Julian's son Josh. From © Mary Kaldor's private collection.

Olly, Andrew Packard and Susanna Linhart, Mary, Julian and Josh. From © Mary Kaldor's private collection.

4.6 Julian with His Children and Grandchildren

Julian and son Ralph ('Charlie'). From © Mary Kaldor's private collection.

Julian Mary and son Josh. From © Mary Kaldor's private collection.

Julian and son Josh. From © Mary Kaldor's private collection.

Julian and son Olly. From © Mary Kaldor's private collection.

Julian and son Ralph ('Charlie') with a camera. From © Mary Kaldor's private collection.

Julian and son Olly in car. From © Mary Kaldor's private collection.

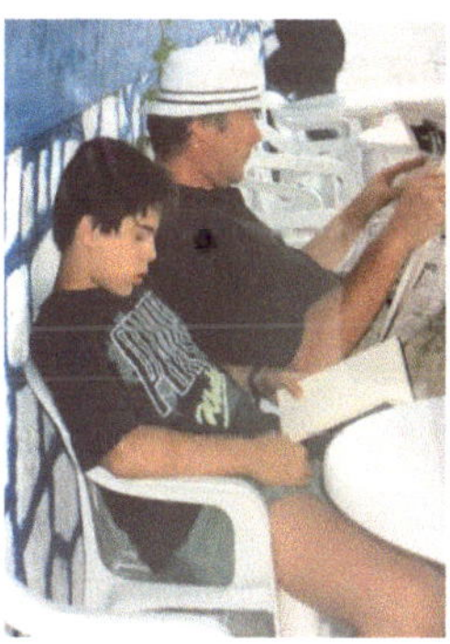

Julian and son Olly reading together. From © Mary Kaldor's private collection.

Julian and granddaughters Cecily and Agnes. From © Mary Kaldor's private collection.

Julian snuggling with grandson Tom. From © Mary Kaldor's private collection.

Julian with grandchildren Felicity and Arthur. From © Mary Kaldor's private collection.

Julian with grandson Tom in Brighton. From © Mary Kaldor's private collection.

Mary and Julian with grandson Arthur in London shop. From © Mary Kaldor's private collection.

Julian's sons Josh Kaldor-Robinson, Olly Robinson and Ralph Perry Robinson ('Charlie').
From © Mary Kaldor's private collection.

Part III
Writings

Chapter 5
Introduction to Part III

Henrietta Wilson, Mary Kaldor, Josh Kaldor-Robinson, Peter Pringle, Alexander Ghionis, and Richard Guthrie

Part III presents an assortment of Julian's work in Chaps. 7 and 8, followed by a bibliography of his chemical and biological warfare (CBW) writings in Chap. 9.[1] These substantive chapters are framed by Peter Pringle's Chap. 6 which reflects on and celebrates Julian's writing style.

We made the selection of papers included in Chaps. 7 and 8 to showcase Julian's diverse interests, style and intellectual contributions. The choice was influenced by several factors. First, we consulted friends, colleagues and associates of Julian, and asked them which papers they thought we should include.[2] These nominations gave

[1] Julian published under different names – including for example Julian Perry Robinson; J. P. Perry Robinson; Julian Robinson – sometimes to facilitate an appropriate ordering of co-authorship.

[2] We are enormously grateful to the people that helped us to select these papers, including Malcolm Dando, Alastair Hay, Kai Ilchmann, Roy MacLeod, Susan Martin, Bob Mathews, Caitríona McLeish, James Revill, Jean Pascal Zanders, and the participants in the Julian Perry Robinson Annual Seminars that took place in November 2022 and November 2023.

H. Wilson (✉)
King's College London, Bristol and London, UK

M. Kaldor
London School of Economics and Political Science (LSE), Brighton and London, UK

J. Kaldor-Robinson
Brighton, UK

P. Pringle
New York, USA

A. Ghionis
Harvard Sussex Program, Science Policy Research Unit (SPRU), University of Sussex, Falmer, Brighton, UK

R. Guthrie
CBW Events, Bradford on Avon, UK

H. Wilson et al. (eds.), *Julian Perry Robinson: Pioneer in the Prohibition of Chemical and Biological Warfare*, Pioneers in Arts, Humanities, Science, Engineering, Practice 32, https://doi.org/10.1007/978-3-032-04064-0_5

us a “longlist” of suggestions, and we used this as the starting point for deciding on a set of papers that provides wide and deep insights into the richness of Julian’s work.

Our selection was also steered by practical constraints. We wanted a collection that was a manageable size – i.e., that could introduce readers to a range of Julian’s writing whilst remaining accessible. At times, this led us to extract passages from longer texts. We also avoided works that were likely to have complicated copyright situations, e.g., publications where it was difficult to find the copyright owner or secure their permission to republish. Luckily, the quantity and diversity of Julian’s writings meant we had many options, including different versions of similar papers, so that papers likely to have copyright issues could be easily substituted for preprints or other alternatives.

Some of the papers made it into the collection because we particularly liked them, some because they had a tangible impact on policy-making and policy-shaping, and some because they demonstrate ways in which Julian’s thinking developed. The selection includes Julian’s testimony to US Congress, as well as his papers and speaking notes for academics, policy-makers and activists – all intended to support the development and implementation of effective policies to prevent CBW through providing robust, fact-based analysis, and communicating honest and expert assessment of contested or ambiguous material.

Chapter 7 presents a selection of Julian’s writing on CBW. The first paper introduces Julian’s thinking about the broad category of CBW and its regulation; after that, articles are presented in chronological order, ranging through analyses of alleged CBW use to obituaries of CBW experts from governmental and non-governmental sectors, and ending with his analysis of chemical weapons use in the Syrian civil war, which he was working on until his death. He co-authored several of the papers we have included here as indicated in the titles and references; we are extremely grateful to these co-authors for allowing us to use these texts. The shorter Chap. 8 presents two examples of Julian’s work on non-CBW matters: first, an extract from his detailed study on the effects of weapons on ecosystems, and second a paper describing conventional weapons’ trends.

Except for correcting obvious typos, we have retained the original spellings and punctuation in the texts and their footnotes, including, for example, by keeping Julian’s preference for spelling the Organisation for the Prohibition of Chemical Weapons with a ‘z’, as this is how it is written in the Chemical Weapons Convention. We have, however, renumbered all the headings and tables to avoid confusion.

For all the breadth of the collection, there are inevitable gaps. People will be able to access additional works within the complete set of Julian’s publications that the Harvard Sussex Program is currently digitizing to make available online.

We hope the following papers – together with the overviews in the Introduction and Part I, as well as Part II’s photographs – will be a stimulating and enjoyable introduction to Julian’s work. We also hope they will guide future scholars, activists and policy-makers in how, and how not, to understand and guard against CBW.

Chapter 6
A Scientist Who Could Write: The Elegant Language of Julian Perry Robinson

Peter Pringle

Elegant is a word often used to describe a ground-breaking scientific experiment, a brilliant insight into a complex question that others had been puzzling over for decades, but elegant is rarely used to describe the way scientists write about their experiments and observations. Typically, scientific papers feature dry, ponderous language, rigidly formatted. There are exceptions, of course, Charles Darwin being one of them, but generally speaking when it comes to the written word scientists are very dull.

Julian's writing is a shining exception. Over the years, he chronicled the minutiae of the scientific and political history of chemical and biological warfare (CBW) – not subjects easily given to elegant interpretations. His papers, reports and essays were sprinkled with original, fluent, carefully crafted, witty insights that are joy to read. Even when the subject was especially complicated, he produced flourishes that endeared and enlightened his readers.

Close followers of his work affectionately refer to these literary nuggets – sometimes phrases, sometimes a paragraph or two – as "Julianisms". They appeared, often unexpectedly, in a wide variety of his work, starting with his early reports of international meetings where he would single out the telling moment others were thinking about but had not recorded, or he would add a defining footnote. Later, as he delved deeper into the politics of CBW, he offered advice on how to untangle the obfuscations of government spin doctors, or expose the falsehoods of a disinformation campaign. Occasionally, he would express the frustration of experts like himself with ill-informed and politically motivated media reports.

He was also not beyond directing his editorial spotlight onto colleagues. In one passage much-admired by his readers, Julian brilliantly analysed a central dilemma

Peter Pringle, Journalist, author, foreign correspondent, New York, United States of America.

P. Pringle (✉)
New York, NY, USA

H. Wilson et al. (eds.), *Julian Perry Robinson: Pioneer in the Prohibition of Chemical and Biological Warfare*, Pioneers in Arts, Humanities, Science, Engineering, Practice 32, https://doi.org/10.1007/978-3-032-04064-0_6

for scientists who advise government on weapons of war: their double loyalty to their science and to their country.

Those literary gems, always scholarly and easily accessible, reflect Julian's endless curiosity, his devotion to humanity, and his insistence on the truth, or as near as he could get to it. They also give a hint of his brief training as a patent lawyer, sometimes sounding, in their judicial fairness, like a legal brief. Here are a few examples.

6.1 On International Meetings and the Telling Footnote

Julian's 1998 description of the first Pugwash[1] meeting in July 1957 began with the agreed details: "Fifteen physicists, two chemists, one biophysicist, a biologist, two physicians and one lawyer… from academic institutions in ten countries (Australia, Austria, Canada, China, France, Japan, Poland, the Soviet Union, the United Kingdom and the United States). The meetings were conducted in English and Russian, the USSR participants being accompanied by an official, who acted as interpreter".

But then Julian added, in a personal footnote, "The late Dr. Vladimir P. Pavlichenko, Assistant General Scientific Secretary of the USSR Academy of Sciences… It was generally assumed throughout the Pugwash movement that Pavlichenko was an officer of the KGB, an assumption of which he must have been aware but did nothing to change".[2]

6.2 On the Issue of Dual-Loyalty of Scientists in War with Reference to the Origins of the Pugwash Movement

"The question remains as to what actually it was, at the level of the individual, that inspired Pugwash and then kept its people participating year after year, in growing numbers. There are, after all no obvious rewards for the foot-soldiers of Pugwash beyond the pleasure of seeing old friends and familiar faces every so often. Perhaps it was the following. Scientists have always had to contend with what has been called their 'double loyalty': a sense of duty not only to their country but also to

[1] The Pugwash Conferences on Science and World Affairs was established to coordinate scientists working across ideological divides to ban nuclear weapons, following the Russell-Einstein Manifesto. See 'About Pugwash', https://pugwash.org/about-pugwash/ [accessed 6 July 2024] and 'Russell-Einstein Manifesto, 1955', Atomic Heritage Foundation, https://www.atomicheritage.org/key-documents/russell-einstein-manifesto [accessed 6 July 2024].

[2] J.P.P. Robinson. 'Contribution of the Pugwash movement of the international regime against chemical and biological weapons'. Pugwash meeting no. 242, Geneva, Switzerland. 28–29 November 1998. P. 6. This is listed in Chap. 9 as item 336 Rev.7.

their science. In some disciplines, this duality could be conflictual, and, during those early years of the 'national security state', it was indeed for more and more people. Loyalty to science is an abstraction not easily described or understandable in its own world. In some scientists it is non-existent, but in others it is passionate, overriding. Maybe it has to do with desire to protect newborn knowledge from deformation, from distraction, from loss, from waste. Maybe it also has to do with belief that science is for the common good. In nuclear physics and the other disciplines that produced the atomic bomb, the double loyalty clearly became a source of much personal stress. What could be more antithetical to belief in science for the good of humanity than use of science to make weapons of mass destruction? Yet what could be more disloyal to one's country than to disregard the military potentials of the atomic nucleus? Some scientists reacted by asserting the primacy of one or the other loyalty, notwithstanding the risk in so doing of becoming branded by their fellows, as some were either as enemy dupes or as bigoted patriots. Most of the rest preferred to ignore their predicament, seeking refuge in the notion that these were 'political matters' which therefore had no place in the laboratory; a notion propagated, one might think, for that very purpose. But there were a few who chose neither the one retreat nor the other: scientists who accepted that, if science were really to be beneficial, the double loyalty could neither be fragmented nor evaded. It had instead to be recognized and somehow accommodated. These were the people who set and kept the Pugwash movement in motion".[3]

6.3 On Western Mainstream Media Reports on Soviet CBW in the Late 1970s

"If popular writing on CBW continues as the domain of buffoons, dupes and hysterics, as is seemingly the case, it still will remain wide open to exploitation by propagandists of all colours. The latter can hardly have failed to play a part in the allegations examined here".[4]

6.4 On the Risk of Publicly Criticizing the Mainstream Media

"Such is the climate of the times that to cast doubt on reports of Soviet malignity is to risk accusations of fellow travelling, or worse. The allegations that have been examined in this paper are entirely in keeping with the image of the USSR that

---

[3] Ibid. p. 5.

[4] 'Chemical and Biological Warfare: Analysis of Recent Reports Concerning the Soviet Union and Vietnam'. March 1980. *ADIU Occasional Paper No 1*. J P Perry Robinson, Science Policy Research Unit, University of Sussex, p. 3. This is listed in Chap. 9 in item 103.

is now being resurrected from the Cold War, and no doubt it is because of this that they are widely assumed to be correct...Yet whatever the prevailing climate of opinion, credulity is a most dangerous factor in questions touching on national security, whether it takes the form of a propensity to believe only the best about potential enemies, or only the worst. Against the dangers of dismissing the allegations out of hand or ignoring them, so preserving a sense of security that may be false, there are the opposite dangers of over-reacting to them, thereby perhaps transforming them into self-fulfilling prophecy. It could be argued that the media have a special responsibility for steering a careful course between these two extremes, and that many of the reports noted in this paper should therefore never have been published without a great deal more substantiation. But information purveyed by the media is as much a reflection of demand as of supply. For an audience that will pay to receive regular frissons of outrage of the awfulness of others or titillating leaks from the more secret corners of government, CBW allegations are an appropriate commodity".[5]

6.5 On What Julian Called the "Aggravated Casualty Data" in the Iran-Iraq War

"No government of a country at war can be expected to be completely candid or completely guileless in its releases of information about the casualties its subjects have suffered at the hands of the enemy, for, to the latter, such weapons-effectiveness data could constitute valuable military intelligence, especially when they relate to individual attacks. That is not to say that all such information that is released is bound to be valueless. On the contrary, there may well be countervailing pressures acting upon the government in favour of candour. Given the political background of the Gulf War, such countervailing practices are likely to have been substantial, though perhaps not always dominant, in the special case of the CW casualties inflicted upon Iran. In the information about them which the Iranian government has chosen to release – and the very-tight news media control exercised by Tehran means that other sources of information are rare, save perhaps the intelligence-gathering assets of other governments – both exaggeration and attenuation have no doubt been interspersed. That, in aggregate, the two may have balanced out is not a completely unreasonable supposition".[6]

[5] Ibid., p. 40.

[6] SIPRI Year Book 1985 includes Julian's Analysis of Iraqi Chemical Warfare Against Iran 1980–84.

6.6 On Omissions and Falsehoods in Government Information and What to Do About Them

Julian analysed the disputed reports of the use of chemical weapons in Syria in 2013, especially in the Ghouta agricultural belt around the eastern and southern periphery. In 2016 he wrote "The available information base is at present too contaminated with error, falsehood and spin to be a dependable and decisive tool. Certainly, we cannot yet say with justifiable confidence that we know what happened at Ghouta and who, then, should be brought to account.

What follows in the present study addresses the difficulty in a systematic fashion, looking for correctives. A palliative for the presence of *error* and *spin* in the information base is the critical opinion of experts, which always needs to be sought. On *omission* and *falsehood* the remedy can only be further research, and it is this on which the present paper is focused".[7]

On omissions:

> "…gaps in public information can be due to the secretive customs and hidden motives of people within the intelligence and security communities, who tend to seem unmindful of any need for countervailing transparency in their professional behaviour. This could well mean that civil society cannot expect to fill any more than the most obvious gaps in public understanding of the Ghouta event".[8]

On falsehoods—or deliberate fabrications:

> "As students of chemical warfare know very well, disinformation – not simply misinformation – can be an obstacle to proper understanding of episodes in which the use of chemical weapons has been alleged. Such is the emotiveness of chemical warfare, especially poison-gas warfare, that such allegations have long contributed to psychological warfare, to 'black' propaganda, and to the vilification of enemies. And in wartime, as Winston Churchill told Joseph Stalin in December 1943, 'Truth is so precious that she should always be attended by a bodyguard of lies'. In a civil war increasingly unconstrained by national borders or by international humanitarian law, the likelihood of fabricated information about the Ghouta even seems peculiarly great".

"There are no rules for detecting disinformation", he writes, but gives some guidelines: "Focus not only on the information but also on the character of the publisher/propagator. In other words, separate the *credibility* of information from the *reliability* of its source.

- Certain writers have reputations as professional purveyors of disinformation.
- Check out each of the otherwise dubious sources that might be purveying disinformation.
- Keep a firm eye nevertheless, on the credibility of the information – its technical, engineering and scientific credibility. There are clear instances in which statements or arguments simple do not stand up to any sort of technical scrutiny.

---

[7] Julian Perry Robinson. 'Gaps, lies and information in the Ghouta event'. March 2016. p. 9. A version of this is listed in Chap. 9 in item 483.

[8] Ibid., pp. 21–3.

- Make the assumption that all sources of information can be manipulated into vehicles of disinformation".[9]

Julian could also turn his editorial spotlight on himself. Appearing as an expert witness at a 1982 US Senate Committee hearing on America's proposed production of a new generation of chemical weapons, Julian, ever self-effacing, introduced himself.

> "My outlook is that of a professional research worker who has been studying chemical warfare matters for some 15 years now, mostly within the European community…I cannot, of course, pretend to portray the debate in its entirety, still less to offer a view that is typical of the 315 million NATO Europeans…The problem is that the number of professional chemical-warfare analysts to whom people in government or politics can turn for informed opinion is rather small, and most of the opinions on offer are likely to be coloured by institutional and other affiliations. The number of real experts on the subject is essentially zero, as there has been no actual experience of poison gas warfare between technologically-advanced belligerents for more than sixty years. The basis of theory underlying current appraisals of the utility of chemical weapons is now too far removed from practice to be much more than conjecture".[10]

Julian's "outlook of a professional researcher", as he modestly described his work to the US Senate Committee, would eventually cover more than half a century of microscopic examination of open sources on CBW and high-level contacts with government scientists and policy-makers. The result is a collection of invaluable and eminently readable guides to interpreting the past and of indispensable tools for understanding the future. He was a real expert.

[9] Ibid, pp. 23–25 (edited).

[10] Testimony of Julian Perry Robinson before US Senate Appropriations Committee, May 5, 1982, p. 7. This is listed in Chap. 9 as item 143.

Matthew Meselson, Julian, and Peter Pringle. From © Mary Kaldor's private collection.

Chapter 7
Selection of Julian's Writings on Chemical and Biological Warfare

Julian Perry Robinson

7.1 Near-Term Development of the Governance Regime for Biological and Chemical Weapons[1]

Editors' note. This paper presents Julian's core thinking on the distinct characteristics of chemical and biological weapons alongside his description of the key elements of the global governance regime against them that were in place at the time he was writing (2007). It also outlines his assessments of the challenges facing this regime and recommends actions that governments, civil society and scholars could take to address these challenges.

This is one of a series of project[2] papers on relationships between non-state actors and technology applicable in chemical/biological weapons (CBW).[3] In presenting a synopsis of CBW, of ways in which these weapons threaten our security, and of

[1] Editors' note. This was first published as Julian P. Perry Robinson. "Near-term development of the governance regime for biological and chemical weapons", 14 September 2006, and is listed in Chap. 9 as item 456. In the original version, Julian included this explanatory note: "Item 456, a 'rolling text' first circulated for private comment on 14 September 2006. An abbreviated version has since been submitted for publication in Nathan Busch & Daniel H. Joyner (editors), *Combating Weapons of Mass Destruction: The Future of International Non-Proliferation Policy*, University of Georgia Press. Another version has been submitted as a discussion paper for the 25th Workshop of the Pugwash Study Group on Implementation of the Chemical and Biological Weapons Conventions, Geneva, Switzerland, 18 19 November 2006". Unless otherwise indicated, the remaining footnotes for this paper are Julian's originals. Apart from correcting obvious typos and renumbering headings and tables in keeping with the requirements of this book, we have retained Julian's formatting and spelling throughout the paper and its notes.

[2] The SPRU project entitled *Non-State Actors and the Globalization of CBW Technologies* that is funded by Phase 2 of the *New Security Challenges* programme of the UK Economic & Social Research Council.

[3] Editors' note. This use of the acronym CBW differs from that in most of the rest of the.

J. P. Robinson (✉)
Brighton, UK

H. Wilson et al. (eds.), *Julian Perry Robinson: Pioneer in the Prohibition of Chemical and Biological Warfare*, Pioneers in Arts, Humanities, Science, Engineering, Practice 32, https://doi.org/10.1007/978-3-032-04064-0_7

international efforts to suppress them, it pays particular regard to one category of non-state actor – elements of civil society active in CBW arms control – and there is also attention to other categories, including terrorists.

The paper starts with an account of what sets CBW apart from other weapons. It then explains the concept of "CBW governance regime" it uses to frame the policy issues, with the main elements of this regime being described one by one in the *Appendix*. Challenges to the regime are then characterized: the emergence of new utilities for CBW; proliferation; and the "creeping legitimization" of certain CBW. Finally, having regard to the imminence of the next review conferences for the 1972 Biological and Toxin Weapons Convention (BWC) and the 1993 Chemical Weapons Convention (CWC),[4] the paper considers possible ways of strengthening the regime, principally measures that would enhance the Organization for the Prohibition of Chemical Weapons (OPCW) and the norm that underpins the two Conventions, such measures including an initiative for the use of international criminal law to build sanctions against non-compliance. The paper closes with a listing of topics for work within the policy-research community that could complement governmental and civil-society action.

7.1.1 The Peculiarities of CBW

Let us be clear, first, about what exactly 'CBW' are here being taken to mean. The defining characteristics are *toxicity* and *infectivity*: CBW are weapons whose intended means for causing harm is either the toxicity of chemicals, including naturally occurring inanimate substances, or the infectivity of pathogenic micro-organisms, including viruses, prions and other such biological agents that are not, in the strictest meaning of the expression, living organisms. CBW are accordingly being seen as 'disease weapons'.[5] There are other categories of chemical and biological agent that, although highly aggressive in their properties, are not useful in either toxic or infective weapons and which are therefore disregarded in the present paper: napalm, white phosphorus, smoke agents, malodorants, high explosives, propellants, bacteria that can damage inanimate materials, and many more besides. These other categories all lie beyond the scope, too, of the CBW disarmament treaties.[6] Article II of the CWC limits the chemicals covered by the treaty to "toxic chemicals and

---

[4] The texts of these two treaties are available in many places, including the website of the Harvard Sussex Program on Chemical and Biological Weapons, www.sussex.ac.uk/Units/spru/hsp, which is cited here because some of the literature referenced later in this paper is posted on it, including the HSP quarterly journal, The CBW Conventions Bulletin.

[5] On the understanding that *disease* is not synonymous with *infectious disease*. Toxicity is a cause of non-infectious disease, as are such other (non-weaponizable) factors as genetic disorder and stress. Some infectious diseases, all of which are caused by the infectivity of pathogens, may be *contagious* in the sense of person-to-person transmissibility.

[6] The assertion here that malodorants and white phosphorus lie outside the scope of the CWC, an assertion that rests on the meaning of "chemical action on life processes", is contentious.

their precursors" qualified by a criterion of general purpose that reads "except when intended for purposes not prohibited under this Convention, as long as the types and quantities are consistent with such purposes".[7] Article II goes on to define a toxic chemical as "Any chemical which through its chemical action on life processes can cause death, temporary incapacitation or permanent harm to humans or animals". The corresponding provision in Article I of the BWC is worded less tightly but also uses a general purpose criterion: it extends to all "Microbial or other biological agents, or toxins whatever their origin or method of production, of types and in quantities that have no justification for prophylactic, protective or other peaceful purposes". There is nothing in BWC Article I about infective agents; the infectivity criterion that the present paper is using to characterize biological weapons (BW), analogous to the toxicity criterion of the CWC, follows instead from the *travaux préparatoires* of the BWC and from three authoritative international texts on CBW – the 1969 report of the UN Secretary-General,[8] which heralded the BWC negotiation,[9] and those of the World Health Organization initially produced as input into the Secretary-General's report,[10] later rewritten for the post-Cold-War post-Nine-Eleven situation.[11]

Readers needing detailed descriptive information about CBW that is both reliable and independent of individual governments and sufficiently extensive to go to the heart of the problems that the weapons present for security – international, national

[7] The "purposes not prohibited under this Convention" are defined later in Article II: "(a) Industrial, agricultural, research, medical, pharmaceutical or other peaceful purposes; (b) Protective purposes, namely those purposes directly related to protection against toxic chemicals and to protection against chemical weapons; (c) Military purposes not connected with the use of chemical weapons and not dependent on the use of the toxic properties of chemicals as a method of warfare; (d) Law enforcement including domestic riot control purposes."

[8] *Chemical and bacteriological (biological) weapons and the effects of their possible use: report of the Secretary-General.* New York: United Nations, 1969.

[9] The immediate impetus for the BWC had come from the UK where, in the wake of the NPT negotiation, the Labour government was looking for a disarmament initiative to propose on the international stage. At an August 1968 meeting of the Eighteen Nation Disarmament Committee in Geneva, in the context of an ongoing inter-governmental debate on strengthening the 1925 Geneva Protocol that had been initiated in the UN General Assembly some two years previously under the stimulus of the Vietnam War, the UK suggested "early conclusion of a new Convention for the Prohibition of Microbiological Methods of Warfare". A year later – immediately after the UN Secretary-General had published his report on CBW – the UK then tabled what became the initial draft of the BWC, thereby separating BW out of what had hitherto been a CBW debate. Such separation was at first welcomed only within the Western Group, but in March 1971 the USSR unexpectedly accepted the idea of a separate BWC, for which it tabled its own draft. The draft was similar to the UK draft, but lacked the latter's prohibition of BW research and of BW use, which included provision for investigation of use-allegations by the UN Secretary-General. The USA at that time was keen to improve the international climate by means of an agreement ahead of President Nixon's visit to Moscow in May 1972. It decided not to push the USSR on the divergences between the UK and Soviet drafts, and in August of that year the USA and the USSR tabled identical drafts for what speedily became the BWC.

[10] *Health aspects of chemical and biological weapons: report of a WHO group of consultants.* Geneva: World Health Organization, 1970.

[11] *Public health response to biological and chemical weapons: WHO guidance.* Geneva: World Health Organization, 2004.

and human security – are well advised to turn first to these three texts. The 2004 WHO volume, especially, allows one to understand what makes CBW so peculiar today and why the security and health impacts of CBW armament could be so heavy. Three features stand out in particular.

First and foremost, CBW may resemble other categories of weapon in that they can attack life, killing their victims no less dead than can bullets or bayonets, but they may also be targeted to disrupt individual processes that contribute to life, which other weapons cannot do save by accident, not design. The nerve gases, for example, target nerve-signal transmission; the blood gases, cellular respiration mediated through the blood. Advances in the life sciences, coupled with those allied technologies that allow the analysis and construction of complex biologically active molecules, could eventually make it possible to design a CBW agent that will interfere with *any* life process that can be understood in molecular terms, whether it be the process of development, inheritance, reproduction, locomotion, sensation, cognition or indeed any other process that keeps us functioning properly, according to expectations. The potential is there, inasmuch as it has not materialized already, for inducing many different forms of malfunction, maybe even ones that discriminate between ethnic groups of human beings. It is this potential for manipulating at will our very humanity, in pursuit of who-knows-what strategy of adversary subjugation, repression or coercion, that makes CBW especially menacing.[12] Science has a way to go yet before the full horrors can be upon us, but the writing is clearly on the wall. For example, the most obvious motivation that can be imputed to Iraq's programme in the 1980s to weaponize aflatoxin[13] is a desire to harm through latent liver cancer a subsequent generation of anticipated adversaries. As Dutch UNSCOM [United Nations Special Commission] commissioner Jack Ooms put it, the objective must have been a weapon of delayed genocide.[14] Artillery shells charged with aflatoxin are said to have been used during the suppression of the Shi'ite uprising in March 1991.[15] This has not been confirmed.

The second outstanding characteristic is that, as destroyers of life on a large scale, some CBW can rightly be called 'weapons of mass destruction' (WMD). When first used, on 22 and 23 April 1915, the device that brought chemical warfare out from its prehistory, namely massed cylinders of liquefied chlorine gas that could simultaneously be opened into the wind, reportedly asphyxiated five thousand French and Canadian troops at Ypres in Belgium and harmed a further 15,000.[16] Similar numbers of people are said to have fallen victim to Iraqi mustard and nerve gas in

---

[12] See, further, Matthew Meselson, "Averting the hostile exploitation of biotechnology", *The CBW Conventions Bulletin* no 48 (June 2000) pp 16–19.

[13] See the entry for 25 August 1995 in *The CBW Conventions Bulletin* no 30 (December 1995) p 17.

[14] See the entry for 6 May 1997 in *The CBW Conventions Bulletin* no 37 (September 1997) p 16.

[15] See the entries for 7 and 7–20 March 1991 and 12 March 2001 in *Chemical Weapons Convention Bulletin* (later *The CBW Conventions Bulletin*) no 12 (June 1991) pp 13–14.

[16] But the records are poor and these commonly and widely quoted numbers have been described as far too large, initially only by German commentators. For a recent account, see Gordon Corrigan, *Mud, Blood and Poppycock: Britain and the First World War* (London: Cassell, 2004) at pp 164–65,

the Kurdish town of Halabja after chemical air raids during 16–18 March 1988.[17] For biological weapons there has been no comparable experience, but during 1964–68 the United States conducted unprecedentedly large field-trials over open sea of aircraft germ weapons, each of which was found capable of laying down a cross-wind line source of pathogenic microbial aerosol tens of kilometres long that was able to infect experimental animals at sea level over a distance of several tens of kilometres downwind.[18] That could translate into an infective hazard to every person living within an area on the order of thousands, even tens of thousands, of square kilometres. In other words, it appeared from experimental data that some biological weapons would be capable of producing effects comparable in their magnitude to the life-destroying potential of nuclear weapons. Within NATO at that time, defence scientists were also anticipating a new generation of *chemical* weapons having comparable area-effectiveness.[19] Yet the actual historical record of CBW employment is not at all dominated by episodes of mass destruction (though in fact for any sort of biological warfare the record is so sparse that adequate definition of the BW threat requires quite other considerations[20]). Published military doctrine shows that most of the military and other utilities for which user-services once valued possession of CBW depended on aggressive properties other than mass killing – such as ability to deny terrain or materiel through contamination (this creating an overt *threat* of mass casualties), or ability to terrorize through fear of imperceptible danger, or ability to degrade the combat efficiency of enemy forces by obliging them to assume the encumbrance of anti-CBW protection, or ability to disconcert opposing forces or hostile mobs through a toxicity that inflicts immediate pain or disablement with relatively little risk of death. One may view the available target effects of CBW as lying along spectra that have highly localized, say, or low-casualty effects at one end and large-area or mass-casualty effects at the other. Where along such a spectrum a given chemical

---

whose view is that, in reality, the chlorine casualties probably numbered around 1500, with 200 deaths.

[17] See the entries for 18 March 1988 and 11 July 2002 in *Chemical Weapons Convention Bulletin* (later *The CBW Conventions Bulletin*), no 1 (Summer 1988) p 11, and no 57 (September 2002) p 55.

[18] Citing the original field-trial report, Ed Regis, in his *The Biology of Doom* (New York, 1999), writes that in 1964, during Operation Shady Grove off Johnston Atoll in the Pacific, Marine Corps *Skyhawks* laid down demonstrably infective clouds of tularaemia and Q-fever bacteria over a sea-level area of nearly 5000 km^2 per aircraft.

[19] North Atlantic Treaty Organization, Standing Group, von Kármán Committee. *Future developments in chemical warfare*, from the report of Working Group X on Chemical, Biological and Radiological Defence, March 1961, as distributed to the UK Ministry of Defence Advisory Council on Scientific Research and Technical Development, paper no SAC 1928, 11 February 1969, in The National Archive, Kew, file WO195/16864. Regis, *op. cit.*, citing another US field-trial report, writes that in 1968, during sea trials off Eniwetok Atoll, Air Force *Phantoms* discharging dry-agent spraytanks loaded with staphylococcal enterotoxin B over caged monkeys and other animals were estimated to have secured 30 percent casualty rates over 2400 square kilometres per aircraft.

[20] On which see, especially, Caitríona McLeish *Accommodating bio-disarmament to bio-technological change: the issue of dual use*, D. Phil dissertation, University of Sussex, December 2002, pp 72–125.

or biological weapon would manifest its effects is determined by the characteristics of the toxic/infective agent being used (such as the contagiousness of any disease it can cause) and the manner of its use, and by the vulnerability of the threatened population, this reflecting such factors as the health status of the population and degree of preparedness for protecting itself against disseminated agent. It remains the case today that, in the design of CBW, increasingly severe technological constraint sets in as the mass-destruction end of the spectrum is approached: the greater and more assured the area-effectiveness sought for the weapon, the greater the practical difficulties of achieving it. This is why the notion of mass-destruction terrorism using CBW is less plausible than its portrayals have often suggested. It is one of the reasons why a recent international study of the vulnerability of EU societies to radiological and CBW terrorism has concluded that any such acts of terrorism today or in the immediate future are unlikely to achieve more than localized nuisance.[21] That said, however, risk-assessment in this area remains very poorly developed indeed, at least in public, and much more research is surely merited, given the magnitude of conceivable consequences.

The third outstanding characteristic of CBW is the existence of a uniquely wide array of societal constraints on CBW armament. This array has been building up over time in the law and custom of nations, so that damaging opprobrium is likely to fasten on any future users of CBW. Only nuclear weapons share this feature, but they have not yet attracted anything approaching the proscription of CBW. Some people even speak with pride about their country's nuclear weapons.

From this third characteristic of CBW there stem elements of state practice and widely accepted law tending to suppress biological and chemical warfare. This is the "CBW governance regime" referred to earlier: an accretion of norms, rules and procedures, both national and international, the whole constituting an entity ('regime') that individual states, through their various internal political processes, may judge themselves to be better off inside than outside.[22] Its emergence can be explained in different ways. One of them is in terms of an ancient cross-cultural taboo, evident in different literatures over the millennia,[23] against weapons that exploit disease. One may well imagine this taboo having become transformed into a norm of international behaviour, providing a basis for the further development of international law that is uncommonly strong in this peculiar field of armament. Weapons of any

[21] See the *Interim Report* dated 17 October 2006 from European Commission Framework 6 project 502476, *Assessment of the vulnerabilities of modern societies to terrorist acts employing radiological, biological or chemical agents with the view to assist in developing preventive and suppressive crisis management strategies* (ASSRBCVUL). Some particulars are posted on the European Union website at http://ec.europa.eu/research/fp6/ssp/assrbcvul_en.htm.

[22] This particular concept of 'regime' is set out in Julian Perry Robinson, "Chemical warfare arms control: a framework for considering policy alternatives", *SIPRI Chemical & Biological Warfare Studies* no 2 (1985). Editors' note. This paper is listed in Chap. 9 as item 181.

[23] A powerful impression of the diversity and antiquity of the codes of professional military behaviour that forbid the weaponization of disease, including poison and infectious disease, can be gained from M. A. Marin, "The evolution and present status of the Laws of War", *Academie de Droit International: Receuil des Cours* 92(2):633–749, 1957. His examples are from the Manu laws of India, Roman law and early commentaries on the Koran, as well as from more recent texts.

type are designed to harm, and there is no obvious reason for regarding one type of weaponized harm as more (or less) reprehensible than another. Is it worse to be the victim of unnatural disease than, say, to be shattered by shell fragments? One may ask this and then be surprised that so strong an obloquy should nevertheless have attached to disease weapons and not to other weapons. There is an irrationality here whose very strength and depth is characteristic of taboo.[24] It is surely a phenomenon that people engaged in CBW policy, whether as practitioners or as researchers, should study more systematically than has yet been evident.

7.1.2 Challenges to the CBW Governance Regime

The formal elements of today's CBW governance regime comprise multilateral, plurilateral and unilateral instruments. Its ten principal components, identified in the table above (Table 7.1), are described in turn in the *Appendix* to this paper. The underlying international norm is CBW non-armament and, in the case of former CBW possessor states, CBW disarmament. A principle of the regime is that it be non-discriminatory, making no distinction between the 'haves' and 'have-nots' among its member states.

On the chemical side, a respected international organization, the OPCW, now exists that oversees CWC implementation and compliance procedures – hampered though it is by some remaining gaps in its legal foundation and by the ever-present possibility of governments (or their voters) losing awareness or appreciation of the OPCW's *raison d'être*.[25] On the biological side, in contrast, there is no such implementing organization to put into effect the terms of the BWC, which on its face is not much more than a statement of the norm plus a few norm-promoting obligations that states parties have not always observed. To say this is not at all to disparage the BWC or to suggest that the biological side of the regime is feeble; on the contrary, the norm

[24] A biological explanation for the taboo was proposed by Michael Mandelbaum in his *The Nuclear Revolution: International Politics Before and After Hiroshima* (Cambridge University Press, 1981), at pp 29–40. This explanation has been disputed, notably by Richard Price, *The Chemical Weapons Taboo* (Cornell University Press, 1997). He and other commentators have looked, for example, to discourse theory in preference to concepts of taboo; see especially Jean Pascal Zanders, "International norms against chemical and biological warfare: an ambiguous legacy", *Journal of Conflict and Security Law* vol 8 (2003) pp 391–410. Zanders does not dispute the existence of a taboo but questions the universality sometimes ascribed to it, both in space and in time.

[25] For a detailed account of the formation of the OPCW and of how the first eight years of its existence proceeded, see the quarterly reviews prepared by the HSP Researcher seconded to OPCW Headquarters and published as "Progress in The Hague" from issue 19 (March 1993) of *Chemical Weapons Convention Bulletin* through issue 67 (March 2005) of *The CBW Conventions Bulletin*. HSP, in the persons of Ian Kenyon (who, as Chief Executive Officer of the OPCW Provisional Technical Secretariat, was in charge of transforming the CWC treaty text into a functioning intergovernmental organization) and Daniel Feakes, is currently producing an edited volume on the Preparatory Commission and the early years of the OPCW, with contributions from other main actors.

Table 7.1 Chief components of today's CBW governance regime[26]

Since 1928	The 1925 Geneva Protocol
Since 1975	The 1972 Biological and Toxin Weapons Convention
Since 1978	The 1977 Environmental Modification Convention
Since 1985	The Australia Group
Since 1987	New wave of unilateral renunciations of CW and/or BW (Romania, Iraq, Russia, South Africa and Libya)
Since 1987	Re-empowerment of the UN Secretary-General to investigate reports of CBW use
Since 1997	The 1993 Chemical Weapons Convention
Since 2002	The Global Partnership against the Spread of Weapons and Materials of Mass Destruction
Since 2003	The Proliferation Security Initiative
Since 2004	UN Security Council resolution 1540 on non- proliferation of WMD

has clearly drawn much strength from state practice and the support of internationally organised civil society.[27] As Ambassador Donald Mahley, former chief US CBW negotiator, said in Geneva during his valedictory statement on 28 April 2006, civilization, norm and law are now such that there would be a massive reaction against the use of biological weapons by anyone, whether state or terrorist, in the future.[28] But the BWC regime suffers, as its critics rightly observe, from an 'institutional deficit' – the absence of an 'OPBW'[29] or anything like it charged with those routine activities of information exchange, assistance networking and shared international procedure that

[26] Editors' note. This table was not numbered in the original.

[27] On which note, especially, the work of Pugwash and its various international CBW 'study groups' over the past 45 years, which is reviewed in J. P. Perry Robinson, "The impact of Pugwash on the debates over chemical and biological weapons" in Allison L. C. de Cerreño and Alexander Keynan, editors, *Scientific cooperation, state conflict: the role of scientists in mitigating international discord* (New York: NY Academy of Sciences, 1998). Note also the CBW-related work of the International Network of Engineers and Scientists for Global Responsibility (INES), www.mindspring.com/lines/default.htm, and of the BioWeapons Prevention Project, www.bwpp.org. See, further, Daniel Feakes, "Global civil society and biological and chemical weapons", in Mary Kaldor, Helmut Anheier and Marlies Glasius, editors, *Global Civil Society 2003* (Oxford University Press, 2003), pp 87–117.

[28] Graham Pearson, "Report from Geneva: the Preparatory Committee for the Sixth BWC Review Conference", *The CBW Conventions Bulletin* no 71 (May 2006) pp 6–15, at p 14.

[29] Editors' note. This refers to the absence of a body attached to the BWC and equivalent to the Organisation for the Prohibition of Chemical Weapons (OPCW).

would entrench responsibility for the norm and the obligations within the bureaucracies and legislatures of BWC states parties. Supranational threats to security, such as these weapons pose, require global solutions involving active international cooperation, as the UK government observed in 2002 in its Green Paper on strengthening the BWC.[30] Nothing less can suffice. Surprisingly successful though the regime has been, it still needs enlightened policy-making for its further development, not least in view of the challenges that are now emerging.

A development or change that causes a state to question its continuing allegiance to the CBW governance regime is by definition, a challenge to that regime. The challenge may be to the regime as a whole or to particular parts of it. If major or many states start such questioning, the challenge is serious, requiring a collective response if the regime is to remain in good order, properly adapted to its ever-changing environment of international relations. For each state party the constant question is whether the benefits flowing from the regime continue to outweigh the attendant costs and also any penalties there may be to the national interest – are we still better off inside the regime than outside it? With that as a framing concept, particular challenges can be identified. Three big ones stand out at present: the emergence of new utilities for CBW; proliferation in a variety of forms; and the creeping legitimization of CBW that are not regarded as WMD.

7.1.2.1 New Utilities for CBW

Disarmament, especially WMD disarmament, is an objective widely seen as beneficial, but armament also can bring benefit, by preserving security. Under some circumstances that could include CBW armament. Military options forgone through renunciation of CBW capabilities might then be significant on the cost side of remaining within the CBW governance regime. The taboo seemingly associated with disease weapons appears to mean that most states are content with CBW disarmament – except that circumstances may now be creating utilities for CBW not previously considered.

At least three types of new utility can now be discerned, and examples of all three seem evident in recent conflicts or in preparedness for them. The first is a consequence of wider changes in the nature of warfare, rather as the shift from 'massive retaliation' to 'limited war' doctrine towards the end of the 1950s elevated the status of CBW in Western military thinking, causing new utilities to be seen for the weapons, especially in Third World settings. Today, a new type of organised violence is taking the place of, for example, those confrontations between highly disciplined and technologically advanced armed forces that characterized the later Cold War. Conflicts these past two decades in the Balkans, the Caucasus, the horn of Africa, Rwanda, Liberia, Sierra Leone, Angola, Afghanistan and Iraq have eroded formerly clear distinctions between war, organized crime and large-scale violation of human rights. These new wars are fought by seeking political control through

[30] *Countering the Threat from Biological Weapons*, Cm 5484, released 29 April 2002.

the displacement or worse of civilian populations and through the sowing of fear and hatred.[31] Because CBW can lend themselves particularly effectively to such objectives, they may have a greater affinity to the new wars than they did to the old, so, notwithstanding the CBW governance regime, they may have an expanding future.[32] This is additional reason why the latest CBW-use allegations, emanating from Sudanese, Israeli, Palestinian, Baluchi and Lebanese sources, should not remain uninvestigated.

The CWC, though not the BWC, provides for a compliance-verification system run by an intergovernmental organization including an international inspectorate that ought in principle to countervail the new-utility challenge. But the routines of that system were designed against Cold-War-period conceptions of utility, meaning that the lists of chemicals and types of industrial facility that the OPCW now has under its immediate surveillance are dictated by the types of chemical weapon that fitted old-war, not new-war, requirements. Basically that meant focussing on toxic chemicals that were so intensely aggressive in their effects that weapons disseminating them would be competitive, in casualty-producing terms, with conventional weapons. Not a great many such toxicants exist,[33] so their coverage in the CWC schedules that govern routine OPCW verification allowed people to suppose that the main threats had thereby been brought under control. In the new wars, however, it is not so much relative aggressivity that determines the utility of CW but rather such other factors as accessibility or availability and terrorizing potential. A whole host of toxic industrial chemicals (TICs) and other chemicals not hitherto regarded as CW agents might thus find application in new-war contexts. The fact that most of these chemicals are not listed in the CWC control schedules does not mean that their use for CW purposes is permitted, nor that the CWC is unavailing against them. It means only that the routine international verification procedures run by the OPCW may not see them. The general purpose criterion that the CWC uses to define its scope obliges the national authorities of states parties to ensure that no toxic chemical is abused.[34] The

---

[31] See especially Mary Kaldor, *New & Old Wars: Organized Violence in a Global Era* (second edition, Cambridge: Polity, 2006).

[32] This future seems already to have begun. Instances of 'new' chemical warfare include episodes in Iraqi Kurdistan, in southern Africa, in Bosnia and perhaps in Chechnya; see J. P. Perry Robinson, "The General Purpose Criterion and the new utility of toxicants as weapons", unpublished paper presented at the 15th workshop of the Pugwash Study Group on Implementation of the CBW Conventions, Oegstgeest, the Netherlands, 23–24 June 2001. Editors' note. This paper is listed in Chap. 9 as item 396.

[33] For discussion of this point, see J. P. Perry Robinson, "The chemical industry and chemical warfare disarmament: Categorizing chemicals for the purposes of the projected Chemical Weapons Convention", *SIPRI Chemical & Biological Warfare Studies* no 4 (1986) pp 55–104. Editors' note. This paper is listed in Chap. 9 as item 196.

[34] CWC Article VI.2 opens thus: "Each State Party shall adopt the necessary measures to ensure that toxic chemicals and their precursors are only developed, produced, otherwise acquired, retained, transferred, or used within its territory or in any other place under its jurisdiction or control for purposes not prohibited under this Convention." Known as the 'Molander chapeau', this obligation prefaces the main provisions of the CWC for its industry control regime.

challenge to the regime therefore lies in the degree to which such national controls may fail to exert a constraining effect.

A second major source of new utility for CBW is the propensity of knowledge newly gained in the life sciences for suggesting novel modes of attack that could be the basis for militarily or politically attractive new forms of weapon. For example, if a hitherto unknown molecular pathway serving a process of life comes to be identified, chemical or biological agents capable of interfering with that pathway might also become identifiable and thus form the basis for a novel weapon. Of course many considerations other than novelty of effect determine the usefulness of a new weapon, so the new science is not itself the challenge to the regime that is here suggested. But it would be a step towards it; and many such can be envisaged.[35] This prospect is not necessarily remote. We should not, for example, disregard the recent statement reliably attributed to a "former high-level Defense Department official" commenting on the feasibility of US attack on Iranian underground facilities: "We can do things on the ground, too, but it's difficult and very dangerous – put bad stuff in ventilator shafts and put them to sleep".[36] Again, it is the general purpose criterion as used in the BWC and in the CWC that is the safeguard provided by the CBW governance regime against this challenge. But it is a safeguard only if it can be activated, and this requires continual monitoring of scientific and technological change for any new development that might challenge the regime. This is a task, it should be noted here, that cannot reliably be left to security authorities alone, simply because their surveillance of new science will always be insufficient: the scientific community at large must also contribute.

A third type of novel utility now becoming manifest is the emerging role of CBW, not in the hands of terrorists or other new-war aggressors as in the first novel utility, but for purposes of counter-terrorism. This utility has demonstrably become a stimulus to rich-country questioning of the regime. It is rooted in past counter-insurgency applications of toxic chemicals, which reach back through the Vietnam War to British, French, Italian and Spanish use of toxic chemicals in colonial situations – a utility that the CWC was intended to suppress. Its re-emergence in counter-terrorist guise is perhaps to be seen in the proliferation of weapons based on Agent CR, evident each year in that part of the OPCW Annual Report addressing declarations of 'riot control agents', for the potency and other properties of CR have caused it to be widely rejected as suited to civil police use. The utility is also evident in the vigorous advocacy to be heard in some quarters for the arming of counterterrorist forces with more advanced types of 'non lethal' CBW. The readiness with which the US Marine

[35] A particularly rich recent source of information on advances in technology that may be applicable to CBW is the so-called 'Fink follow-on' report: Institute of Medicine and National Research Council of the US National Academies, Committee on Advances in Technology and the Prevention of Their Application to Next Generation Biowarfare Threats (Co-Chairs: Stanley M Lemon and David A Relman), *Globalization, Biosecurity, and the Future of the Life Sciences*, Washington, DC: The National Academies Press, 2006.

[36] Seymour M Hersh, "The Iran plans: would President Bush go to war to stop Tehran from getting the bomb?", *New Yorker*, 17 April 2006. It is not clear whether it was a literal or a euphemistic 'sleep' that was meant.

Corps has taken to toxin weapons of this type – devices disseminating Agent OC[37] – is further indication. So is the absence of any serious criticism of the Russian government for having authorized use of toxic chemicals other than riot-control agents by the *spetznaz* forces that, on 26 October 2002, liberated 634 of the theatre-goers taken hostage by Chechen separatists in Moscow. The other 129 hostages were killed by the toxicant used, which is said to have been an agent "based on derivatives of fentanyl" that had been developed by USSR special services.[38] Comparable in some respects, if very different in others, is a counterterrorist utility for CBW that Israel, for example, has demonstrated in its espousal, following the Munich Olympics outrage in August 1972, of assassination as a major tool in counterterrorism.[39]

7.1.2.2 Proliferation of CBW

Nowadays when people speak of CBW proliferation it is not always clear what they are talking about. The concerns expressed most commonly are probably about clandestine acquisition of CBW by non-state entities, such concerns having supplanted or perhaps augmented earlier concerns about 'rogue states'. The characteristics that have caused states to be described as rogues include their supposed interest in acquiring CBW, so the concern could have been at root a questioning of the efficacy of the CBW governance regime, in turn becoming stimulus to the various plurilateral initiatives that have since contributed to the regime. Latterly these initiatives have been directed also against non-state entities. Such entities include business corporations heedlessly serving a lucrative marketplace, criminal organizations feeding a black market, and terrorist groups seeking new weapons.

Proliferation of CBW among terrorists is, for the largely technological reasons addressed earlier, probably myth. While it is known that certain terrorist groups have indeed looked at CBW options, such intent as they have had to acquire CBW has not been translated into evident capability. Thus far, other means for terrorist violence have proved more attractive or more accessible to them and, since the aberration of Aum Shinrikyo in the 1990s, only the most footling terrorist attempts to acquire CBW have been observed.[40] This is not to say that it could not happen: the lesson to draw from the still-unresolved 'anthrax letters' affair in the USA during late 2001 is that biological weapons can put potential for great harm into the hands of

[37] See the entry for 22 March 1996 in *The CBW Conventions Bulletin* no 32 (June 1996) p 27.

[38] See the entry for 26 October 2002 in J. P. Perry Robinson, "Disabling Chemical Weapons: a Documented Chronology of Events", 1 November 2003, pp 139–42.

[39] See Aaron J. Klein, *Striking Back: The 1972 Munich Olympics Massacre and Israel's Deadly Response* (New York: Random House, 2005) pp 104–11 and, describing the CBW assassination of Wadi Haddad in 1978, pp 205–8. On the attempted assassination of Khalid Mish'al, see the entries for 25 September 1997 and 19 February 1998 in *The CBW Conventions Bulletin* no 38 (December 1997) p 29 and no 40 (June 1998) p 23.

[40] For an informed review of the evidence of interest in CBW by actual terrorist groups today, see Milton Leitenberg, *Assessing the Biological Weapons and Bioterrorism Threat,* US Army War College Strategic Studies Institute monograph, December 2005, at pp 21–42.

technically competent and skilled individuals.[41] A similar lesson can also be drawn from two other instances in which pathogenic microbes were deliberately released for purposes, not of terrorism (as normally understood), but for criminal or other purposes: in the Mato Grosso of Brazil during 1957–63, when smallpox and other diseases were introduced into Indian tribal populations so as to facilitate large-scale takeovers of land; and in a small town in Oregon, USA, in 1984, when Rajneeshee cultists sought to influence local elections by infecting voters with salmonellosis through contamination of salad bars in neighbourhood restaurants. The CWC and BWC in fact place useful tools at the disposal of guardians against such acts, for they oblige states parties to introduce and enforce effective controls on access and use of toxic and infective materials by all persons within their jurisdictions. Whether the CBW governance regime is more or less applicable in counter-terrorism than it is in counter-proliferation is now a matter for particular inquiry by security scholars. Yet it still seems to be at the state level that CBW proliferation is, on the available evidence, a more serious challenge to the regime.

Here it is necessary to differentiate 'vertical' and 'horizontal' proliferation. Vertical proliferation describes the process whereby states possessing CBW weapons expand or upgrade existing CBW capabilities. It thus refers solely to states that have opted to remain outside the regime and, for those that have done so in order to preserve CBW capability (as opposed to lack of interest or even awareness), the challenge to the regime is serious on several counts. Above all, it strengthens pro-CBW constituencies in both national and bureaucratic politics, thereby increasing the political costs to the country's leadership if it were to advocate joining the regime. Most probably this process is operating in the treaty-holdout states[42] of Egypt, Israel and Syria, where it has been greatly complicated by nuclear weapons. Syria has been uniquely forthcoming about its avowed CW capability, starting in November 1996 when its ambassador to Egypt told a lecture audience in Alexandria that Syria would retaliate with chemical weapons against Israeli nuclear attack or threat[43]; and President Bashar Assad has spoken recently about Syrian rights to possession of CBW as safeguard against "Israeli aggression".[44] Nor have Israeli officials been reticent about their view of Syrian CW[45] and, by implication, the counter-CW value of Israeli nuclear weapons.

Horizontal CBW proliferation presents an altogether different kind of challenge to the CBW governance regime, one that is rooted in the otherwise mostly beneficial tendency of industrial and other technologies to diffuse rapidly around the world

[41] See, further, Martin Rees, *Our Final Century: Will the Human Race Survive the Twenty-first Century?* (London: Heinemann, 2003); and Milton Leitenberg, *The Problem of Biological Weapons* (Stockholm: The Swedish National Defence College, 2004), pp 137–55.

[42] See *Appendix* below.

[43] See the entry for 25 November 1996 in *The CBW Conventions Bulletin* no 35 (March 1997) p 24.

[44] See the entry for 6 January 2004 in *The CBW Conventions Bulletin* no 63 (March 2004) p 37.

[45] See the entries for 29 April 1997, 27 May 1997, 19 June 1997, 22 June 1998, 26 November 2001, 29 July 2003 and 25 July 2004 in *The CBW Conventions Bulletin* no 36 (June 1997) p 31, no 37 (September 1997) p 19 and p 23, no 41 (September 1998) p 30, no 55 (March 2002) p 18, no 61 (September 2003) p 40, and no 65 (September 2004) p 48.

under the various influences of globalization. The capability of individual states to acquire CBW weapons, if they so choose, is thereby enhanced, and, if they still need specialized assistance, clandestine procurement networks have now gained increasingly dense cover within which to operate. That networks of this type can indeed spring up to meet demand was clearly shown by the UNSCOM/UNMOVIC [United Nations Special Commission/ United Nations Monitoring, Verification and Inspection Commission] investigation of Iraqi CBW acquisition[46] and by the Libyan CW programme. At the level of know-how and other intangible technology, comparable assistance is increasingly available from another quarter as well. A great proliferation of competence in technological matters relating to CBW is currently underway in the United States as thousands of millions of dollars are dedicated annually to countermeasures against bioterrorism,[47] a competence that may by now surpass whatever still remains available from the Soviet biological-weapons programme. Even states that have no immediate wish to acquire CBW may nevertheless move to take advantage of these various possibilities as a hedge against circumstances changing – by, for example, building what could serve as 'break-out capacity' into their industrial infrastructure rather as the USSR created 'mobilization capability' for manufacturing BW within its biotech industry during 1973–92. Iran, victim on a terrible scale to Iraqi chemical weapons during 1983–88, very probably falls into this category[48] while at the same time being amongst the most vocal supporters of the CBW governance regime and, especially on the medical side, a proactive participant in its international procedures. Nor is there any clear impropriety in such a position, for all the industrial powers, including ones that menace Iran, have manufacturing industries to which they could turn at short notice for CBW agents (whose full weaponization would, however, be more demanding). The fact nevertheless remains that horizontal proliferation of this type is a threat to confidence in the regime and therefore a serious challenge to it for as long as the problem of 'dual use' persists. If this challenge is to be contained, it must first be understood, and this requires a continuing policy-research effort directed at the nature, processes and drivers of the diffusion of 'dual use' technologies applicable to CBW.[49]

---

[46] On UNSCOM investigations, see especially: Tim Trevan, *Saddam's Secrets: The Hunt for Iraq's Hidden Weapons* (London: HarperCollins, 1999); Graham Pearson, *The UNSCOM Saga: Chemical and Biological Weapons Non-Proliferation* (London: Macmillan, 1999) and *The Search for Iraq's Weapons of Mass Destruction: Inspection, Verification and Non-Proliferation* (Basingstoke and New York: Palgrave Macmillan, 2005); and Scott Ritter, *Iraq Confidential* (New York: Nation Books, 2005)

[47] For detailed expenditure figures, see Stephanie Chang and Alan Pearson, *Federal funding for biological weapons prevention and defense, Fiscal Years 2001 to 2007*, Washington, DC: The Center for Arms Control and Non-Proliferation, June 2006.

[48] See J. P. Perry Robinson, "Dual technology and perceptions of Iranian chemical and biological weapons", draft of 20 July 2005. Editors' note. This paper is listed in Chap. 9 as item 450]; and International Institute for Strategic Studies, IISS Strategic Dossier. *Iran's Strategic Weapons Programmes: A Net Assessment* (London: Routledge, 2005)

[49] For recent research see: Caitríona McLeish, "A background note on dual use technologies: can technology studies inform the dual use debate?", and Emmanuelle Tuerlings and Caitríona McLeish, "Is risk assessment a useful method to govern dual use research?", papers presented at the 21st

7.1.2.3 Creeping Legitimization of Certain CBW

In the ability of CBW agents to target themselves on particular life processes, there is growing scope to 'tailor' the nature or severity of their effects to a strategic or tactical objective. In that such tailoring could, as has been seen, open the way to weapons suited to hugely malign purpose, CBW present a long-term danger that demands an alert and strong CBW governance regime. That same tailoring can, however, provide weapons of an altogether more acceptable character, including ones having effects gentler than most other means of violence. Examples include the 'tear gas' of police forces; the psychochemical weapons that, according to past US Army teaching,[50] would cause the enemy to "linger in overpowering reverie"; and the entirely mythical knock-out agents of "war without death" that have figured in science fiction since the Nineteenth Century. Add to these chemicals the various infective agents that can induce highly debilitating diseases of low mortality, and a category of CBW is created whose features seem far distant from those of WMD, whose possession may appear desirable, and whose constraint by the regime may therefore come to seem a liability, notwithstanding the abyss into which the tailoring could also take us.

A rather wide variety of commercial, political and military interests stand to benefit from exclusion of some or all of these non-WMD CBW from the governance regime. *Sub rosa* campaigning to that end has long been under way, most notably during the last months of the CWC negotiation in mid-1992, when the protagonists of what was then starting to be called 'Non Lethal Warfare' (NLW) came up against governmental officials charged with securing consensus on those parts of the CWC text that dealt with 'riot control agents' (RCA).[51] The issue turned then on whether RCA fell within the definition of 'toxic chemicals', subject, thereby, to the general purpose criterion that would serve to regulate the duality of their application either in warfare (prohibited) or in law-enforcement (permitted). For reasons that remain unclear to this day, the USA favoured exclusion but, finding itself isolated in this position within the Western Group, secured a compromise[52] in which the CWC expressly prohibited use of RCA "as a method of warfare" but remained silent on the toxic character of RCAs, thus perpetuating a semblance of ambiguity on whether

Workshop of the Pugwash Study Group on Implementation of the CBW Conventions, Geneva, 4-5 December 2004; and Caitríona McLeish and Paul Nightingale, "The impact of dual use controls on UK science: results from a pilot study", *SPRU Electronic Working Paper Series*, paper no 132, April 2005.

[50] US Army Chemical Center and School, Fort McClellan, "New chemical agents and incapacitating agents", Lesson Plan LP6075, undated [*ca* 1965].

[51] A close account of these and related events is to be found in J. P. Perry Robinson, "Solving the problem of 'law enforcement", a paper presented at the 19th Workshop of the Pugwash Study Group on Implementation of the CBW Conventions, Oegstgeest, the Netherlands, 26–27 April 2003, especially in the version currently being revised for publication as "Law enforcement, 'non lethal' warfare and the norm against biological and chemical armament", 10 May 2003. Editors' note. The latter paper is listed in Chap. 9 as item 421.

[52] Proposed not in fact by the USA but by a group of eleven Neutral and Non-Aligned States looking also, and in sharp contrast to the USA, for a CWC prohibition of herbicide warfare.

the toxicity criterion fundamental to the CWC did or did not capture RCA. The way became open for determined NLW protagonists to argue that, if tear gas was not proscribed by the CWC, then neither should the more modern varieties, for which they coined the category-label 'Advanced RCA Technology' (ARCAT). Subsequent ARCAT development projects funded by the US government have included work on the fentanyls and other such intensely supertoxic chemicals. The process that can be seen here is a surreptitious equation of toxicity with lethal toxicity. In this attempt to loosen the CWC constraint on the weaponization of other forms of toxicity we have started to see a creeping legitimization of non-WMD CBW, which is a most serious challenge to the regime. A situation in which some types of toxic weapon are allowed but not others is certain to be unstable.

The instruments of creeping legitimization include, not only 'public diplomacy' and other more hidden pressures for exemption, but also national legislation. In the USA the 'Ensign Amendment' of the 2006 Defense Authorization Act asserts that "riot control agents are not chemical weapons".[53] Fortunately no other state party to the CWC has adopted such a position, nor even commented publicly on what the USA has done. It remains to be seen whether the United States can keep the topic off the agenda of the Second CWC Review, which is where it clearly ought to be.

7.1.3 Strengthening the CBW Governance Regime

Two features stand as bulwarks against the challenges to the CBW governance regime. The first is an OPCW that, despite severe budgetary constraint and opaque operating methods but having, in general, constructive US support, is now an effective multilateral institution that is actively suppressing chemical-warfare armament in most of the world. Second, in regard to disease-warfare more generally, modern customary and conventional international law has transformed an ancient taboo into an enforceable norm of international behaviour. Together these are the principal reasons why chemical and biological warfare are rare occurrences even in today's conflict-ridden world. Because of what CBW could become, they are treasure that must not be frittered away.

Loose talk of CBW "non-proliferation" is a step in that direction, for it carries the implication – no doubt unintended by most of those who use the term – that dominant

[53] In its section 1232, the US National Defense Authorization Act for Fiscal Year 2006 states: "It is the policy of the United States that riot control agents are not chemical weapons and that the President may authorize their use as legitimate, legal, and non-lethal alternatives to the use of force that, as provided in Executive Order 11850 (40 Fed.Reg.16187) and consistent with the resolution of ratification of the Chemical Weapons Convention, may be employed by members of the Armed Forces in war in defensive military modes to save lives, including the illustrative purposes cited in Executive Order 11850." On 27 September 2006, in evidence to the Senate Armed Services Subcommittee on Readiness and Management Support, which is chaired by Senator John Ensign, the Defense Department in the person of Joseph Benkert, Acting Principal Deputy Assistant Secretary of Defense for International Security Policy, testified that the "Administration agrees with [this] statement".

powers have a right to possess disease weapons; and there are indeed certain states in which some leadership figures actually seem to believe this to be so, at least in the case of disabling chemical weapons having counterterrorist application. Yet as a step in the wrong direction this one is surely far outpaced by the growing practice of treating CBW as though they are a subset of WMD. For while some CBW can properly be treated as WMD, all CBW are WMD only when eccentric meanings are given to these two technical terms. The original 1947/48 United Nations definition of WMD,[54] for example, some of which is followed in present-day US legislation on the subject,[55] embraces only "lethal" chemical and biological weapons, whereas the CBW disarmament treaties of 1972 and 1993 have no such limitation, lethality being a criterion of scope in neither of them. The hazard here is not only the suggestion that non-WMD CBW are acceptable – that as long as CBW do not kill us in large numbers they are legitimate instruments of security. Nor is it only the danger that the fuzzy concept of WMD is now supplanting that long-standing precept of the laws of war whereby weapons-employment that does not discriminate between combatant and non-combatant is impermissible, on whatever scale it may occur. No less pernicious is a third implication: that, being in the same category, nuclear weapons are a legitimate, even a necessary, counter to CBW, whether as deterrent instrument of retaliation or as means for destroying CBW capability. We can see this implication displayed in the current pressure in the USA for acquisition of 'bunker busting' nuclear weapons and in the entry into the declared national policy of some of the nuclear-weapon states (China is in fact the only clear exception) of stated willingness to use nuclear weapons in response to CBW attack. There is further reason beyond the legitimation it seems to offer for nuclear armament why this last is so dangerous a policy. Any decision on such nuclear release would have to rest on what recent events have shown to be very uncertain ground indeed: intelligence about supposed adversary CBW behaviour. Ambiguities, uncertainties and other peculiarities attach to CBW to such a degree that the methods of intelligence assessment that seem to work

[54] The Charter of the United Nations, in Article 26, requires the Security Council to formulate plans for a "system for the regulation of armaments". An early step in this direction was agreement to subdivide the problem into categories of armament that would, because of their inherent character, require categorically different approaches. "Weapons of mass destruction" (WMD), in contradistinction to "conventional weapons", were seen as one such category, though there was no immediate consensus on what exactly the category should comprise. During the latter part of 1947, the United States proposed in the Security Council that WMD be defined as "atomic explosive weapons, radioactive material weapons, lethal chemical and biological weapons, and any weapons developed in the future which have characteristics comparable in destructive effect to those of the atomic bomb or other weapons mentioned above"; see UN Security Council document S/C.3/SC.3/7/Rev.1 dated 8 September 1947. It was this definition that the United Nations subsequently adopted and, as it has never formally been changed, is presumably what has been meant in the frequent UN references down the decades to WMD.

[55] Section 1403 paragraph (1) of the 1997 Defense Authorization Act defines a WMD as "any weapon or device that is intended, or has the capability, to cause death or serious bodily injury to a significant number of people through the release, dissemination, or impact of (a) toxic or poisonous chemicals or their precursors; (b) a disease organism; or (c) radiation or radioactivity". See, further, W Seth Carus, "Defining 'Weapons of Mass Destruction' ", National Defense University Center for the Study of Weapons of Mass Destruction *Occasional Paper* no 4, January 2006.

adequately outside the CBW field do less well within it. The oddities of CBW, which are rooted mainly in the 'dual use' nature of much of their contributing technology, are too obtrusive to be swept under a collective WMD carpet. Professional CBW intelligence analysts of course recognise this; it is at the interface between them and their political masters that the attendant misapprehensions are liable to become dangerously misleading or, if not that, the uncertainties then become subordinated to the political imperatives of the moment. The failure to find CBW in Iraq after the last coalition invasion; the cruise-missile bombardment of Al-Shifa Pharmaceutical Industries in the Sudan before that[56]; and, from still earlier times, the Yellow Rain fiasco in which President Reagan's administration mistook beeshit for BW[57]; all of these and more are cautionary tales.

It is presumably ideas of common cause that have promoted the conflation of CBW and nuclear weapons. Maybe, for example, the current alarms over bioterrorism can be turned to advantage by invigorating the quest for nuclear-weapons non-proliferation, or *vice versa.* Common cause imbues the current policy of the European Union of negotiating a WMD "non-proliferation clause" into the Association Agreements it is concluding with non-EU states and other entities,[58] a policy that started with Syria.[59] At another level, the machinery of international security cooperation and shared procedures so laboriously worked out and put into effect by the IAEA and the OPCW may have seemed at first glance also to be applicable *mutatis mutandis* to biosecurity, and therefore good reason for bringing N, C and B together under a single precautionary roof. There may indeed be advantage in such joint approaches, but is it sufficient to risk damaging what exists already? No, one may well think. More scrutiny, including historical research, needs at least to be given to the question of functional links between CBW on the one hand and nuclear weapons on the other – an area strangely neglected by scholars of international security.

The impending Sixth BWC Review Conference (20 November to 8 December 2006) has stimulated much work on how best to strengthen the biological side of the present CBW governance regime, and the Second CWC Review Conference (7–18 April 2008) has begun to do the same for the chemical side. (Work at the biological-chemical interface, which could well be important for the overall CBW regime, now seems to rest, if only by default, with civil society.) The US government, which has faced criticism for its seemingly hostile attitude towards the BWC since mid-2001,[60]

---

[56] See the entries for 20 and 24 August and 17 September 1998, 3 May and 12 October 1999, and 11 February 2001 in *The CBW Conventions Bulletin* no 42 (December 1998) pp 24–25 and 30–31, no 45 (September 1999) pp 17–18, no 46 (December 1999) p 37, and no 52 (June 2001) p 37.

[57] Julian Robinson, Jeanne Guillemin and Matthew Meselson, "Yellow Rain: the story collapses", *Foreign Policy* no 68 (Fall 1987) pp 100–17.

[58] Council of the European Union, note from the General Secretariat of the EU to the Delegations, "Fight against the proliferation of weapons of mass destruction: Mainstreaming non-proliferation policies into the EU's wider relations with third countries", 14997/03, Brussels, 19 November 2003.

[59] See the entry for 19 October 2004 in *The CBW Conventions Bulletin* no 66 (December 2004) p 56.

[60] Which was when the United States effectively brought the BWC Protocol negotiations (see below) to a close; see Daniel Feakes and Jez Littlewood, "Hope and ambition turn to dismay and neglect:

has put forward suggestions, albeit chiefly of a plurilateral rather than multilateral kind.[61] The often contrasting recommendations of the WMD Commission chaired by Hans Blix of Sweden also merit the most careful attention.[62] Ordering principles for these and the many other proposed ways forward[63] are needed and, for this, an attractive logic is to prioritize around the two just-identified bulwarks of the regime: to put most effort into (a) enhancing the OPCW and (b) strengthening the norm. In comparison, the various other sorts of proposal, such as codes of conduct for scientists, seem mere tinkering at the edges, useful though they are as topics for international discourse at times when real progress is politically difficult.

Enhancing the OPCW means soliciting maximal support for the Organization in national politics and as much freedom for it to get on with its CWC-assigned tasks as the international system can countenance. Enhancing the OPCW also means ensuring that, in its necessary pursuit of adaptation to changing international relations, it is protected from having to compromise on principles or procedures that lie at its heart. Undoubtedly the best judge of how to do this is the OPCW Technical Secretariat. But its institutional memory is now severely challenged by its having to implement an employment policy that admits only a seven-year job tenure. For this reason, and especially for matters that impinge on political sensitivities and on which the Secretariat may therefore be unable to develop its own recommendations, 'scape-goat' study by outside specialists may become essential. A clear case in point is implementation of the primary safeguard that the CWC provides against the emergence of new utilities for CBW and the associated pressures for their legitimization. That safeguard is the comprehensive nature of the CWC's prohibitions as embodied in the general purpose criterion that is also protection against adverse technological change and against abuse of dual technology.[64] Especially in regard to NLW, this is

---

the Biological and Toxin Weapons Convention in 2001", *Medicine, Conflict and Survival*, vol 18 (2002) pp 161–74.

[61] But in regard to the BWC specifically, the USA has put forward no proposals since five years ago: see the entries for 25 July, 1 November and 19 November 2001 in *The CBW Conventions Bulletin* no 53 (September 2001) pp 48–49 and no 55 (March 2002) pp 12 & 16–17.

[62] The Weapons of Mass Destruction Commission, final report, *Weapons of Terror: Freeing the World of Nuclear, Biological and Chemical Arms*, Stockholm, Sweden, 1 June 2006.

[63] Some additional governments (notably Canada) and several nongovernmental bodies have put forward constructive and detailed recommendations for the Sixth BWC Review Conference. See, especially: The Geneva Forum, *Meeting the challenges of reviewing the Biological & Toxin Weapons Convention,* March 2006, www.geneva-forum.org/Reports/20060309-10.pdf; Andrew J Grotto and Jonathan B Tucker, *Biosecurity: A Comprehensive Action Plan* Washington, DC: Center for American Progress, June 2006, at pp 14-19; Graham S Pearson, Nicholas A Sims and Malcolm R Dando, editors, *Strengthening the Biological Weapons Convention: Key Points for the Sixth Review Conference* University of Bradford Department of Peace Studies, September 2006; and VERTIC, 'A new strategy: strengthening the biological weapons regime through modular mechanisms', *Verification Matters* [London: VERTIC] no. 6, October 2006.

[64] For a uniquely strong governmental statement on the importance of implementing the general purpose criterion properly, see UK Department of Trade & Industry, *Operation of the Chemical Weapons Act 1996: Annual Report 2004*, July 2005, pp 26–28. This statutory annual report of the UK CWC National Authority is available on www.dti.gov.uk/non-proliferation/cwcna.

an area that civil-society friends of the OPCW will expect the Second CWC Review to examine most carefully. Meanwhile the OPCW's own projects for strengthening the norm should receive the widest possible outside assistance, including the active support of civil society: the action plans for promoting universality of the CWC and for improving implementation of CWC Article VII on national measures.

Strengthening the norm means, first, not weakening the norm and, second, enhancing it—as by promoting mechanisms for ensuring that the norm is not disregarded at the national level and is more enforceable than it is at present at the international level. How is it possible to weaken a norm that seems to rest on so strong a taboo? Perhaps by fragmentation: by treating BW and CW as though they were entirely different from one another instead of variations on the same theme, that of disease weapons; or by trying to differentiate CBW according to bogus criteria of lethality or mass-destructiveness. Sometimes, certainly, different types of CBW will require different practical arrangements from the regime; but, in the interests of norm integrity, the so-to-say default position must surely be common treatment wherever possible. Nor is that an unnatural default. As scientific disciplines, chemistry and biology are converging. And the BWC and the CWC anyway overlap strongly in their scope: toxins, for example, are covered by both treaties; and it is not unreasonable to contemplate the routine international procedures of the OPCW being extended into the domain of toxins and other inanimate biological agents, which is another important and now-pressing topic for policy research.

Beyond that, two different forms of enhancement may now be judged worthwhile. The first is the mechanism noted earlier: machinery for entrenching responsibility for the norm and rules of the BWC within the bureaucracies and legislatures of BWC states parties. For such machinery the CWC has a demonstrably workable model in the obligation it has placed upon its states parties to designate or establish National Authorities for liaising with other states parties, and with the OPCW, empowered by penal legislation to implement the provisions of the treaty within their jurisdictions. A number of OPCW member-states have gone further and created formal lines of communication between their National Authorities, their legislatures, their industries and their civil society.[65] Some of the CWC National Authorities already include BWC cells. The necessary precursor of the mechanism being proposed here is an international campaign to ensure that BWC states parties take Article IV more seriously than many seem to do at present.[66] That article contains the obligation upon

---

[65] As in the UK, which has a formal CWC National Authority Advisory Committee and where the National Authority must report annually to Parliament.

[66] On which see especially: Angela Woodward, *Time to lay down the law: national legislation to enforce the BWC*, a VERTIC report, London, October 2003; and Treasa Dunworth, Robert J Mathews and Timothy L. H. McCormack, "National implementation of the Biological Weapons Convention", *Journal of Conflict and Security Law* vol 11 no 1 (2006) pp 93–118.

each state party to "take any necessary measures to prohibit and prevent" all BWC-violative activity within its jurisdiction. The EU has already agreed a Joint Action[67] to this end and, in regard to non-state entities, further pressure is emanating from UN Security Council resolution 1673 (2006) that has just extended 1540 (2004). There is thus some expectation, at this writing, that the Sixth BWC Review Conference will agree an Article IV action plan similar to the OPCW's on CWC Article VII. All that would then be lacking to complete the norm-strengthening mechanism would be some form of standing international secretariat for the BWC capable of networking with the 'BWC national authorities'. Though there has in the past been resistance to creating any such body, times have now changed, and the BWC institutional deficit has become a disgrace to the states that created it. Their affirmations of support for the BWC would no longer be credible were they to persist in that stance. Civil society would then have a 'naming and shaming' duty to perform.

The second, and complementary, means for strengthening the norm is machinery for better enforcement. Sovereign states do not readily commit themselves in advance to punitive or other reactive sanctions against one another, so a possible way forward is a mechanism for holding *individuals* personally accountable for any such violations. Such an individual-responsibility approach is already discernible in international affairs. Individuals, or at least sub-state entities, in their identity as persons involved in WMD proliferation, are the main target of UN Security Council resolution 1540 (2004). The former president of Iraq, Saddam Hussein, is now on trial facing charges of genocide arising out of the CW attacks on Halabja and elsewhere in Iraqi Kurdistan. From among the several categories of stakeholder in the development of biotechnology, individual scientists are being made responsible for impeding access to 'dual use' materials judged helpful to bioterrorists. All of this is surely good for the norm. Yet the defining feature of a universal norm is not that it should dictate how particular individuals or groups behave, but rather how *everyone* should behave, high or low, head of state or simple citizen. All of us are responsible for the continued well-being of the norm against the weaponization of disease, and the onus is therefore upon us all, individually, to uphold it. It is all very well saying (as many people continued to do at the time of the 2005 BWC intersessional meetings) that scientists need to think more about conceivable end-applications of their work and be taught about the international treaty that is meant to suppress BW. That treaty, the BWC, codifies the norm, but like the Geneva Protocol and the CWC, it places its primary constraints on the behaviour of states, not of individuals. If the norm is to be strengthened, the overall regime must be developed so that its relevance to the individual as well as to the state becomes clear to all. And unless this is done in step with the current moves towards codes of conduct and principles of practice for scientists, those moves will merely become empty acts of discrimination, unpopular and ignored. A more compelling way of inserting the sanction of individual accountability into the regime is needed.

---

[67] "Council Joint Action 2006/184/CFSP of 27 February 2006 in support of the Biological and Toxin Weapons Convention, in the framework of the EU Strategy against the Proliferation of Weapons of Mass Destruction", *OJ* [*Official Journal of the European Union*] L65, 7.3.2006, p 51.

Now that international criminal law is emerging from the constraints of the Cold War, just such a mechanism has become possible. It is a very simple one: a new international convention that would confer on national courts jurisdiction over individuals present in their national territory, regardless of their nationality or official position, who order, direct or knowingly lend substantial assistance to the acquisition, production or use of CBW anywhere. Such a convention might take various forms. One possible text, prepared with advice from an international group of eminent jurists, is the Harvard Sussex Draft Convention.[68] This has already attracted favourable notice from a number of governments and now constitutes the basis for a possible further initiative in the WMD area by the European Union.

If support could indeed be mustered for the several norm-strengthening and organization-enhancing measures just described, the CBW governance regime would surely be better able to resist the challenges that now face it, even the ultimate challenge of emergent CBW that can change the human identity.

7.1.4 *Conclusion: An Agenda for the Near-Term Future*

So from this synopsis we now have several pointers to what ought to be done in order to develop the CBW governance regime. Some are recommendations for action by governments, perhaps matters for exploration during the 2006 and 2008 Review Conferences. Others are tasks for civil society. Others still are topics for attention within the scholarly research community: areas of CBW and the CBW governance regime where knowledge still seems insufficient to direct worthwhile policy and in which, therefore, policy-orientated research would be valuable. All three are itemized below.

7.1.4.1 Governmental Action

- Improve national implementation of the 'Molander chapeau' in CWC Article VI.2.
- Ensure that both the general purpose criterion and the toxicity criterion that together bound the scope of the CWC are adequately addressed during the Second CWC Review.
- Promote agreement on an action plan to secure proper implementation of BWC Article IV (national measures) by all states parties.

[68] Matthew Meselson and Julian Robinson, "A Draft Convention to Prohibit Biological and Chemical Weapons under International Law" in R. Yepes-Enriquez & L. Tabassi (eds), *Treaty Enforcement and International Cooperation in Criminal Matters* (The Hague: TMC Asser, 2002) pp 457–69. Editors' note. This paper is listed in Chap. 9 as item 406. Another account of the HSP draft convention is reproduced as '7.10 A Draft Convention to Prohibit Biological and Chemical Weapons under International Criminal Law' in Chap. 7 of this book.

- Each state party to designate or establish a national authority to ensure that its obligations under the BWC and any associated agreements are fulfilled, and to liaise with other such national authorities, including liaison for assistance purposes.
- Establish a standing international secretariat for the BWC charged with routine activities of information exchange, assistance provision, and other such networking of the national authorities.
- Start exploratory intergovernmental talks on a new international convention that would criminalize acts by any person that the BWC and CWC prohibit to states and would require either prosecution or extradition of such persons.

7.1.4.2 Civil-Society Action

- Campaign actively for each of the six governmental actions outlined above, and be alert to governmental backsliding.
- Both nationally and internationally, promote current awareness and understanding of the OPCW, its purpose and functions, the challenges it faces, and opportunities for advancement.
- Provide support where possible for OPCW action plans and research.
- Watch for allegations of CBW employment (or other BWC/CWC violation) and, if they arise, use media or other contacts to press for investigation.
- Enter into the research programme outlined below and into the dissemination and communication of its research findings.

7.1.4.3 Research Questions

- How should the risks presented by the possibility of terrorist resort to CBW best be assessed?
- Can concepts of CBW taboo offer policy guidance?
- How should scientific and technological change best be monitored for early warning of any consequent challenges to the CBW governance regime?
- Could counter-terrorism applications of the CBW governance regime interfere with the counter-proliferation applications?
- How should risks inherent in the diffusion of 'dual use' technology applicable to CBW best be understood and assessed?
- Why have CWC states parties had such problems with the general purpose criterion, and what does that say about how best to implement it in the future so as to contain the 'new utility' and 'creeping legitimization' challenges to the CBW governance regime?
- What does historical experience of links between nuclear weapons and CBW teach us about possible CBW stimuli to nuclear proliferation or about other CBW-related influences on nuclear-weapons policy today?

- How useful and how feasible might it be to extend the routine procedures of the OPCW verification system further into the domain of toxins and other inanimate biological agents?

7.1.4.4 And Finally

If the norm against the weaponization of disease is to remain strong, it could be a grave mistake to fragment it, so there is argument for always considering chemical and biological weapons together, except when there is unquestionable reason for treating them separately, and for avoiding, wherever possible, equation of CBW with WMD.

To frame the policy issues it raises, this paper has used a concept of CBW governance regime. It is a concept not without drawbacks, most notably the focus it imposes upon states as primary actors. Although the concept does not exclude the increasingly important roles of civil society, other framing concepts might be more accommodating towards them and therefore more worthwhile. One such concept is that of human security.[69] How it could be applied to CBW issues is a subject that might productively form part of any programme of study that includes the research topics listed above.

7.1.5 Appendix: The Existing CBW Governance Regime

The chief components of what this paper has termed the 'CBW governance regime' are here identified and described in turn, first the major multilateral elements of the regime, then developments that have not, in contrast, rested on wide international consensus, following instead from 'plurilateral initiatives' and from a sequence of unilateral renunciations. Some of these latter elements are not always in harmony with the former, multilateral, elements – not least as regards membership and the coercive roles some of them assign to their leaders. Their legal status in relation to the multilateral elements is therefore contentious. They are all, nonetheless, a reality of current international politics.

The 1925 Geneva Protocol, which is an international treaty whose states parties have agreed among themselves not to use CBW against one another, is the bedrock of the existing CBW governance regime. More properly known as the Protocol for

[69] *A Human Security Doctrine for Europe: The Barcelona Report of the Study Group on Europe's Security Capabilities*, presented to EU High Representative for Common Foreign and Security Policy Javier Solana, 15 September 2004, posted at www.lse.ac.uk/Depts/global/5publications3.htm; Helsinki Process Secretariat, "Working Papers of the Human Security Track of the Helsinki Process on Democracy and Globalization", *Helsinki Process Papers on Human Security* (Finland: Foreign Ministry's Publications, 2005); Mary Kaldor, Mary Martin and Sabine Selchow, "Human security: a new strategic narrative", paper presented at the Human Security Workshop, Berlin, 16–17 October 2006.

the Prohibition of the Use in War of Asphyxiating, Poisonous or Other Gases, and of Bacteriological Methods of Warfare, it was negotiated during a League of Nations conference in Geneva on the arms trade and was signed on 17 June 1925. Being drafted as a contract between its parties, it is in effect a no-first-use agreement, which is a characteristic that some states emphasized in reservations they attached to their instruments of ratification. According to a communication in October 2002 from the Ministry of Foreign Affairs of France, which is the depositary of the treaty, the Protocol has 130 states parties (not counting Taiwan) including the five permanent members of the UN Security Council, and one additional signatory state.[70] The Protocol builds upon earlier international agreements, such as ones from the 1899 and 1907 peace conferences in The Hague, and is now widely considered to have entered customary international law, thereby becoming binding upon all states whether they have or have not formally joined the treaty.[71] The fact that the definition of war crimes in the Rome Statute of the International Criminal Court extends to use of chemical but not biological weapons[72] is a reflection less of the state of customary law in 1998 than of a North-South political deal cut during the negotiation then on account of nuclear weapons.

The 1972 Biological Weapons Convention is an international treaty ratified by 155 of the 171 states that have now signed it[73] whereby they renounce germ weapons in order to "exclude completely" the possibility of such weapons being used against human beings, other animals or plants.[74] It reflects the renunciation of biological weapons by the defeated Axis powers after the Second World War, as in the 1954 Revised Brussels Treaty, as well as the subsequent unilateral renunciations by other states noted below. The BWC extended the regime of CBW no-first-use established by the Geneva Protocol by explicitly outlawing development, production and stockpiling of biological and toxin weapons, but it has rather few of the ancillary provisions that are nowadays often thought essential: means of monitoring, even enforcing, compliance with its prohibitions; and prescriptions of procedure for implementing its other rules, or of international organization for assisting states parties to discharge their obligations. Opportunities for extending the regime during the review conferences for which the Convention provides were, however, soon taken up, as with the

---

[70] See pages 109–110 and 335–40 of the 2004 *WHO Guidance* (footnote 11 above).

[71] For a strong descriptive and analytical account of the birth of the 1925 Geneva Protocol, see Rodney McElroy, "The Geneva Protocol of 1925" in Michael Krepon and Dan Caldwell, editors, *The Politics of Arms Control Treaty Ratification* (New York: St Martin's Press, 1991) pp 125–66. For masterly discussion of the scope of the treaty, see R. R. Baxter and Thomas Buergenthal, "Legal aspects of the Geneva Protocol of 1925", *American Journal of International Law* vol 64 (1970): 853–79. On the present state of customary law in this area, see Jean-Marie Henckaerts and Louise Doswald-Beck, editors, *Customary International Humanitarian Law* (ICRC and Cambridge University Press, 2005).

[72] See the entry for 30 June 2000 in *The CBW Conventions Bulletin* no 49 (September 2000) p 38.

[73] These numbers reflect the situation as of September 2006.

[74] On the BWC, see especially Nicholas Sims, *The Diplomacy of Biological Disarmament: Vicissitudes of a Treaty in Force, 1975–85* (London: Macmillan Press, 1988) and *The Evolution of Biological Disarmament* (Oxford University Press, 2001).

institution of a consultative procedure in the event of problems arising, and also a variety of 'confidence-building measures' characterized as "politically binding". But subsequently confirmed reports of gross violation of the BWC by the USSR, as well as intelligence on Iraqi BW, promoted belief that such measures could never in themselves be sufficient. This led, in 1995, to the opening of negotiations for an agreement among the states parties on a legally (as opposed to 'politically') binding instrument that would strengthen the treaty by establishing verification or other compliance-promoting procedures. These 'Compliance Protocol' negotiations collapsed in 2001.[75] In their aftermath, an "intersessional process" ran during 2003-2005, the outcome of which is due for assessment during the impending Sixth BWC Review Conference.

The 1977 EnMod Treaty, as the Convention on the Prohibition of Military or Any Other Hostile Use of Environmental Modification Techniques is commonly known, among other things prohibits warfare with chemicals toxic to plants "having widespread, long-lasting or severe effects". This treaty entered into force in 1978, but is at present subscribed to by less than a majority of the world's states, though among the permanent members of the UN Security Council only France is a non-party.[76]

The 1993 Chemical Weapons Convention, like the BWC, originated in intergovernmental talks on CBW that commenced in 1968. It prohibits development, production and stockpiling of weapons toxic to human beings or other animals, or assistance in acquiring such weapons, and obliges parties to the treaty not only to institute domestic compliance-assuring measures, including penal legislation, but also to participate in a verification system operated by an international agency that the treaty established in The Hague, the Organization for the Prohibition of Chemical Weapons.[77] The OPCW includes a trained international inspectorate of 150–200 people. The treaty further extends the 1925 Geneva Protocol by including among its provisions an express prohibition of any use of toxic chemical weapons, including the uses in reprisal and in retaliation-in-kind that the Protocol could not outlaw. Together with the BWC, the CWC is the core of today's CBW governance regime. Initially inspired by the Vietnam War and heavily marked by the Cold War, both treaties are, in terms of membership, nearing universality, but with important holdouts, most conspicuously North Korea, Egypt, Israel and Syria.[78] Both treaties seek to preclude international

---

[75] For the definitive account of these negotiations, see Jez Littlewood, *The Biological Weapons Convention: A Failed Revolution* (Aldershot, UK: Ashgate, 2005).

[76] On the EnMod Treaty, see especially: Lawrence Juda, "Negotiating a treaty on environmental modification warfare", *International Organization* vol 32 no 4 (Autumn 1978) pp 975–91; and Susana Pimiento Chamorro and Edward Hammond, "Addressing environmental modification in post cold war conflict", Edmonds Institute occasional paper, 2001.

[77] For an unrivalled description of the CWC, see Walter Krutzsch and Ralf Trapp, *A Commentary on the Chemical Weapons Convention* (Dordrecht: Nijhoff, 1994).

[78] North Korea is party to the BWC but not the CWC. Both Egypt and Syria have signed but not yet ratified the BWC, while for the CWC they have done neither. Israel has signed but not yet ratified the CWC; for the BWC it has done neither. All four countries are understood to have developed CW and/or BW capability.

transfers of prohibited items, these being defined by their purpose because of the general purpose criteria that the two treaties used to set their scope – principally in order not to interfere with 'dual use' goods meant for unprohibited purposes. Such an approach to the governance of 'dual use' technology had become more fully developed by the time of the CWC, so much so that the CWC expressly outlaws factories for chemical weapons. The BWC had made no similar provision, its negotiators having been nervous of inhibiting vaccine production and commerce in products of biotechnology.[79] Nor, in contrast to the CWC,[80] does the BWC require disclosure of past biological-weapons programmes, though that is a provision of one of the confidence-building measures adopted in 1991.[81]

Empowerment of the UN Secretary-General to investigate use-allegations. In November 1987 the UN General Assembly reaffirmed the powers of the UN Secretary-General to investigate allegations of CBW use.[82] Such investigations took place on several occasions in Asia and Africa during 1981–93.[83] Since April 1997, the OPCW has been able, at the request of member states, to investigate allegations of chemical warfare. It has not yet formally done so. What happens if a UN investigation verifies an allegation is left for decision by the processes of international politics. Express provision for sanctions is largely absent from the overall regime.

The Australia Group, which began work in May 1985 at the Australian embassy in Brussels after a year of preparatory activity there and in other Western capitals, seeks to harmonize supply-side controls on dual-use technology applicable to CBW

[79] See Emmanuelle Tuerlings, *Dual-use biotechnology: Prospects for governance through arms control*, D. Phil dissertation, University of Sussex, January 2003.

[80] Of the 180 states parties to the CWC (as of 20 October 2006), 12 have declared post-1945 possession of factories for making chemical weapons. They are Bosnia & Hercegovina, China, France, India, Iran, Japan, Libya, Russia, Serbia & Montenegro, South Korea, the UK and the USA. Of these 12, five plus one other have declared possession of CW stocks that total in all about 70,000 tonnes of CW agent, which are currently being destroyed under strict international verification. These six possessor states are Albania (declaring 16 agent-tonnes to the OPCW), India (1044 agent-tonnes, maybe somewhat more), Libya (24 agent-tonnes), Russia (40,000 agent-tonnes), South Korea (1056 agent-tonnes, maybe somewhat less) and the USA (27,800 agent-tonnes). Like the (China-supplied) Albanian stockpile, the Libyan one appears far too small to have had much military significance, but Libya also declared that it possessed thousands of tons of precursor chemicals – a supply sufficient to provide a stockpile of chemical weapons exceeding those of India and South Korea together. The amount of mustard and nerve gas used by Iraq during its 1980-88 war with Iran is believed to have totalled 2540 agent-tonnes. The UK had destroyed its cold-war stockpile of some 62,000 agent-tonnes well before the CWC had been agreed, as had other erstwhile possessors.

[81] On which see, especially, Nicolas Isla, "Transparency in past offensive biological weapon programmes: an analysis of Confidence Building Measure Form F 1992–2003", Hamburg Centre for Biological Arms Control *Occasional Paper* no 1, June 2006. The BWC states parties that have used Form F to declare such programmes are Canada, France, Iraq, Russia, the UK and the USA.

[82] UN General Assembly resolution 42/37C. On the present status and possible futures of machinery for investigating BW use, see especially Jez Littlewood, "Investigating allegations of use of chemical and biological weapons: the role of the United Nations Secretary-General", draft of 5 April 2006.

[83] For references see page 129 of the 2004 WHO volume, *supra*.

by promoting common standards for the formation and implementation of national export-control policies.[84] The AG is one of the earlier anti-proliferation plurilateral initiatives. It was inspired by the discovery that the chemical weapons used by Iraq in its war with Iran were not USSR-supplied but home-made, put together from 'dual use' commodities and know-how imported from the global marketplace. Its membership and range of activities have expanded over the years, most notably in the early 1990s, when it took on biological as well as chemical export controls and also agreed controls on specific dual-use equipment applicable to the production and dissemination of CBW agents. As the BWC and the CWC advance closer towards universality of membership, the AG will presumably come to serve mainly as a safety-net against non-compliance with the technology-transfer provisions of the two treaties.

A new wave of unilateral renunciation. Unilateral actions on chemical and/or biological weapons taken since 1987 by Romania, Iraq, Russia, South Africa and Libya represent a different category of regime element, all having occurred outside the contemporary framework of the CBW governance regime. In Russia's closure of its biological-weapons programme in 1992,[85] and perhaps also for the ending of the South African CBW programme in 1993,[86] the renunciations were made in order to put right actual violation of the BWC. The Iraqi renunciation in 1991 was conjunct with ratification by Iraq of the BWC, as was the Libyan WMD renunciation in 2003 with accession to the CWC.[87] Romania decided against proceeding with production of chemical weapons in 1987, having instituted a development programme in 1968 that it closed down altogether in 1990, after the fall of Ceausescu, subsequently

---

[84] For detailed accounts of the AG, see: Julian Perry Robinson, "The Australia Group: a description and assessment", in H. G. Brauch, H. J. van der Graaf, J. Grin and W. A. Smit, editors, *Controlling the Development and Spread of Military Technology* (Amsterdam: VU University Press, 1992); Amy Smithson "Separating fact from fiction: the Australia Group and the Chemical Weapons Convention", Henry L. Stimson Center Occasional Paper, Washington, DC, March 1997; and Robert J Mathews, "The development of the Australia Group export control lists of biological pathogens, toxins and dual- use equipment", *The CBW Conventions Bulletin* no 66 (December 2004) pp 1–4. Recently the AG, hitherto a secretive organization, has established its own website, at http://www.australiagroup.net/.

[85] For recent additions to the literature on the biological-weapons programme of the erstwhile Soviet Union, see especially: John Hart, "The Soviet biological weapons program' in Mark Wheelis, Lajos Rózsa and Malcolm Dando, editors, *Deadly Cultures: Biological Weapons since 1945* (Harvard University Press, 2006); Jan T Knoph and Kristina S. Westerdahl, "Re-evaluating Russia's biological weapons policy, as reflected in the Criminal Code and official admissions: insubordination leading to a president's subordination", *Critical Reviews in Microbiology* vol 32 (2006) pp 1–13; and I. V. Domaradskij and W. Orent, "Achievements of the Soviet biological weapons programme and implications for the future", *Revue scientifique et technique de l'Office international des Epizooties* vol 25 (2006) no 1, pp 153–61. In preparation for several years, the investigative book on the subject by the US scholars Milton Leitenberg and Ray Zilinskas is still awaited.

[86] See Chandre Gould and Peter Folb, *Project Coast: Apartheid's Chemical and Biological Warfare Programme* (Geneva: UNIDIR, 2002).

[87] See Matthew Meselson and Julian Perry Robinson, "Libya and 'dual use' ", *The CBW Conventions Bulletin* no 65 (September 2004) pp 1–3.

asking the USA to verify the termination of its programme, this happening in 1994.[88] Precedent for the new wave of unilateral actions existed in the US renunciation of biological weapons in 1969 and of toxin weapons in 1970,[89] which set the way to multilateral agreement on the BWC, and in the renunciation of biological weapons in 1972 by France, which until 1984 did not consider the BWC likely to be worth joining on account of the absence of international procedures for compliance-verification. All of these unilateral actions may in turn be regarded as further steps on the path of CBW deproliferation[90] that the UK had (unintentionally) set in the late 1950s – a path that later largely disappeared from public sight, though not from actuality, amidst the furors about CBW proliferation on the part of 'rogue' states. For the Iraqi renunciation, outside coercion was the primary motor, renunciation being a condition of the ceasefire that ended the Kuwait War, as to a degree was coercion partly responsible for the Libyan renunciation also, although in fact internal Libyan factors seem to have been the primary motivation.[91] Nevertheless, the intergovernmental agreements, the acts of national legislation, the increasing eschewal of CBW in the custom and behaviour of states, and the political climate associated with the renunciations have made them substantial elements of today's overall CBW governance regime and one that may yet find resonance in the security policies of Egypt, Israel, North Korea and Syria. At root they reflect assessments that the security benefits of possessing CBW are not worth much when set against the benefits of being inside the regime and against the costs of remaining outside or in violation of it. Possession or non-possession of nuclear weapons has been a prominent part of this calculus,[92] as when the UK decided, in 1956, not to proceed with large-scale manufacture of nerve gas because nuclear weapons now offered "the major deterrent to war".[93]

The Global Partnership against the Spread of Weapons and Materials of Mass Destruction is another of the 'plurilateral initiatives', this one set going at

---

[88] See the entries for 13 February 1990, 30 June 1994, 22–26 August 1994 and 21 June 2001 in *Chemical Weapons Convention Bulletin* (later *The CBW Conventions Bulletin*), no 8 (June 1990) p 8, no 25 (September 1994) pp 23–26, no 26 (December 1994) p 17 and no 53 (September 2001) p 40.

[89] On which see, especially, Jeanne Guillemin, *Biological Weapons: From the Invention of State-Sponsored Programs to Contemporary Bioterrorism* (New York: Columbia University Press, 2004) at pp 112–30.

[90] On chemical-weapons deproliferation, see Julian Perry Robinson, "Chemical-weapons proliferation in the Middle East", in Efraim Karsh, Martin S Navias and Philip Sabin, editors, *Non-Conventional- Weapons Proliferation in the Middle East*, Oxford: Clarendon Press, 1993, pp 69–98.

[91] Dafna Hochman, "Rehabilitating a rogue: Libya's WMD reversal and lessons for US policy", *Parameters* [US Army War College] vol 36 no 1, April 2006.

[92] See "Analysis and implications" in Mark Wheelis, Lajos Rózsa and Malcolm Dando, editors, *Deadly Cultures: Biological Weapons Since 1945* (Harvard University Press, 2006).

[93] Brian Balmer, whose *Britain and Biological Warfare: Expert Advice and Science Policy, 1930–65* (Palgrave, 2001) explores relationships between BW and nuclear weapons as evident in UK state papers, is now engaged in a comparable study of CW.

Kananaskis, Canada, in June 2002 by the G7/P8 Group of Major Industrialized Countries.[94] It served initially to attract and provide a framework for international financing for the destruction of former Soviet WMD capabilities, but is now broadening its ambition into the development of measures for "international non-proliferation and disarmament assistance", including biosecurity projects and supportive activities in states that have renounced WMD.

The Proliferation Security Initiative of 2003 was not so much a plurilateral initiative as a coalition inspired and led by the USA of states undertaking to interdict, by armed force if they so chose, international shipment of goods thought destined for WMD programmes considered illegal by the coalition.[95] Notwithstanding its assertions of legitimacy, the PSI coalition includes states not party to the UN Convention on the Law of the Sea, although in October 2005 states parties to the Convention for the Suppression of Unlawful Acts against the Safety of Maritime Navigation agreed on extensions of that treaty which would, in effect, legalize maritime PSI activities directed against materials related to CBW or nuclear weapons.

UN Security Council resolution 1540 (2004) and its implementing machinery are elements of the international response to perceptions of impending WMD terrorism.[96] As regards CBW, this Chapter VII resolution seeks to universalize parts of national BWC and CWC implementing legislation (including transfer provisions) in regard to non-state entities. Even states not parties to the treaties, being bound as all UN member-states are by Security Council resolutions, are now under obligation to enact and enforce such legislation. Its implementation is promoted through the '1540 Committee' that it established, the lifetime of which was extended until 27 April 2008 by resolution 1673 (2006).

---

[94] On this relatively little known element of the CBW governance regime, see especially the website of the Strengthening the Global Partnership Project, which is an international consortium of Asian, European and North American research institutes led by the Center for Strategic and International Studies in Washington DC: www.sgpproject.org.

[95] See, especially, Christer Ahlström, "The Proliferation Security Initiative: international law aspects of the Statement of Interdiction Principles", *SIPRI Yearbook 2005: Armaments, Disarmament and International Security* (Oxford University Press, 2005) pp 741–65.

[96] See Ben Steyn, "Understanding the implications of UN Security Council resolution 1540", *African Security Review*, vol 14 (2005) no 1, pp 85–91; Bruno Demeyere, "The proliferation of international nuclear law's actors: Resolution 1540 and the Security Council's fight against weapons of mass destruction falling into terrorists' hands", *Nuclear Law Bulletin* [OECD] supplement 75, pp 35–61, 2005; Elizabeth Rindskopf Parker and Bryan Pate, "Implementing UN Security Council resolution 1540 to combat the proliferation of biological weapons", *Biosecurity and Bioterrorism*, vol 3 no 2 (June 2005) pp 166–73

7.2 Extract – The Negotiations on Chemical-Warfare Arms Control[97]

Editors' note. Published in 1980, this extract assesses the possibilities for a chemical weapons convention. When Julian wrote the paper, such a convention was the subject of 'exploratory discussions' in the multilateral negotiating body then called the Geneva Disarmament Committee.[98] After considering the scope of a future treaty, the extract considers how a convention could be verified, pointing out the political and technical difficulties involved in this, as well as similarities and differences in states' positions in the discussions.

7.2.1 *Scope: State of Agreement*

One way of thinking about the proper scope of a chemical-weapons convention is to start by classifying the types of chemical, associated hardware and associated activities which together would constitute a full capability for waging chemical warfare.[99] The more of these components of capability the convention proscribes, the more likely will it be to exclude the possibility of chemicals ever being used as weapons. But, wholly or in part, most of the components may in many countries have *raisons d'être* other than offensive chemical warfare – for example, supplies of toxic chemicals needed as intermediates in industrial manufactures (the so-called 'dual-purpose chemicals') or research facilities for studying methods of protection against CW attack. A convention that seeks to eliminate or preclude them completely will therefore be unacceptable: exceptions will have to be made. But any such exception would inevitably complicate verification since the latter would then hinge, not on the relatively straightforward matter of possession or nonpossession, but on intention behind possession. With respect to these components of capability, therefore,

---

[97] Editors' note. This extract is from pages 36–38 and 40–43 of J. P. Perry Robinson, "The Negotiations on Chemical-Warfare", *Arms Control (London)* 1(1): 30–52, May 1980. The paper is listed in Chap. 9 as item 098. Unless otherwise indicated, the remaining footnotes for this paper are Julian's originals. Apart from correcting obvious typos and renumbering headings and tables in keeping with the requirements of this book, we have retained Julian's formatting and spelling throughout the paper and its notes.

[98] Robert J. Mathews (2004), "The Verification of Dual Use Chemicals under the Chemical Weapons Conventions through Open Source Research: The Pugwash-SIPRI Thiodiglycol Project", Henrietta Wilson, Olamide Samuel and Dan Plesch (eds), *Open Source Investigations in the Age of Google.* World Scientific Publishing, p 226. https://doi.org/10.1142/9781800614079_0012.

[99] One such categorization of the components of CW capability is set out in J. P. Perry Robinson, 'Chemical-warfare capabilities of the Warsaw and North Atlantic Treaty Organizations: an overview from open sources' in SIPRI, *Chemical Weapons: Destruction and Conversion* (London: Taylor & Francis, 1980). [Editors' note. This is listed in Chapter 9 as item 96]. Certain matters which that paper neglects, particularly military-staff requirements, are considered by S. J. Lundin in a contribution to *World Armaments and Disarmament: SIPRI Yearbook 1979* (London: Taylor & Francis, 1979), pp.470–89, 'Stockpiles of chemical weapons and their destruction', and in his paper, 'Chemical weapons: too late for disarmament?', in *Bulletin of the Atomic Scientists*, vol.35 (10); pp.33–7 (1979).

it is logically impossible for any state-party to gain or provide complete assurance of preclusion. Potential states-party must therefore judge how much uncertainty on this score they are able to accept. With respect to other components of capability, purely practical reasons may limit verification of preclusion or elimination, in which case further judgements about acceptable uncertainty must be made. This is what is meant by the somewhat contradictory concept of 'adequate verification'. Clearly there will be differences in the magnitude of risk entailed in incomplete verification as between all the different elements in the scope of a chemical weapons convention. Some components of capability will be more critical than others. Components whose possession by potential adversaries would the more greatly reduce the amount of time or resources needed to build up a fully operational CW capability must be fully accommodated within the scope of the treaty. Stockpiles of chemical weapons are one obvious example; 'dedicated' production facilities another.

Under circumstances where the degree of mutual trust prevailing among potential states-party is low, a case can be made for restricting the scope of the convention only to those proscriptions on which a high degree of assurance can be gained about their implementation. A broader scope could be counterproductive since mistrust and uncertainties about implementation could feed upon one another to the point where confidence in the treaty as a whole disappeared. But to take verifiability alone as a determinant of what precisely should or should not be banned might have the no less counterproductive effect of locking the treaty into the present, denying it that forward-looking character which, as is noted above, may well be the primary requirement of many potential states-party. As the pertinent technologies advance, the currently irreducible uncertainties about implementation could recede; and as international relations fluctuate, so too may the acceptability of uncertainties. In this particular linkage of scope and verification, the twin objectives of elimination and preclusion are in conflict with one another. If a solution is to be achieved, it may perhaps most likely be found in a conjunction of specific verification and confidence-building procedures. Such an approach has been under study for several years within the Pugwash Chemical Warfare Study Group[100] and, as is described later, is now receiving attention within intergovernmental forums.

[100] The Pugwash CW Study Group is a body of scientists loosely organised via the Executive Committee of the Pugwash Conference on Science and World Affairs from a still-growing number of countries (now 22), each of its members having a particular interest or specialisation in aspects of chemical-warfare defence or arms limitation. Though it includes officials and chemical-industry representatives from East, West and nonaligned countries, the PCWSG is a private body in the sense that its members participate in their personal capacity. Through a series of workshops and roundtables that began in 1974, the PCWSG has in effect come to serve as an informal international specialist forum complementing the Geneva disarmament committee. Confidentiality is maintained on the detailed proceedings of PCWSG workshops so as to encourage full and frank discussion, but from each one a unanimously agreed final report is published, and participants are free to release any papers they may themselves have contributed. Further particulars, including literature references, are available in a paper by the present author, 'Current state of chemical-warfare arms limitation', presented at the 6th Pugwash CW Workshop, Kansas City, May 1978. Editors' note. This is listed in Chap. 9 as item 90.

7.2.2 *Verification: State of Agreement*

In the process of gaining assurance of the implementation of treaty-provisions to a degree commensurate with security needs, three distinct contributory processes can be differentiated: monitoring, assessment and evaluation. *Monitoring* concerns the collection of data originating either in elimination activities, such as stockpile destruction, or in activities that might conceivably be proscribed ones. The data may be of many varieties: the rate of flow of munitions into a demilitarization/detoxification plant, for example, or the relative intensities of different infra-red wavelengths in light reflected from the surface of a weapons proving ground, or the amount of product acrylonitriles leaving a factory that uses hydrogen cyanide as a starting material. *Assessment* concerns the interpretation of such data: the process, first, of inferring information about the nature of the activities from which the data have emanated and, second, of judging from such inferences and any other pertinent information the probabilities of the activities under observation actually being – wholly or in part – what they appeared to be. The likelihood of any such interpretation resulting in an assessment of probability approaching unity would depend upon the types of data monitored and the capacity available for monitoring them. The highest-grade data would generally be accessible only on site, so that if on-site monitoring were not possible the probabilities thus assessed could be far short of unity, meaning a high degree of uncertainty about the activity being monitored. *Evaluation* concerns the appraisal of the broader implications which the assessed probabilities may carry when taken together. Uncertainty, should it arise, always means a possibility of abrogation. In the light of prevailing perceptions of the general intentions of the state in question, *evaluation* asks: is it a real possibility? And if abrogation were happening to the extent that the degree of uncertainty could admit, would it seriously endanger overall security, having due regard to any countermeasures already available? The latter include anti-chemical-warfare protective measures; the more extensive these are, therefore, the easier it may be to satisfy overall verification requirements.[101]

Technical considerations are obviously very important in questions of verification. They set the limits within which political accommodations must be reached on what is to be accepted as adequate verification. These technical limits do not reside solely in the capacities of available monitoring and assessment techniques. For example, in the particular case of verifying elimination, the types of data available for monitoring and the degree of access needed to monitor them will depend upon the particular methods adopted for elimination. These methods will vary in their economic, public-safety, environmental-impact and other such characteristics, so that some will be more attractive than others on practical grounds; and there will be variations in this respect from country to country. For this reason the final choice of elimination method must be left to the countries concerned; but, in making their choices, verifiability could be an additional criterion. The possible need for negotiated agreement in this respect

---

[101] This point is discussed further in: M.S. Meselson and J. P. Perry Robinson, 'Chemical warfare and chemical arms limitation', *Scientific American*, forthcoming [sic]. Editors' note. This corresponds to item number 102 in Chap. 9.

has long been recognised in the CW talks. It has led to much discussion of the technicalities of detoxifying chemicals and of past experience in the demilitarization of CW munitions. This has imposed further demands on the already high degree of technical understanding and general numeracy required of the CW negotiators.

Of the three component processes in verification, that of evaluation would necessarily be conducted by each state-party in respect of potentially hostile states. For this purpose each state-party would have a capacity of its own, through what are commonly termed 'national means', for performing at least some monitoring and assessment. But unless these 'national means' amounted to an extraordinarily large and technologically sophisticated intelligence apparatus, one that could readily be focussed on foreign activities of chemical-warfare concern, there might well be critical uncertainties which they could never resolve to an adequate degree. Supplementary aids in the form of specific treaty provisions for verification therefore appear to be necessary. This much has long been accepted in the intergovernmental talks, but the form which the aids should take remains deeply controversial.

There is general agreement that the formal control machinery to be established should have both national and international components. This concept of complementarity was originally put forward as a basis for negotiation by the group of nonaligned states during the first phase of the CW talks,[102] and subsequently enabled constructive dialogue gradually to be joined on verification questions between Eastern and Western delegations. The former envisaged national control organs playing the dominant role in verification; the latter, an international organ. Over the years both sides have moved towards an intermediate position, though exactly how far they have moved is at present concealed by the secrecy under which the bilateral US-Soviet negotiations are conducted. A step of particular significance seems to have occurred early in the seventh round when the USSR accepted in principle the idea of 'challenge' on-site inspections.[103] Where this has left the original rather detailed Socialist-country proposals on the constitution and function of national control organs remains unclear. It is hard to envisage states-party formally creating special control bodies, though in most countries there probably already exist organisations of one kind or

[102] CCD/310 of 25 Aug. 1970.

[103] Some 18 months were to elapse before this development was noted in a joint communiqué from the bilateral negotiations. The first public news came from a press briefing given by the senior US negotiator: see *The Times*, 19 Jan. 1978. The bilateral statement of 31 July 1979 (see note above) contains the following passage:

A convention should include provisions in accordance with which any party should have the right ... to request from another party with respect to which suspicions have arisen that it is acting in violation of obligations under the convention relevant information on the actual state of affairs, as well as to request investigation of the actual state of affairs on site, providing appropriate reasons in support of the necessity of such an investigation. A party may agree to such an on-site investigation or decide otherwise, providing appropriate explanation. [Editors' note. Endnote 16 in the original document reads: "The joint statement of 31 July 1979, read to the CD [Conference on Disarmament] by Ambassador V. L. Issraelyan of the USSR, is the most detailed progress report yet to have emerged in public from the bilateral talks. It is available as a CD Document, no. CD/48 of &. Aug. 1979."].

another to which control duties could reasonably be assigned. Within the CD [Conference on Disarmament][104] the present consensus seems, from the formal records, to be that there should be some sort of international body functioning as a contact-point between national organs but having some responsibilities of its own. Whether routine monitoring tasks should be included among these further responsibilities and, if so, what types, are especially contentious questions, primarily because the type of monitoring most productive for verification purposes would require the presence of foreign personnel at sensitive locations.

Solutions to this problem of intrusiveness and derogation of national sovereignty are currently being sought in focussed systems of verification that minimise the role of international on-site inspection. The focussing device is to require states-party to declare any stocks of chemical weapons they may possess and the facilities that they had formerly assigned to chemical-weapons activities – production is the activity mainly being considered in this regard – and then to accept intrusive techniques for monitoring their elimination. The latest US-Soviet joint statement outlines a system of this type, but makes it clear that there is no agreement yet on the use of on-site inspection in anything other than challenge procedures. Moreover, following SALT [Strategic Arms Limitation Talks] precedents, the statement places express emphasis on 'national technical means of verification', specifying that states-party should not impede the application of such means by 'deliberate concealment measures'.

Soviet objections to mandatory international on-site inspection have been expressed, as in other arms-control negotiations, in terms of the potential for abuse – for industrial or military espionage. In the particular case of verifying stockpile elimination, the Soviet delegation has argued, further, that on-site inspections by foreign personnel would carry a serious risk of disseminating information about CW agents that could promote proliferation. However, since it is not seriously believed in the West that the USSR possesses CW agents of a kind not already described in the open literature, the basic untenability of this particular argument may be expected to lead, sooner or later, to its abandonment. As for the verification of nonproduction, it is undoubtedly the case that site inspections would carry some potential for abuse, though it is now apparent from the international site visits to commercial chemical factories organised during March 1979 by the West German and British Governments, that the sensitivities of industrial corporations about commercial secrecy are not as acute as had previously been supposed.[105] Moreover, the United States now says that routine inspection of civilian chemical production plant, with the exception

[104] Editors' note. The Conference on Disarmament is the international forum for negotiating disarmament. See "Conference on Disarmament" [online] Available at: https://disarmament.unoda.org/conference-on-disarmament/ [Accessed 25 January 2025].

[105] The initiative for these site visits, which were organised through the CD [Conference on Disarmament], was taken by Chancellor Schmidt during his address to the UN General Assembly Special Session on Disarmament on 25 May 1978, followed up in a *note verbale* to the UN Secretariat (see UN document A/S-10/AC.1/13 of 6 June 1978). The Federal German Government had been stimulated by the successful international site visits to German factories and laboratories organised by the Pugwash CW Study Group during the previous year; subsequent PCWSG site visits were accepted by a US chemical corporation and by the US Army (to a CW agent/munition detoxification/demilitarization facility) in 1978, and by a Swedish corporation in 1979. For particulars of the

of facilities handling certain 'dual-purpose' chemicals, is not necessary for verification purposes. It is not yet clear whether, as has been suggested,[106] this reflects the existence of a US technical capacity for detecting 'single-purpose' CW-agent production by 'national means'. In the US view, systematic and mandatory on-site inspection is necessary only for verifying the elimination of stockpiles and of production facilities hitherto dedicated to CW agents and munitions. In summary, then, it would seem that US and Soviet positions on on-site inspection are separated by a gulf which, though large, is neither as wide as is commonly supposed nor intrinsically unbridgeable.

7.3 Chemical Arms Control and the Assimilation of Chemical Weapons[107]

Editors' note. This paper introduces Julian's innovative work on weapons assimilation within his broader assessment of chemical warfare prohibitions. Julian demonstrated that, to be effective, new technologies must be *assimilated* – that is, military social structures and behaviours, and associated decision making, must change if novel weapons are to be fully operationalized. Showing that weapons' utility goes beyond technical research and development, the concept of assimilation provides a richer understanding about the ways in which some weapons become normalized while others don't, and thereby gives scholars, practitioners and policy-makers additional tools to build effective constraints against weapons systems.

CD site visits, see CD/15 of 24 April 1979 on the British exercise and CD/37 of 12 July 1979 on the German one. The next PCWSG site visit is to take place, by invitation, at a chemical factory in East Germany: one that has been alleged, in certain West German publications, to be a location for nerve-gas production.

[106] Colonel C. H. Bay, 'The other gas crisis: chemical weapons', US Army War College Strategic Studies Institute special report no. ACN 79022, 1 Aug. 1979. Colonel Bay suggests, moreover, that the technique to which he refers is applicable, also, to the low-toxicity non-commercial chemicals that would be needed for the 'binary' munitions with which the US armed services are currently planning to equip themselves in modernization of their CW-weapons capability. The now-demonstrated feasibility of binaries, which would be loaded not with CW agent but with chemicals that react together to produce CW agent en route to the target, has commonly been supposed to have greatly aggravated the problem of nonproduction verification. For a detailed account, see J. P. Perry Robinson, 'Binary nerve-gas weapons' in SIPRI, *Chemical Disarmament: New Weapons for Old* (Stockholm: Almqvist & Wicksell, 1975), pp. 21–99, and 'The United States binary nerve-gas programme: national and international implications', *ISIO Monographs* (University of Sussex: Institute for the Study of International Organisation) no.10, 1975.

[107] Editors' note. This was first published as J. P. Perry Robinson. 'Chemical arms control and the assimilation of chemical weapons'. February 1981. *International Journal* (Toronto) 36(3): 3–5, July 1981. It is listed in Chap. 9 as item 117-published. Unless otherwise indicated, the remaining footnotes for this paper are Julian's originals. Apart from correcting obvious typos and renumbering headings and tables in keeping with the requirements of this book, we have retained Julian's formatting and spelling throughout the paper and its notes.

Chemical weapons are treated in one of the older extant international agreements which are now regarded as arms control: the 1925 Geneva Protocol. Parties to this treaty have renounced 'the use in war of asphyxiating, poisonous or other gases, and of all analogous liquids, materials or devices'. The protocol also outlaws 'bacteriological methods of warfare'. There are no provisions for verification. About a hundred states are parties, including all members of the North Atlantic Treaty and Warsaw Treaty alliances. On the evidence, in particular of numerous resolutions of the United Nations General Assembly over the past fifteen years, the prohibitions expressed in the protocol have entered customary international law and are now binding upon all states.

Although chemical warfare may thus appear to be well and truly banned, the search for supplementary control of these arms has been on the active agenda of the principal international negotiating body since 1968. In part this has been a natural consequence of the manner in which the United Nations has approached the task of disarmament: in 1947 the organization defined 'weapons of mass destruction' to include at least some types of chemical weapon whose elimination from national armaments thereby became, in the words of General Assembly resolution 41(1), an 'urgent objective' of the United Nations, one that has dominated subsequent disarmament effort. But in part also the continuing quest for additional chemical arms control has been stimulated by doubts about the existing régime. It would be remarkable indeed if a treaty formulated during the comparative calm of the mid-1920s remained fully adequate under the conditions of great political and technological change that have characterized the intervening period.

That there is room for further chemical arms control is obvious enough; but what exactly it should comprise, having due regard to what is feasible, is less obvious, despite the past twelve years of intergovernmental discussion. A clarification of basic principles, then, is the aim of the present essay. The focus is not on the negotiations for more chemical arms control[108] but on the strengths and weaknesses of the control that we have already. The theme is the possible existence of generally operating de facto constraints on national chemical-warfare armament, discernible in the history of the relationships between the prevailing legal norms on the one hand and state practice on the other. It is on the bedrock of these constraints that further measures of chemical arms control might most efficaciously be built.

I

The growth of chemical technology during the latter part of the nineteenth century carried clear practical implications for the nature of future warfare, most conspicuous in the rapidity of innovation in the field of explosives and propellants. That toxic chemicals might come to contribute to weaponry more extensively than in the past was also appreciated, so much so that at intergovernmental conferences at Brussels (1874) and The Hague (1899 and 1907) specific measures of chemical arms control were included among the agreements reached. Military and naval establishments began to consider the possibilities seriously during the rearmament period preceding

[108] Detailed chronological accounts of these negotiations are most easily found in successive volumes of The United Nations disarmament yearbook and its antecedents.

World War I. By the end of 1914 primitive chemical weapons had been introduced into the fighting by both French and German forces and others were under development in Britain. All of these weapons were, however, carefully designed not to contravene the letter of the existing prohibitions.

The initial resort to chemical warfare on a large scale was by Germany, in April 1915. By exploiting the area-effectiveness and permeating characteristics of airborne clouds of poison discharged from cylinders of liquefied poison gas (the production technology for which only Germany had acquired on any scale at that time), the protagonists of this new mode of warfare hoped to provide a means both for breaking the stranglehold of trench warfare and for reducing demand for high explosives, the supply of raw materials for which was threatened by the British naval blockade. As it happened, the German high command did not take kindly to this innovation and did not follow through its initial success; and anti-gas protective measures were soon introduced by the French and allied armies which took much of the edge off the new weapon. But despite protracted resistance from many field commanders, poison gas came into increasing use on both sides, with artillery delivery soon predominating, and it eventually claimed well over a million casualties.[109] The ammunition procurement plans for 1918 and 1919 of the principal belligerents suggest that by the end of the war chemical weapons were fast becoming assimilated as conventional armament.

Yet in the great majority of wars that have been fought since, including World War II, there has been no use of chemical weapons. To this extent the 1925 Geneva Protocol, negotiated essentially as a symbol of outrage against the carnage of the Great War, may be regarded as effective arms control. The important question to ask, however, is whether there might not have been the same general abstention from chemical warfare even in the absence of the Geneva Protocol.

At the time of World War II, more than a dozen of the belligerent states possessed significant stocks of chemical-warfare agents, the total supply by the end of the war in fact greatly exceeding the total consumption during World War I. This could be taken as evidence of disregard for the protocol were it not for the device of belligerent reprisals that has long been entrenched within the law of warfare as a sanction lending force to the law. In 1936 Italy had sought to justify its use of chemical weapons in Ethiopia as reprisal against Ethiopian illegalities.[110] But a capability for exacting reprisals was not the rationale underlying the World War II stocks in the policies of their possessors. In those British, American, and German policy documents that are now available for inspection it is apparent that a capability for retaliation, not reprisal,

---

[109] A particularly informative account of the chemical warfare of World War I is by Dorothy Kneeland Clark, *Effectiveness of chemical weapons in WWI* (Bethesda MD: The Johns Hopkins University Operations Research Office, November 1959, Tactics Division Stall Paper ORO-SP-88).

[110] For an account of the use of chemical weapons during the Italo-Ethiopian War drawing from Italian state archives, see Alberto Sbacchi, 'Legacy of bitterness: poison gas and atrocities in the Italo-Ethiopian War 1935-1936.' in *Geneva-Africa* (Geneva: Institut d'Etudes du Développement) XIII (1974), 30–53. This paper also records the consideration given by Mussolini and his then commander-in-chief in East Africa, General Badoglio, to the use of biological weapons, the former proposing, the latter opposing.

was the dominant consideration, and that this was explicitly valued for its in-kind deterrent properties. Churchill, Roosevelt, and Hitler all advertised their chemical weapons in just such terms. Herein, then, lies an explanation for the abstention from chemical warfare during World War II that is independent of the existence of the Geneva Protocol; and it is not inconceivable that, in the many other wars where chemical weapons have not been (and are not being) used, circumstances favouring use have simply not obtained. It is not as though the abstention has been complete: there have been allegations of the use of chemical weapons in at least 29 of the 200 or so wars of the past 55 years.[111]

Chemical weapons have three major characteristics which set them apart from more conventional weapons. In the first place, like biological and radiological weapons, most of them belong to the category of 'biospecific' weapons, being adapted only for assaults on life.[112] Killing or disabling large numbers of enemy people (or other living organisms of value to the enemy) has sometimes been a primary objective of military operations, but it has rarely been the sole objective, so much so that a fighting force carrying biospecific weapons as part of its basic load will have a restricted military utility. Secondly, because individuals can be protected relatively easily against many of the effects of chemical weapons without at the same time preventing their pursuit of other activities, it may be unusually difficult for a military commander to predict the likely outcome of any chemical-warfare operation he may mount, or to estimate the expenditure of chemical munitions likely to be needed. Thirdly, the effects of chemical weapons manifest themselves over a time-period that is very different indeed from that of conventional weapons, being characterized by a delayed onset, a prolonged duration, and a dependence on weather and other environmental conditions. This, too, limits predictability and requires abnormal skills for exploitation. It may also so weaken controllability in both space and time that, in battlefield operations, the consequent hazard to neighbouring friendly forces or noncombatants may restrict use of the weapons. Taken together these three features mean that, even on purely military considerations, the circumstances favouring chemical warfare may be rare. And it is rare, also, to find any serving military officer who speaks favourably of chemical warfare.

Seen as constraints on the use of chemical weapons, these features are not necessarily over-riding or irreducible. With sufficient technical competence, field intelligence, logistical flexibility, and advance planning, the military benefits of chemical weapons could no doubt often be realized without excessive military cost. Yet

[111] On the evidence currently available, few of these allegations can be taken as authentic. For particulars of allegations prior to 1971 see Stockholm International Peace Research Institute (SIPRI), *The problem of chemical and biological warfare* (6 vols; Stockholm and New York 1971-5). 1, 125–214.

[112] There exists in international law a category of chemical weapon that may be termed 'anti-matériel', such as the 'equipment or apparatus expressly designed to use, for military purposes, the ... anti-lubricating or catalysing properties of any chemical substance' specified in the 1954 revised Brussels Treaty (which established the Western European Union and placed restrictions on West German armament). So far as is known, no country has equipped itself with such weapons; they remain a curiosity of ·World War II weapons research.

this would require much prior preparation: appropriate organization of combat and support units, special training of personnel, procurement of special equipment, and other such mission-specific allocations of manpower and material resources. If sought on anything other than a trivial scale, these allocations would be at clear cost to other programmes of military preparedness and would entail significant organizational changes, the more so since chemical warfare lies well outside the mainstream of military theory and practice. The necessary preparations would therefore have to overcome not only the inertia of any military bureaucracy but also the active opposition of its disaffected components. Mechanisms can thus be described whereby the various operational constraints on use of chemical weapons become transformed into institutional constraints on the acquisition of a chemical-warfare capability.

It is at this level, arguably, that the Geneva Protocol exerts its greatest influence, for the existence of explicit and well-known prohibitions cannot fail to aggravate the institutional constraints.[113] There is a reinforcing mechanism as well. Because of the international and domestic political ramifications of chemical-warfare armament that stem from its legal status and peculiar emotiveness in public opinion, all but the most routine matters for decision tend to be pushed up into the highest reaches of intragovernmental politics; where, for a variety of reasons, neglect, deferment, or carry-on-as-before are more likely outcomes than positive action necessitating institutional change. In short, the Geneva Protocol, far from being an irrelevancy, may be seen as a principal factor imposing political considerations on purely military ones in chemical-warfare decision-making. It has thus constituted a major obstacle to the assimilation of chemical weapons into national force postures.

Of course, the weight carried by political considerations and their general complexion will be governed by the nature of the body politic, whether one is talking of national or of governmental politics. The strength of the institutional constraints, residing ultimately in the inherent tendency of any organization to resist change that may affect its basic structure and way of doing things, will vary from country to country; so too will the strength of the reinforcing political constraints. How great the variations may be is a matter for debate. Some commentators contend, for example, that because the assimilation of chemical weapons into the forces of the USSR cannot have been inhibited by factors of a type that have operated strongly in the United States (such as the opposition of an alert public opinion mediated through the Congress), the Russians have now assimilated chemical weapons whereas the Americans have not.[114] But such contentions are partial: they do not admit the possible operation of inhibiting factors peculiar to the Soviet system; nor do they admit that

[113] See, further, Abram Chayes, 'An inquiry into the workings of arms control agreements,' *Harvard Law Review* LXXXV (March 1972), 905-69.

[114] See, especially, William M. Carpenter *et al*, *Evaluation of chemical warfare policy alternatives 1980-1990*, prepared for the United States Office of the Assistant Secretary of Defense for International Security Affairs, under contract MDA903-76-C-0383 (Stanford Research Institute Strategic Studies Center, February 1977), III, 6–8. For further discussion of asymmetric assimilation see my *Should NATO keep chemical weapons? A framework for considering policy alternatives*, SPRU Occasional Paper no 4 (Brighton: University of Sussex, Science Policy Research Unit, August 1977). Editors' note; this is listed in Chap. 9 as item 87.

the important constraints may be ones that are common in kind, if not necessarily degree, to any system of government. Outside the USSR very little is known for certain about present Soviet chemical-weapons capabilities and intentions, though much has been surmised.[115]

The relationship between assimilation and the prevailing legal norms is necessarily a continuing and reciprocal one. Assimilation may be expected to proceed most rapidly during periods of general rearmament: when the total resources available for war-preparedness programmes expand so as to diminish at least the more sensitive opportunity costs of chemical-warfare allocations. The particular types of new chemical-warfare capability then acquired are at first likely to be those that can be accommodated the most readily within existing force posture. Thus, aircraft munitions (bombs and spraytanks) predominated in the procurement of chemical ammunition by the major powers during the onset of World War II.[116] This circumstance, coupled with the disdain with which most of the senior army commanders of the period regarded chemical warfare, served to concentrate contingency planning for use of chemical weapons more at the strategic than at the tactical level, thereby consolidating the emergent posture-led policies of in-kind deterrence. Battlefield chemical weapons, though available in increasing numbers, did not become assimilated to the extent that any belligerent could realistically contemplate their use. Such incentives as there were to do so (for Germany, for example, when facing the Allied landings on the Normandy beaches) fell foul of the realization that even minor tactical use was likely to be met with terrorizing chemical air raids on civilian populations, including those of allied countries. But as the tide of war lifted this constraint from some of the belligerents, the latter did begin to consider exploiting their stocks of chemical weapons (as during United States planning for the invasion of the Japanese home islands). For them, assimilation had thus advanced significantly, even though it had not done so to the point where there was sufficient organizational and logistical readiness to support resort to chemical warfare.[117]

State practice had thus been brought close to a fundamental departure from the legal norms. As it was, the widespread adoption of in-kind deterrence policies had in effect already diminished the absolute ban on chemical warfare to a disavowal only of first use of chemical weapons. For some states (led by France and soon followed by the Soviet Union and Britain, though not by Germany) this custom

---

[115] Open Western publications on this subject are of most uneven quality, but see Charles J. Dick, 'Soviet chemical warfare capabilities,' *International Defense Review* XIV (1981), 31·8, and his contribution to the forthcoming issue of *Soviet Army Review Annual*. A discussion of the limitations of the available literature is contained in my 'Chemical warfare capabilities of the Warsaw and North Atlantic Treaty Organizations: an overview from open sources,' in SIPRI, *Chemical weapons: destruction and conversion* (London 1980), 9–56. Editors' note. This is listed in Chap. 9 as item 96.

[116] See, for example: Frederic J. Brown, *Chemical warfare: a study in restraints* (Princeton 1968); Olav Groehler, *Der lautlose Tod* (East Berlin 1978); and Paul Harris, 'British preparations for offensive chemical warfare 1935–39.' *Journal of the Royal United Services Institute* CXXV (June 1980), 56–62.

[117] Brown, *Chemical warfare*.

had been foreshadowed in the reservations of the right to retaliate in kind entered when ratifying the Geneva Protocol. As of today, about one-third of the parties have made (or succeeded to) this reservation, including all five permanent members of the United Nations Security Council; and the Geneva Protocol is often held up as a possible model in proposals for a no-first-use agreement on nuclear weapons. With a view to redressing this situation, recommendations have been made in recent years that those states that have made reservations should now withdraw them. A few have actually done so.

A general proposition may now be stated. The strength of the chemical arms-control régime resides in the power of the prevailing legal norms on chemical warfare to inhibit assimilation of chemical weapons; but as assimilation proceeds, the norms are threatened. An effective route to a strengthened régime should therefore lie in the creation of specific new constraints on assimilation. New arms control measures that do not constrain assimilation are unlikely to survive.

Support for this general proposition may be drawn from the more recent assimilation history of chemical weapons in the United States (which is the only country to have opened pertinent state papers). The seminal policy document for American chemical-warfare programmes during the Cold War rearmament is the 1950 Stevenson Report.[118] Its basic recommendation, 'that the necessary steps be taken to make the United States capable of effectively employing toxic chemical agents at the onset of a war', was accepted by the defense secretary and the joint chiefs of staff, and implementing directives setting out responsibilities and schedules were duly issued. But another of its recommendations, 'that chemical, biological and radiological weapons not be restricted by a policy of use in retaliation only and that the present interim policy regarding chemical warfare be changed accordingly', was not accepted. It survived, however, as a bone of contention; and when the chemical-warfare preparedness programme persistently failed to keep up to schedule, the restrictive use policy was commonly portrayed as the primary obstruction, despite exhortation from the secretary of defense that 'the policy must not be allowed to deter achievement of preparedness in chemical warfare'.[119] The argument was, essentially, that the services could not be expected to give the necessary priorities to development of doctrine and combat readiness for use of a weapon that might never be permitted them. By 1956 a compromise policy on chemical (and biological) weapons was being considered by the joint chiefs, promulgated in March 1958 as 'preparedness for use at the discretion of the President'. Thus, to accelerate assimilation, the United States government was prepared to abandon its no-first-use policy. The way was then open, a few years later, for the adoption of three of about a dozen different applications of chemical and biological warfare proposed by the Army Chemical Corps as its contribution to the growing American counterinsurgency effort in Indochina.[120] As

[118] *Report of the Secretary of Defense's Ad Hoc Committee on Chemical, Biological and Radiological Warfare* (chairman, Earl P. Stevenson), 30 June 1950.This committee had been convened in October 1949.

[119] United States, Department of Defense, directive no 200.01-ITS, 21 December 1951.

[120] In all, about 100,000 tons of chemical-warfare agents were employed during the Vietnam War by United States and allied forces. For particulars (derived from a comprehensive review of United

it happened, new obstacles to assimilation were to be generated by the Vietnam War, so much so that in 1969 President Nixon reinstated the no-first-use policy.

Now, twelve years later, the United States is again on the verge of a new round of chemical-warfare rearmament, in the form of the multibillion-dollar 'binary-munitions' programme for which President Reagan has sought facilitating appropriations in his supplemental defence budget for the fiscal year 1981. If this programme goes ahead, can one expect reaction against the no-first-use policy to be entirely absent from the consequent assimilation processes?

II

Rearmament and war are not the only circumstances that may stimulate assimilation of chemical weapons. Another requirement is some form of institution dedicated to chemical warfare standing in readiness to promote chemical weapons when circumstances are favourable. One may, indeed, describe the assimilation history of chemical weapons in the United States in terms of the changing weight of the Army Chemical Corps within the United States defence bureaucracy; and no doubt the same could also be done for the USSR, were corresponding data available, in terms of the influence enjoyed by the Chemical Directorate of the Soviet defence ministry.[121]

Technological change, transmitted through such institutions, stands out as a further powerful influence on assimilation. Those forms of change which may be seen to have been especially influential fall into two broad categories. One category comprises innovations that have loosened the military-operational constraints on the use of chemical weapons. The pre-eminent example is the advent of the 'nerve gases': organophosphorus compounds that exploit the dependence of parts of the human nervous system on an enzyme which they can reach and inhibit with extraordinary specificity. They appeared, initially in a German industrial laboratory concerned with pesticides, soon after the great opening up of organophosphorus chemistry in the 1930s. Exemplified by poisons such as Tabun (1936), Sarin (1938), Soman (1944), and the V agents (1952), the nerve gases are much more amenable to use in mobile ground warfare than their predecessors because of the rapidity of their action combined with the smallness of their effective doses. Their significance became evident not so much during World War II (when only Germany possessed nerve gas on any scale, and that only Tabun, which was not so striking an advance as its successors) as during the Cold War rearmament. By that time nuclear weapons had greatly diminished the status of chemical warfare, not least by usurping roles chemical weapons had previously possessed as instruments of strategic attack or retaliation. The nerve gases, with the increased tactical utility they appeared to confer on chemical warfare, offered some compensation. Yet to capitalize on that utility, even

States government and other sources), see Arthur H. Westing for SIPRI, *Ecological consequences of the second Indochina war* (Stockholm 1976). About 90 percent of this tonnage comprised herbicides used for crop destruction and defoliation, the remainder being the irritant agent CS used as an adjunct to firepower and manoeuvre. There were allegations of the use of chemical anti-personnel casualty agents by both sides, but they remain unsubstantiated.

[121] On which see United States Army Foreign Science and Technology Center, *Chemical warfare capabilities – Warsaw Pact countries*, AST-1620S-104-79 (31 October 1979), 16.

for in-kind deterrence only, would demand close integration of chemical weapons into force structure and military doctrine; and the institutional constraints on such assimilation were unlikely to have been weaker than they had been during World War II. It seems that only the United States, the Soviet Union, and France considered it worthwhile equipping their forces with nerve-gas weapons, though for a time other states (such as Britain) maintained residual stocks of World War II chemical weapons.

The technological change represented by the nerve gases preserved the World War I mode of chemical warfare from the oblivion to which the abstention during World War II seemed likely to consign it. In the second category of influential technological change are the developments which have opened up other modes of chemical warfare, ones to which resort may be less constrained by public opinion. Thus, during World War II chemicals were discovered in Britain and the United States which, in terms of application densities, are as deleterious to plants as nerve gases are to people. These chemicals were soon to appear in commerce as selective weed killers (such as 2,4-D), but military applications of more concentrated formulations were under close study as well. The British chiefs of staff noted 'their possible use for purposes of internal security within the Empire, e.g. for the destruction of food supplies of dissident tribes in order to control an area'.[122] This was a notion that would later be put into practice in Malaya, Algeria, Vietnam, Portuguese Africa, and (if recent reports are true) Eritrea.[123] Again, the discovery of such intensely psychotropic drugs as LSD (lysergic acid diethylamide) during the 1940s and 1950s opened up the possibility of chemical weapons which could cause large numbers of casualties without actually killing people. The associated vision of war without death was widely advertised in the United States during the late 1950s and did much to persuade Congress to increase chemical-warfare appropriations at that time. So far, however, this notion of an 'incapacitating' chemical casualty agent has not (so far as is known) been reduced to practice in the form of weapons having sufficient military attractions to remain in today's active operational inventories; a few have come and gone; the search continues. But what the notion has done is to promote the use in war of those short-term incapacitants used by police forces as 'tear gases': irritant agents, such as CS gas, that harass rather than cause casualties and which can enhance the effectiveness of conventional firepower and tactical manoeuvre. For the United States during World War II and the Korean War, and apparently for Britain during the Greek civil war, no distinction was seen between such applications and other forms of antipersonnel chemical warfare, and the occasional proposals that came up for exploiting them during the fighting then were accordingly rejected. But the practice was put to wide use during the Vietnam War and may well be what lies behind (if anything does) the recent allegations of chemical warfare in Laos, Kampuchea, and Afghanistan.[124]

---

[122] United Kingdom, Chiefs of Staff Committee, Joint Technical Warfare Committee, 'Potentialities of weapons of war during the next ten years', T.W.C. (45)42, 12 November 1945.

[123] The military or any other hostile use of herbicides 'having widespread, longlasting or severe effects' is now accepted, at least by the United States, as prohibited under the terms of the 1977 Environmental Modification Convention.

[124] Hard evidence on what, if any, chemical agents have been used in these countries is scarce and contradictory. The most substantial body comprises the records of interviews with H'Mong refugees

These lesser forms of chemical warfare present a peculiarly insidious threat to the chemical arms-control régime. Their quite widespread use has familiarized many military personnel with the operational intricacies which herbicides and irritants share with all other types of chemical-warfare agent, thereby loosening at least one set of constraints on assimilation. Because those states that have used them have felt obliged to maintain that the weapons are not subject to the same proscriptions as other chemical weapons, the scope of the prevailing legal norms has been called into question. And since no unambiguous distinction can be drawn between chemical irritants or herbicides, on the one hand, and 'poisonous ... materials', on the other, the entire body of law is thereby impugned.

Further examples of both these categories of technological change may be expected to appear in the future. It would be invidious to speculate on them in any detail here. It may be noted, however, that the science embodied in the latest chemical weapons (at least of Western arsenals) is that of the early 1950s, and a great deal has happened in the laboratories of the life and molecular scientists since then. The chemical industry is, moreover, beginning to shift away from its petroleum base; were investment to increase substantially in the biotechnologies, as now seems to be happening, a whole range of novel or currently obscure supertoxic substances might then become accessible on a large scale. Pronounced military attractions can be seen for an amenable casualty agent with greater (about two orders of magnitude) toxicity, lethal or incapacitating, than the nerve gases.[125]

Technological change may also act, it should be observed, to retard assimilation of chemical weapons. Currently within the realm of technical feasibility are ways of reducing the cumbrousness and increasing the efficiency of anti-gas protection. Existing protection already goes a long way towards negating the utility of chemical weapons. Again, the general trend in non-nuclear weapons design is towards greater precision of effects, greater single-round effectiveness, and reduced collateral damage: characteristics which are often the very opposite of those of chemical weapons. As force structure and doctrine adapt themselves to these new weapons, chemical weapons may become still less assimilable.

III

The closer one studies the history of chemical weapons in particular countries, the more one is led to a view of national levels of chemical armament as having been determined less by demand than by supply. The momentum of past armament mediated through a favourable institutional structure appears at least as responsible as any clear conviction of need for the sporadic, sometimes transient, accumulations of chemical weapons by states during the general rearmaments of the past fifty years. Because poorly assimilated weapons can make no certain contribution to a state's war-fighting capacity, one can see why these accumulations have been justified

---

from Laos published by the United States government. While these records have been interpreted to indicate the use of at least three different agents, including nerve gas and perhaps even an advanced toxin agent, they are not obviously incompatible with heavy use of an irritant such as Adamsite which contains an arsenical moiety capable of persisting in the environment as a lethal hazard.

[125] See SIPRI, *Problem of chemical and biological warfare*, II, 279–81.

largely in terms of their deterrent value: a hedging of bets against an adversary having assimilated them more closely. Those governments that have expressed themselves in public on the subject – which have conspicuously not included, since 1938, that of the USSR – have appeared content to confine their declaratory chemical-weapons policies to retaliation only; and some governments have now expressly renounced chemical weapons altogether. All this has served to reinforce the current chemical arms-control régime of no first use and thus further to constrain assimilation of the weapons.

The existing régime should therefore be capable, one might suppose, of holding in check developments that would bring about peoples' worst fears regarding chemical warfare: the mass destruction of life, especially among noncombatants, which its use could entail, or the dangerous regional instabilities that might follow from the proliferation of chemical weapons out from under their constraining nuclear umbrella.

There are two main reasons for judging this view to be unduly complacent. One is displayed in the current alarm in the West about the tilting of the 'military balance' in favour of potential adversaries, for this has brought forth specific alarms about Soviet chemical-warfare capabilities and intentions which have been exacerbated by an insufficiency of reliable intelligence to the point where otherwise responsible commentators now seem to have no hesitation in portraying Soviet chemical-warfare forces as structured into a first-use capability that is regarded as a conventional means of warfare. Whether correct or not, this perception of complete assimilation of chemical weapons by the USSR must necessarily weaken confidence in the strength of the existing chemical arms-control régime. That confidence has, moreover, been weakened still further by allegations that forces of the USSR and of Soviet client-states have been using chemical weapons in Afghanistan, southeast Asia, and the Horn of Africa. The United Nations secretary-general has been empowered by the General Assembly to investigate the reports,[126] but the political context hardly favours a conclusive outcome. The allegations fester on, heightening mistrust.

The second cause for doubt lies in the fundamental instability of the mechanisms from which the chemical arms-control régime derives its strength. There may be little about a retaliatory chemical-warfare capability to make it inherently distinguishable from an initiatory one, other than the prevailing intentions of its possessors; and these may change as assimilation advances. Stockpiles built up as deterrents may thus have the effect of sustaining, even inducing, the threats they are supposed to deter; and once accepted as necessary components of a state's military posture, they may become enmeshed within the apparently ceaseless spiral of modernization and counter-modernization whereby assimilation must sooner or later result.

The remedy that has been sought since 1968 is that of chemical disarmament. The idea of negotiating a whole category of weapons out of existence is a singularly optimistic one, but events seem to have given it a growing validity. In 1972 an

---

[126] United Nations, General Assembly, resolution 35/144C, adopted after a stormy passage on 12 December 1980 by a vote of 78 to 17 with 36 abstentions. The United Nations has appropriated $653,700 for the investigation.

international convention was opened for signature which outlaws the development, production, and stockpiling of biological and toxin weapons and which commits states-party to continue negotiations 'in good faith' on chemical weapons; it was negotiated within the Geneva Committee on Disarmament under the co-chairmanship of the two superpowers and entered into force in 1975. In the communiqué from the Moscow summit of the summer of 1974, the United States and the USSR announced their intention of considering a joint initiative for a chemical-weapons convention, and in the summer of 1977 bilateral negotiations to this end commenced. By mid-1979 these had progressed to the point of broad agreement on the scope of the convention to be sought and on the use of on-site inspection as an instrument of verification by challenge. This progress was reaffirmed and extended on various points of detail the following year, notwithstanding the turmoil of Afghanistan. During 1980 the Geneva Committee on Disarmament itself began what are virtually parallel negotiations on a chemical convention, building upon the previous eleven years of exploratory talks and detailed technical study within the Committee.[127]

Progress in these endeavours will of course be determined by the degree of common interest the negotiating partners perceive in a treaty. It is certainly possible to identify reasons why chemical disarmament could benefit both the Soviet and the Western blocs. Whatever relative advantage the former may or may not currently possess in chemical-warfare capability, the prospect of the highly developed Western chemical industry finally mobilizing in earnest for chemical warfare can hardly be an attractive one. For the West, currently committed in the NATO area to its flexible-response, forward-defence strategy, there are the disturbing implications of the fact that, in military-operational terms, the more plausible retaliatory applications for chemical weapons, given the widespread deployment of antichemical protection, overlap strongly upon those currently assigned to battlefield and theatre nuclear weapons; chemical weapons thus threaten escalation control at its most crucial juncture, the decision on initial nuclear release.

Yet as the negotiations advance further, the purely practical problems of designing, let alone agreeing upon, a credible chemical disarmament régime have become increasingly apparent. To trade away the security benefits of possessing chemical weapons, or an open option to do so, against the promise of never being menaced by chemical warfare, a state must be able to assure itself that its adversaries too are observing the treaty. Those benefits may not in fact be great, given the narrow limits to the military utility of chemical weapons and the extent to which antichemical protection and the existence of nuclear weapons now vitiate the notion of in-kind chemical-warfare deterrence. Provided antichemical protection is kept up, quite a modest degree of verification should be enough: no more, one might argue, than

[127] For the latest progress reports of these bilateral and multilateral negotiations, see Committee on Disarmament, documents CD/112 of 7 July 1980 and CD/131/rev 1 of 4 August 1980, reproduced in 'Report of the Committee on Disarmament to the 35th General Assembly,' appendix 11, volume 11.

would suffice to diminish the likelihood of mistaken suppositions of abrogation.[128] Unfortunately the base from which a state may draw chemical-weapons capability is diffuse, stretching in principle – now that the apparently successful development of binary munitions[129] has obviated the former need for large factories dedicated to production of supertoxics – right across its military and chemical industrial base. The problem confronting verification is thus that of significantly increasing the transparency of the military-industrial milieux within which illicit activities might be taking place without at the same time exposing other activities that might legitimately be kept secret. The problem is further complicated by asymmetries both in the military-industrial transparency of different countries and in their normal intelligence-gathering capacities, the latter being most obvious as regards satellite reconnaissance. The current sticking point in the negotiations is, ostensibly at least, the role to be given to mandatory routine on-site inspection (as opposed to challenge inspections). The Western position is that this is an essential requirement. One may well wonder, however, whether, even with an army of inspectors having unlimited access, the assurance that could thereby be derived might not be false; and when suspicion and mistrust are high, the number of sites that some states party might require to be inspected, and the frequency of their inspection, could well exceed the limits of practicability.

It is now coming to be appreciated that, just as mistrust may aggravate the demands made of verification, so too may appropriate confidence-building measures diminish those demands. Functional relationships between confidence-building measures and verification methods have thus been proposed, such as early declaration of stocks and production facilities with a view to the formal verification provisions of the treaty then being focussed on their elimination. Other such schemes can be envisaged, more, perhaps, than have yet been studied in detail by the negotiating governments. So peculiarly complex is the verification aspect of chemical arms limitation that, as matters stand at present, these schemes may well offer the only way forward.

It is not easy to be optimistic about the prospects for an acceptable treaty on chemical disarmament. For this reason it could be worthwhile examining the potential of chemical-warfare confidence-building measures, not as adjuncts to a future chemical disarmament treaty, but as means for enhancing the existing chemical arms-control régime. The foregoing analysis suggests two measures in particular. One would be the institution of a standing consultative committee of states party to the Geneva Protocol: an organ that would be able to draw upon the services of experts, including laboratory facilities, to resolve whatever problems about the operation of the protocol a state-party might wish to raise. The second measure would be the institutionalization, possibly through a permanent secretariat of the

---

[128] For the reasoning behind this proposition, see my 'The negotiations on chemical-warfare arms control', *Arms Control* I (May 1980), 30–52. Editors' note. This paper is listed in Chap. 9 as item 098. The current book includes an extract of the paper.

[129] Rather than actual poison gas, a binary munition has separable loadings of less toxic chemicals adapted to react together to generate poison gas after the munition has been activated. For particulars, see my 'Binary nerve-gas munitions' in SIPRI, *Chemical disarmament: new weapons for old* (Stockholm & New York 1975), 21–99. Editors' note. This is listed in Chap. 9 as item 68.

consultative committee, of information exchanges designed to reduce those ambiguities in national chemical-warfare postures which may currently be stimulating destabilizing fears about underlying intentions. The value of the exchanges would depend on reciprocity developing in the volume and detail of the information flow. An obvious subject for the exchanges would be particulars of national research and other programmes in the field of antichemical protection.

In their application to the ills of the present chemical arms-control régime, such confidence-building measures would be palliative rather than remedial, for only indirectly could they retard any continuing assimilation of chemical weapons. But once in operation they could form a more solid basis than exists at present for the pursuit of real remedies: remedies that might range from the ambitious objective of comprehensive chemical disarmament to such narrower goals as the withdrawal of forward-deployed chemical weapons or other regional measures. With poison-gas rearmament now seriously being contemplated once again, time is running out.

7.4 "The Soviet Union and the Biological Weapons Convention" and a Guide to Sources on the Sverdlovsk Incident: Discussion of the Paper by Philip Towle[130]

Editors' note. An unusual outbreak of anthrax in Sverdlovsk (now Yekaterinburg, Russia) in 1979 triggered suspicions that the Soviet Union was violating the 1972 Biological Weapons Convention. While the USSR explained that the disease outbreak was the result of people eating contaminated meat, allegations persisted that it was caused by an accidental release of anthrax spores from an illicit bioweapons programme.

In this extract from 1982, Julian assesses the likely cause of the outbreak by analysing publicly available information. Peter Pringle suggested including the paper in this book, explaining that "Julian dissects the evidence on the Sverdlovsk incident, and uses it to say why further studies by academics are needed – as opposed to press reports – if arms control and good policy is to survive. Also, he poses his perennial question: how good can we expect government bureaucrats to be in coping properly with such evidence?"

After the collapse of the Soviet Union, Jeanne Guillemin, Matthew Meselson and Julian were able to conduct a thorough investigation of the incident and found that the 1979 outbreak was indeed a leak from an illicit bioweapons lab. In 1992, Russian President Yeltsin acknowledged the illicit Soviet biological warfare programme.

[130] Editors' note. This is a preprint of J. P. Perry Robinson, 'The Soviet Union and the Biological Weapons Convention' and a guide to sources on the Sverdlovsk incident'. July 1982. In *Arms Control* (London), vol 3, no 3 [Special Issue on 'The Verification of Arms Control Agreements', edited by Ian Bellany and Coit D Blacker]: December 1982, pp 41–56. It is listed in Chap. 9 as item 148. Unless otherwise indicated, the remaining footnotes for this paper are Julian's originals. Apart from correcting obvious typos and renumbering headings and tables in keeping with the requirements of this book, we have retained Julian's formatting and spelling throughout the paper and its notes.

We cannot doubt that the Sverdlovsk incident weighs heavily and adversely on the whole future of arms control. This was recognised from the start.[131] Indeed, much of the information that has kept the incident in the public eye has stemmed from quarters which equate arms control with a reprehensible pursuit of detente.[132] Philip Towle has now taken us beyond the early commentaries by differentiating aspects of the affair from which may be derived specific practical implications for policy.

The paper rests for the most part on an account of how some parts of the press have portrayed the Sverdlovsk outbreak of anthrax in relation to the obligations which the USSR assumed in 1975 under the 1972 Biological and Toxin Weapons Convention. Because the paper has been limited in its scope to only a narrow band of the available source material, we should not expect to understand from it why the US Government believes, as it apparently does, that the Sverdlovsk incident provides strongly indicative evidence of Soviet noncompliance with the treaty. Still less should we expect to gain an impression of any particular strengths or weaknesses in the assessment on which that belief rests. If we want to inform ourselves fully about these matters, we can do so only through access to secret papers or by talking with the US evaluators under circumstances uninhibited by the security considerations that necessarily attend intelligence affairs. Above all, we cannot assume that the full range of the relevant intelligence data is reflected in the press accounts; and if it were, we could not know it. Yet it is very largely through the press that opinions take root. Press treatment of the Sverdlovsk incident is thus a proper and important subject for study.

Personally, I believe that for those of us in the academic wing of the arms-control community, such study is not only proper and important; it is also obligatory. Supposedly at least, our research is less tyrannised by market demands and pressure of deadlines than that of journalists. We are meant to have well-developed critical and analytical faculties. We sometimes even have access to sources closed to the press.

---

[131] See, in particular: statement of Soviet Ambassador V. L. Issraelyan, 21 March 1980, during the closing session of the Biological Weapons Convention Review Conference; the several *Washington Post* editorials on the subject (22 March 1980, 26 April 1980, 2 July 1980 and 22 November 1981); the statement reviewed and approved by the Council of the Federation of American Scientists, in *FAS Public Interest Report* 33(6): 1–3 June 1980; and the references to the issue in *Proceedings of the 30th Pugwash Conference on Science and World Affairs Breukelen 20–25 August 1980* (London Pugwash Central Office, 1981), pages 16,26 and 223–4. As yet there seem to have been only brief references to the Sverdlovsk implications in the more academic arms control literature: B. M. Blechman, 'Do negotiated arms limitations have a future?', *Foreign Affairs*, Fall 1980, pp. 102–25 at note 4; R. J. Einhorn, 'Treaty compliance', *Foreign Policy* no 45 (Winter 1981/82) pp. 29–47; and Jane M.O. Sharp, 'Restructuring the SALT dialogue', *International Security* 6 (3): 144–76, Winter 1981/82 at p. 151. The last of these has an instructive comparison between the treatments given in the United States to the Sverdlovsk incident and supposed Soviet violations of SALT agreements; for another such comparison, see M. M. Lowenthal, 'Soviet SALT I compliance and the Sverdlovsk CBW incident: a comparison of United States reactions', Congressional Research Service, Library of Congress, 12 June 1980.

[132] For explicit commentary on the Sverdlovsk incident from this standpoint, see, for example, the editorial by W.F. Hahn in *Strategic Review*, Spring 1980; and the allusions to the incident in successive *Wall Street Journal* editorials (especially 22 October 1980, 6 November 1981 am! 20 May 1982).

In our modicum of independence from state apparatus we enjoy a privilege in return for which we are expected, among other things, to maintain an actively questioning stance towards received wisdom. So, for as long as there remains even a small chance that the press and the people who shape and direct security policy have got things wrong about Sverdlovsk, and may be sustaining one another in delusions from which we, as people who judge arms control to be crucial to long-term security, believe catastrophe could result, we have a duty to be critical; and publicly so. We might then appear, it is true, as apologists for Soviet behaviour, but that is an occupational hazard familiar to many of us. Towle's paper has set us on a path which, it seems to me, we ought to follow.

Towle could of course have written a great deal more than he has offered us about the details of the Sverdlovsk incident; and other people could no doubt have offered rather different readings. It is not my intention here to set out my own understanding of the affair, although in the Appendix I do identify and annotate the published sources on which it rests. Instead I will put forward a few detailed reactive comments which may add clarity to the background from which Towle's conclusions have emerged. The comments fall into three groups: ones having to do with the plausibility of the Soviet explanation of the Sverdlovsk anthrax; ones having to do with the evidence fostering doubts about that explanation; and ones having to do with the nature of the Soviet response to those doubts. My remarks close with a discussion of one particular implication, and a conclusion drawn from it.

7.4.1 Plausibility of the Soviet Explanation

From what is known about the epidemiology of anthrax and its history in the Soviet Union, Towle's imputation of a low plausibility to the Soviet tainted-meat explanation seems to me to need qualifying. Anthrax is not a rare disease in temperate regions. Even in Britain there are several hundred cases per year among livestock.[133] In the Sverdlovsk oblast alone some 270 different locations are registered by Soviet anthrax control authorities as 'fixed unsafe areas', and in the Sverdlovsk oblast between 1936 and 1968 there were, apparently, 159 outbreaks of anthrax.[134] Although the incidence of recorded cases of human, as opposed to animal, anthrax in the Soviet Union may now, as Towle says, be only a quarter of what it was in 1950, it was then running at the appalling level of one recorded cases per hundred thousand people,[135] so that today some 700 cases per year would be about the norm. Among those cases,

---

[133] V. Wyatt, 'Anthrax: recipe for a blunt weapon', *New Scientist* 87: 721–2 (1980).

[134] V. M. Popugaylo *et al* in *Pressing Problems 'of Anthrax Prophylaxis in the USSR: Abstracts of Reports* (1974): 50 (in Russian), as cited in I. S. Bezdenezhnykh and V. N. Nikiforov, 'An epidemiological analysis of incidences of anthrax in Sverdlovsk', *Zh. Mikrobiol. Epidemiol. Immunobiol.* 1980 (5): 111–13 (in Russian).

[135] P. N. Burgasov *et al*, *Sibirskaya Yazva* (Moscow: Meditsina, 1970) (in Russian; English-language translation available as JPRS 76944).

the great majority – more than 95 per cent[136] – could be expected to display the disease in its cutaneous form. As to the rare intestinal form (and it may be noted that smoked sausage is a common foodstuff in Sverdlovsk), the Soviet medical literature contains reference to at least ten different outbreaks during the period 1923–40 in different parts of the Soviet Union,[137,138] and to four others during 1946–52[139]: an average approaching five outbreaks of human intestinal anthrax per decade that were both recorded and referred to in open sources. Since the population of Sverdlovsk represents about 2 percent of the total Soviet population inhabiting known anthrax areas, and on the assumption that the ratio of population at risk to total population is much the same across those areas, one may therefore reckon that there is a 5–10 per cent probability of an outbreak of human intestinal anthrax occurring in Sverdlovsk each decade. Twenty-seven cases were the most recorded for any one of the Soviet outbreaks; in other countries, such as Kenya or Uganda, single outbreaks affecting five or ten times that number of people have been common. The Soviet Government has not, therefore, asked us to believe in something extraordinarily unlikely.

7.4.2 Contra-indicative Evidence

Towle says that the Soviet explanation 'is highly unlikely because of the timing, the location and the nature of the victims of the disease'. The epidemiological survey of the Sverdlovsk anthrax outbreak published one year later in the Soviet scientific literature[140] provides some data on these matters but not many, and the US Government has disclosed little of its own intelligence on them. In advancing his opinion, Towle was relying on his earlier observation that the peak incidence of anthrax in a typical year is usually during the summer, not the spring, and that anthrax is most prevalent in low-lying areas. These are both points that have been addressed in Soviet commentary[141]:

> What happened in Sverdlovsk oblast? The winter of 1978/79 happened to be severe, and many weakened animals were found among the sheep and cattle; their susceptibility to infection increased. It must be said that in our oblast anthrax troubles have also been noted earlier. Therefore, despite the yearly preventive inoculations, the danger of disease to livestock grew.

---

[136] ibid., and see also L. M. Marchuk and O. F. Bogatyrev, 'Epidemiology and hygiene', *Feldsher. Akush.* 40(2): 10–15 (1975) (in Russian).

[137] K. K. Aver'yanov, 'Epidemiological outbreaks of anthrax', *Zh. Mikrobiol. Epidemiol. Immunobiol.* 1940 (8): 35 (in Russian); and I. I. Elkin (ed), *A Course in Epidemiology* (1947; published by Pergamon in an English-language translation in 1961).

[138] A. Ya. Zaporozhchenko, 'On the casuistry of cases of intestinal anthrax', *Gigiena i Sanitaria* 1943 (7): 44–50 (in Russian).

[139] A. Ya. Zaporozhchenko, 'The epidemiology and clinical course of intestinal anthrax', Zh. Mikrobiol. Epidemiol. Immunobiol. 32(7): 41–5, 1961 (in Russian).

[140] Bezdenezhnkyk & Nikiforov, *op. cit.*

[141] 'Strict observance of veterinary regulations', *Chelovek i Zakon*, September 1980 (in Russian).

The earlier instances of anthrax in the oblast are, as noted above, a matter of record.

The evidence militating against the Soviet explanation is altogether stronger if Towle's questions of time, location and nature are considered on a smaller scale: days or weeks, rather than seasons; places, down to individual streets, rather than general topography; detailed symptomatology. On the last of these, especially, the contra-indications are impressive. There exist, for example, 'second-hand eyewitness' reports of signs and symptoms which indicate greater pulmonary manifestation of the disease than might be expected were the anthrax bacteria to have invaded their victims solely by way of ingested food. There are even said to be autopsy data that point to anthrax of the lungs, not of the gut. And there is said to be no irrefutable evidence available in the West of one single case of intestinal anthrax at Sverdlovsk. Yet caution is needed before concluding anything definite here, for the intestinal form of anthrax is so sparsely described in the authoritative medical literature, and cases are so rare in Western countries, that there is no consensus on what exactly its manifestations should be. Much uncertainty therefore has to be conceded, however strongly individual analysts may hold to one or another opinion. That uncertainty is compounded, moreover, when following another of Towle's lines of questioning, that of incubation period. The timespan over which people in Sverdlovsk were coming down with the disease is variously reported as between six and ten weeks. This is far too long to be credibly explicable in terms of natural variation in incubation period following exposure to a transitory cloud of airborne anthrax spores. It points instead to a quasi continuous, not instantaneous, source of infection.

Towle finds it difficult to see what sort of evidence could be adduced against the Soviet explanation other than that which has 'already surfaced'. In fact, rather little has surfaced, and almost all of it in the form of conclusions drawn rather than original data. It seems to me that five principal categories of information would be required.

a. Baseline data, in forms more detailed than the aggregated epidemiological data quoted above, with which to assess the probability of anthrax naturally entering food-chains reaching the reported victims at the particular times and locations in question.
b. Detailed symptomatology from which to evaluate the diagnosis of anthrax and attempt a differential diagnosis of inhalation versus intestinal anthrax.
c. Bacteriological data to verify the diagnosis of anthrax (unverifiable without such data).
d. Autopsy data to sharpen the differential diagnosis.
e. Circumstantial data (from meteorological records, satellite observation, accounts of special quarantines or other such departures from the norm, etc.) with which to evaluate several different hypotheses about causation.

This is information of an intricate kind, much of it accessible by outside parties only through channels of a nature barely capable, if at all, of transmitting it reliably in requisite detail. Interpretation of the data thus acquired, in any case demanding high technical competence, would thus be complicated by the necessity of allowing for distortion and other limitations of transmission. It is known that the intelligence-collection efforts of the US and other foreign intelligence services have acquired data from at least four of the five categories above. But we do not know whether the accumulated data have been sufficient to permit confident conclusions, notwithstanding the rather confident conclusions we have heard from US Government policy-level officials. Nor do we know whether those officials are adequately aware of the full range of data that is needed before confident conclusions can be drawn. This last point seems to me to be the real crux of the matter before us. I shall return to it.

In relation to the supposed dissidents' reports, Towle rightly remarks on 'the atmosphere in a town overtaken by a fatal epidemic'. We should not overlook – as the Americans evaluators did until February 1981 – the presence in Sverdlovsk during much of the time in question of a Russianspeaking American physical chemist and his family. Professor Donald E. Ellis of Northwestern University, Illinois, entered Sverdlovsk, where he was to work at the Institute of Chemistry as part of an official US-Soviet scientific exchange, some ten days after the epidemic reportedly began; left one month later to work in a lab in Novosibirsk; and returned to Sverdlovsk about one month after that. He was aware of no special restrictions having been placed on his movements, and knew nothing about the epidemic until after his return to America at the end of the summer.[142]

7.4.3 Soviet Responses

There is one other category of possible explanation for Soviet reticence on the details of the Sverdlovsk anthrax in addition to those which Towle lists: administrative or related blunders during attempts at controlling the epidemic, even civil malpractices exacerbating it. Several possibilities can be envisaged, ranging from the use of faulty vaccine to official complicity in meat black-marketeering.

7.4.4 Two Problems

There are implications in the foregoing comments which bear on the conclusions presented in Towle's paper. In the Sverdlovsk case, as in the Yellow Rain mycotoxin one, the evidence for infringement of the relevant international law is indeed highly technical. For that reason there is, as Towle rightly says, difficulty in convincing the public. I believe, though, that that technicality gives rise to a more fundamental question: How good can we expect governmental bureaucracies to be in coping with it properly? In principle, a government such as the American or British will have at its disposal as good a competence on most scientific and technical matters as it is

[142] L. H. Gelb, 'Keeping an eye on Russia', *New York Times Magazine*, 29 November 1981.

possible to have. That is in principle. In practice, such resources may prove difficult to utilise.

An extreme case illustrates the point. A January 1982 release by the US State Department on Yellow Rain had to do with data gained from analysis of blood samples taken from supposed trichothecene-mycotoxin victims in Cambodia. It contains this passage[143]:

> A trend toward depressed white cell count in the nine individuals exposed to chemical agent was observed. One effect of trichothecenes which persist for several weeks following exposure is a depressed white cell count.

There is a lot of science in this press release: numbers, formulae, peculiar words, &c., &c, all of it creating an impression of diligent scientific inquiry having been at work – as, indeed, there has been. Yet two sentences further on, there is this:

> ... the sampling size was so limited that there was no real statistical difference between control and exposed groups when Student's *t*-test was applied.

This qualification of course negates the 'trend' proclaimed in the first passage, thus rendering negligible as evidence that particular finding in the blood, which was one of the two main messages of the press release. One would like to think that such a display of innumeracy was typical only of those who write the public releases of the State Department. Regrettably, there are other indications which do not allow us to assume this. How far back, then, into the intelligence-evaluating channels may such deficiencies extend? And can we assume that the people in government who have to act on the intelligence appraisals have a proper capacity for distinguishing what may loosely be called a scientific fact from a scientific opinion – for recognising the extent to which the appraisals they are given require qualification in the light of such distinctions? And should it ever come to the launching of public accusations of treaty violation, could we be confident that the decision makers would know that they have sufficient data to do so?

There is a second, but related, problem. In the case of the Sverdlovsk anthrax, the US Government has not taken the position that the available evidence is conclusive of treaty violation; only that it 'points strongly' to such a conclusion. This presumably, although not necessarily, means that some modest probability is still conceded to the Soviet explanation being correct. Back in September 1980, when answering a question about this before the Senate Armed Services Committee, the Director of the US Defense Intelligence Agency said flatly: 'We don't know and [the Soviets] are not sufficiently forthcoming to bridge the gap between what we believe happened and what they say happened'.[144] If that gap is bridgeable in theory – i.e. if the Soviet

---

143 US Department of State, 'Analysis of blood samples from chemical attack victims indicates probable trichothecene exposure', 4 pp., 29 January 1982. This release is reproduced in the report by the Secretary of State to the Congress on 22 March 1982, *Chemical Warfare in Southeast Asia and Afghanistan*, published as Department of State Special Report no 98. However, in the version of the finding reported to the United Nations (UN doc A/37/102 of 25 February 1982), the reference to a 'trend' does not appear.

144 Lt-Gen. E. T. Tighe, in US Senate, Committee on Armed Services, hearing, *Chemical Warfare*, 4 September 1980, at p. 23.

explanation is correct – it may still be unbridgeable in practice: the precise and highly technical nature of the information that is missing may mean, for the reasons discussed earlier, that the information is simply not accessible.

7.4.5 A Conclusion

The information which the US Government would need before its assessment of the probability of the Soviet explanation being correct could be shifted in one direction or the other is apparently of a type that the Russians can hold secure, intentionally or unintentionally. And the information which the Russians would need to release in order to promote acceptance of their explanation is defined by the contra-indicative evidence acquired by the Americans, which the US Government currently seems to be holding secure.

If these suppositions are correct, the only remaining way forward to the truth would seem to lie in private bilateral discussions at a technical level, well away from propagandists: meetings during which data that either side would otherwise not disclose are shared and evaluated by experts in the appropriate scientific disciplines. If so, can it be assumed that the decision-makers on either side appreciate that the technicalities of the situation are such that this form of intercourse may indeed be the last resort: that it may be the only possible way to proceed if the prospects for arms control that have been damaged by the Sverdlovsk incident are to be restored? On the Yellow Rain issue, too, the same question may be asked. And if, despite my contrary imputation earlier, such an assumption is indeed justifiable, would not a failure to institute the bilateral contacts amount to a repudiation of arms control?

7.4.6 Appendix: A Guide to Sources on the Sverdlovsk Incident

There is this passage in the latest annual report to the US Congress by the Secretary of Defense[145]:

> The United States now has very good reasons for believing that the Soviet Union has violated the Biological Weapons Convention – an arms control treaty negotiated, signed, and ratified when the illusions of 'détente' were most prevalent. We have evidence of an inadvertent release of anthrax bacteria from a highly secured military installation in the Soviet city of Sverdlovsk during the Spring of 1979. This incident points strongly, we believe, to biological warfare activities in the Soviet Union that exceed those allowed under the treaty for protective purposes. We regard the explanation offered by the Soviet Government – that the outbreak of anthrax was due to natural causes – as inconsistent with our analysis of the evidence.

---

[145] 1 Caspar W. Weinberger, US Secretary of Defense. *Annual Report to the Congress: Fiscal Year 1983* (8 February 1982), page I–21.

The nature of that evidence and of its inconsistencies with the Soviet explanation can be discerned only dimly in the open literature on the Sverdlovsk incident. While the available source material gives us some factual reason for going part of the way towards Secretary Weinberger's conclusion, to go the rest of the way we must trust to his judgement. Exactly the same is true of the Soviet explanation: fragments of supporting evidence are available, but with great gaps remaining.

For people who are hesitant about accepting on trust what either the Soviet Foreign Ministry or the US Defense Department asserts, the notes which follow identify the major items of source material from which the reader may judge just how great must be the leap of trust in accepting one or the other rendering of events. The notes are also meant to offer some slight guidance through the maze of ancillary problems: of lies – the possibility of a published source in fact being disinformation for purposes of deception or provocation; of rumour purveyed as report; of opinion masquerading as fact, and prejudice or untutored surmise as informed judgement; of secrecy motivated by cover-up, not by justifiable security; of obscuring scientific and other technicalities; of intelligence assessments becoming stripped of their caveats and qualifications during transmission into the public domain; and other such impediments of understanding. But here, too, most must inevitably be left to the reader's personal judgement.

There are four bibliographical sections and one chronological one. The purpose of the last is to aid inquiry by future students of the incident, at such time as documentation currently held secret becomes open to public inspection.

7.4.6.1 An Annotated Bibliography

There are three broad categories of source material available in the public domain which offer details about the Sverdlovsk incident: (I) reports published in the West but purportedly transmitted from within the USSR through samizdat channels; (II) publications that seem to disclose official US intelligence; and (III) Soviet publications. There is also a mass of secondary and tertiary material drawing from these sources. The principal items in each category are these:

Category (I) Sources

N. N., 'Sverdlovsk again faces death', Possev (Frankfurt) 36(1): 7–8, January 1980 (in Russian). This item is datelined 'Sverdlovsk, November 1979'.

Anon., "Reliable information received in Israel ... [reported by] Jews arriving from the Soviet Union", reported in Ma'ariv (Tel Aviv), 21 March 1980.

M. Popovskiy, statement before the Subcommittee on Oversight of the Permanent Select Committee on Intelligence, US House of Representatives, during a public hearing on 29 May 1980 subsequently published as The Sverdlovsk Incident: Soviet Compliance with the Biological Weapons Convention? (Washington, D. C.: USGPO, 1980, pp 18). The source cited by Popovskiy, who emigrated from the USSR in 1977, was "a letter through the underground sent to me from Moscow [by a friend] in January [1980] ... received ... in February".

Anon., 'The bacteriological catastrophe of April 1979 in the city of Sverdlovsk', Russkaya Mysl (Paris) no 3315, 3 July 1980 (in Russian; English translation published on 11 July 1980 by Center for Appeals for Freedom, Freedom House, New York).

In its issue for October 1979, Possev had also carried a short anonymous piece about a bacteriological accident associated with a secret research facility in Novosibirsk that had caused several thousand deaths. In its April 1980 issue (page 7), it reported that this story had been wrong, the Novosibirsk epidemic in fact being the Sverdlovsk one reported in its January issue.

We may now safely conclude that it was the Possev reports, in their prepublication forms, which initiated Western press-reporting of the affair. That coverage began with the sensational Now! (London) cover story of 26 October 1979, 'The great Russian germ war disaster', and an item in the 27 October 1979 issue of the Springer mass-circulation tabloid Bild Zeitung (Hamburg), the details of both of which were identical with the October Possev story. Likewise, the items in the Economist Foreign Report of 30 January 1980 (no 1617, pages 3–4, 'Accident at Soviet germ warfare plant'), the Daily Telegraph (London) of 11 February 1980 (page 6, the column by Robert Moss) and, despite its 'Moscow, 13 February' dateline, Bild Zeitung of 13 February contain nothing substantive that was not also contained, in no less detail, in the January Possev story. It was the translation and publication of the two Bild Zeitung items in a Foreign Broadcast Information Service Daily Report (Washington, D.C.: CIA) for 14 March 1980 which first stimulated widespread public and governmental notice of the affair in the United States – notice which was soon to be enormously enhanced by the State Department's public endorsement of the reports and, later, by the Congressional inquiry to which Mr. Popovskiy gave the evidence cited above, an inquiry which, among other things, was investigating the possibility of a cover-up of the affair by the Carter Administration.

There is some textual evidence, by no means conclusive, to suggest that the Possev and Russkaya Mysl pieces had a common author, very different though these two journals are. Possev, reporting a circulation of about 2000, is the principal organ of the People-Workers Alliance (NTS), a revolutionary anti-Soviet group founded in 1930, reputed to be heavily infiltrated by a variety of national intelligence services, East and West. The US taxpayer has subsidized some of its publication; the KGB has used its pages on occasion as a vehicle for disinformation, provocation or identifying lines of dissident communication. Russkaya Mysl was established by the first post-Revolution Russian emigrant community in France, and still seems basically monarchist. It is said to be unsubsidized by foreign governmental agencies, and reports a circulation of about 8000.

Further commentary on the Category (I) sources is contained in Zh. Medvedev, 'The great Russian germ war fiasco', New Scientist (London), 87: 360–1, 31 July 1980.

Category (II) sources

Leslie H. Gelb, 'Keeping an eye on Russia', New York Times Magazine 29 November 1981, pages 31–3, 59–60, 64, 66, 148 and 172–3. This is the fullest summary yet published of disclosed US intelligence on the Sverdlovsk epidemic. Its author was an official in the

Carter Administration during its initial period, though not one whose duties involved him directly in the Sverdlovsk appreciation.

In this article, Gelb relates information acquired by US Government analysts from two types of intelligence source: satellite imagery ('before and after' pictures of the implicated military installation, located in the southern part of Sverdlovsk city), and a variety of human sources. The latter included 'secondhand witnesses' of the epidemic who had borne testimony out of the Soviet Union "highly regarded" by the CIA, reports that included medical particulars. Some of the detailed assessment of this intelligence made by the US analysts is also reported in the Gelb article (e.g. that the estimates of total fatalities made by different analysts ranged from 20 to 1000); and a few instances are noted where US intelligence conflicted with details related in the Category (I) sources.

The following items are the principal ones among many which also seemingly purvey disclosures or leaks of US intelligence, and which augment substantively the Gelb account:

B. Gwertzman, 'Soviet mishap tied to germ-war plant', New York Times, 19 March 1980, page 1 (humint).

W. Beecher, 'US rejects Soviet anthrax claim', Boston Globe, 29 March 1980 (satellite imagery).

R. Burt, 'New reports on Soviet anthrax convince U.S. of germ-war tie', New York Times, 29 March 1980 (humint).

C. Corddry, 'U.S. insists anthrax cases came from germ weapons', Baltimore Sun, 29 March 1980, page 4 (humint).

F. S. Hoffman for AP, as in 'Anthrax caused deaths in USSR, U.S. asserts', Philadelphia Inquirer, 4 April 1980, page 8C (humint).

R. C. Toth, 'Pall spreads from Soviet anthrax toll', Los Angeles Times, 7 April 1980 (humint and satellite imagery).

R. Evans and R. Novak, 'What did happen at Sverdlovsk?', Washington Post, 28 May 1980, page 19 (humint, and content of a DIA assessment).

M. Hopkins, 'The Sverdlovsk incident', New Leader, 2 June 1980, pages 11–12 (humint).

L. Aspin (Chairman, Oversight Subcommittee of House Intelligence Committee), speech in the House of Representatives, 30 June 1980 (correlations between Popovskiy testimony and US intelligence).

US House of Representatives, Permanent Select Committee on Intelligence. Report of the Subcommittee on Oversight, Soviet Biological Warfare Activities, 30 June 1980, pp 5 (allusion to humint).

L. Garmon, 'What happened at Sverdlovsk?', Science News, 2 August 1980, pages 72 and 78 (correlations between the Russkaya Mysl account and US intelligence).

E. T. Tighe, Lt-Gen., USAF (Director, DIA), statement before the Senate Armed Services Committee, 4 September 1980; see Hearing, Chemical Warfare, page 23 (humint).

N. Wade, 'Death at Sverdlovsk: a critical diagnosis', Science 209: 1501-2, 26 September 1980 (humint).

J. Anderson, 'Soviet breaches of treaty promises', Washington Post, 19 November 1980, page E15 (content of a CIA /sic/ assessment).

Category (III) sources

Of the numerous Soviet publications on the Sverdlovsk anthrax epidemic and on the biological-weapons causation alleged for it, the three following items appear to include all the particulars of the epidemic so far disclosed by the Soviet Government in public; there may have been more in the three private communications to the US Government on the subject. The first item expounds the tainted-meat explanation. A delay of some 48 hours had occurred after the initial US State Department announcement of its suspicions (at one of its daily press briefings) before the public Soviet reports acknowledged that there had been an anthrax epidemic at the time and place in question (although one of the first such rebuttals to have been displayed before the Soviet domestic public – an article in the 26 March issue of Literaturnaya Gazeta – made no special mention of anthrax, carrying only a vague reference to foot-and-mouth disease). The second item, which it states was received for publication on 29 August 1979, is a short technical paper bearing the names of, respectively, the Chief Epidemiologist and the Chief Specialist in Infectious Diseases of the RSFSR Ministry of Health, and which gives rise to more questions than it answers. It was extensively cited in the first item as evidence in support of the tainted-meat explanation, and has been so cited in subsequent official Soviet statements. The third item, by the chief judge of the cognizant District Court, describes the prosecution of two people found guilty on, so it reports, 11 and 18 July 1979 of violating veterinary and public-health regulations in selling meat from three sheep and a calf that had died of anthrax.

Ye. Nikolayev, 'Who is inspiring slander against the Soviet Union?', Izvestia, 15 June 1980, page 4 (in Russian; English translation available in Current Digest of the Soviet Press 32(24): 5, 16 July 1980). The same text had been broadcast in Russian and English from Moscow on 12 June 1980; for text, see BBC Summary of World Broadcasts SU/6445/A1/1 of 14 June 1980.

I. S. Bezdenezhnykh and V. N. Nikiforov, 'An epidemiological analysis of incidences of anthrax in Sverdlovsk', Zhurnal Mikrobiologii, Epidemiologii, i Immunobiologii 1980(5): 111–13 (in Russian).

I. P. Balayev (Representative of the Chkalovsk Municipal Peoples Court of Sverdlovsk), quoted in 'Strict observance of veterinary regulations', Chelovek i Zakon, September 1980 (in Russian).

7.4.6.2 Some Aids to Assessment

The 30-odd publications cited above constitute only a small fraction of the available literature on the Sverdlovsk epidemic. The present writer's files contain some hundreds of additional items, but it is very possible that other noteworthy ones may have been missed, particularly in the German literature. Also, there is a sizeable quantity of supplementary source material that should perhaps be identified as a fourth category of pertinent information: dispatches on the incident by foreign press correspondents based in Moscow. Of these, the following in particular are informative:

R. Beeston, ' "Impudent slander" says Russia', Daily Telegraph (London), 20 March 1980.

UPI dispatch quoted in 'Kremlin confirms epidemic, denies link to weapons', Washington Post, 21 March 1980, page 28 (reporting a telephone interview with the editor of the Sverdlovsk evening newspaper Vecherny Sverdlovsky; on which compare 'Anthrax accident may have killed hundreds', Financial Times, (London), 20 March 1980, and R. Beeston, 'Attack on U.S. "ravings" by Russia', Daily Telegraph 21 March 1980, page 4).

'Reports from Soviet town link anthrax, military site', Baltimore Sun, 27 March 1980, page 4 (a Chicago Sun-Times dispatch reporting rumours reaching Moscow from Sverdlovsk).

D. K. Willis, 'Soviets: US double-crossed us on germ-warfare charges', Christian Science Monitor, 28 March 1980, page 10 (a feature story reporting in detail the previous week's events in Moscow following the US diplomatic demarche and press announcement on the Sverdlovsk epidemic).

D. Satter, 'Support for Soviet explanation of anthrax outbreak', Financial Times (London), 10 June 1980 (reports interviews with people resident near Sverdlovsk).

Reuter dispatch printed as 'Soviet reports lethal centuries-old anthrax', New York Times, 15 January 1982 (reporting an article in Sovetskaya Rossiya, 14 January 1982: A. Lebedinskiy, 'A virus from the Seventh Century. A horrible disease breaks out. The specialists are puzzled: where did it come from?').

However, to judge from conversations with three people involved in the US assessment of the incident, it is probably fair to assume that anyone who reads all the cited items will acquire a good idea both of the pertinent information that is in the public domain and of the types that are not. The following items may be helpful in evaluating it all:

Routine meteorological-data returns to the World Meteorological Organization, from the reporting station at Sverdlovsk airport, especially returns during the period 24 March to 7 April 1979.

For access to the literature on biological warfare in general, and on the BW-utility of anthrax bacteria in particular (including the erstwhile stockpiling of anthrax weapons by, at least, Britain, Japan and the USA):

J. P. Perry Robinson and M. Leitenberg, The Rise of CB Weapons (1971; vol. I of Stockholm International Peace Research Institute, The Problem of Chemical and Biological Warfare, 6 vols, Stockholm: Almqvist & Wicksell, 1971-75).

J. P. Perry Robinson, C.-G Hedén and H. von Schreeb, CB Weapons Today (1973; SIPRI, op. cit., vol. 11).

R. Harris and J. Paxman, A Higher Form of Killing (London: Chatto & Windus, 1982).

For access to the Russian professional literature on the natural occurrence of anthrax in the Sverdlovsk region and other parts of the Soviet Union:

I. S. Bezdenzhnykh and V. N. Nikiforov, op. cit.

P. N. Burgasov, B. K. Cherkasskiy, L. M. Marchuk and Yu. F. Shcherbak, Sibirskaya Yazva (Moscow: Meditsina, 1970; English translation available as Anthrax from FBIS as JPRS 76944 of 5 December 1980).

B. K. Cherkasskiy, A. I. Savinykh and V. Yu. Markov, 'Peculiarities of anthrax epidemiology today in the USSR and ways of improving the prevention of the disease in people through vaccination', Zh. Mikrobiol. Epidemiol. Immunobiol. 1978 (11): 130-4.

B.K. Cherkasskiy, 'Current problems in the prophylaxis against anthrax', Zh. Mikrobiol. Epidemiol. Immunobiol. 1979(8): 3–6.

I. I. Elkin (editor), A Course in Epidemiology (1947; translated from the Russian by C. R. Pringle; Oxford: Pergamon, 1961), pages 491–7.

L. M. Marchuk and O. F. Bogatyrev, 'Epidemiology and hygiene', Feldsher. Akush. 40(2): 10–15, 1975.

E. N. Shlyakhov, E. V. Gruz and V. I. Prisakar, 'Optimization of anthrax vaccination on the basis of rational classification of inhabited localities', Zh. Mikrobiol. Epidemiol. Immunobiol. 1980(4): 91–5.

P. N. Zhurin, 'Diagnostics and epidemiology in two cases of intestinal anthrax in one family', Zh. Mikrobiol. Epidemiol. Immunobiol. 1969(4): 125.

It is instructive to note that, of the 237 citations yielded by a computer-search of the professional scientific and medical literature for publications on anthrax during the period 1966 to April 1980, 84 were Soviet.

On the bacteriology, pathology, epidemiology, clinical manifestations and diagnosis of anthrax, and on variations in the clinical and post mortem picture presented by anthrax according to whether the bacteria enter through the lungs or the gut:

K. K. Aver'yanov, 'Epidemiological outbreaks of anthrax', Zh. Mikrobiol. Epidemiol. Immunobiol. 1940(8): 35.

F. L. Bryan, 'Infections and intoxications caused by other bacteria', in H. Riemann and F. L. Bryan (editors), Food-Borne Infections and Intoxications (2nd ed., New York: Academic Press, 1979), pages 211–97.

P. N. Burgasov et al, op cit.

A. B. Christie, Infectious Diseases: Epidemiology and Clinical Practice (2nd ed, Edinburgh: Churchill Livingstone, 1974), pages 787–815.

D. P. Perl and J. R. Dooley, 'Anthrax', in C. H. Binford and D. H. Connor (eds), Pathology of Tropical and Extraordinary Diseases (Washington, D.C.: Armed Forces Institute of Pathology, 1976), volume I, pages 118–23.

V. Wyatt, 'Anthrax: recipe for a blunt weapon', New Scientist 87: 721–222, 4 September 1980.

For an authoritative exposition of epidemiological principles, especially with regard to food-borne disease, and of their application in identifying contributory and noncontributory factors in the spread of an epidemic (Table 7.2):

F. L. Bryan, 'Epidemiology of food-borne diseases', in Riemann & Bryan op cit, pages 3–69.

Table 7.2 A chronology[146]

Mid 1975	Interpretation of satellite imagery suggests to US analysts that constructions then in progress at several locations in the USSR, one of them Sverdlovsk, may have to do with manufacture or storage of BW weapons[147]
Mid 1976	US Arms Control and Disarmament Agency reportedly[148] urges, unsuccessfully, a comprehensive National Security Council study of the satellite intelligence
ca 2 April 1979	Explosion reportedly occurs at Cantonment no 19, Sverdlovsk, rumoured to have released bacteria that then killed a thousand people[149]
Spring 1979	Outbreak of anthrax among livestock, with secondary cases among people, in the vicinity of Sverdlovsk[150]
July 1979	Verdicts passed down by a Sverdlovsk district court on two people charged with distributing anthracose meat[151]
Early Feb 1980	From a 'secondhand witness', the CIA receives detailed testimony on the Sverdlovsk anthrax epidemic suggesting that the disease had taken the pulmonary form[152]
March 1980	An inter-agency working group on the Sverdlovsk incident is established by the Carter Administration under State Department chairmanship[153]

(continued)

[146] Editors' note. This table was not numbered in the original.

[147] W. Beecher, *Boston Globe*, 28 Sept 1975, page 1; and N. Wade, *Science* 192: 38 & 40, 2 April 1976.

[148] W. Beecher, *Boston Globe*, 15 Oct 1976, page 18.

[149] N. N., *Possev* 36(1): 7–8, January 1980.

[150] Ye. Nikolayev, *Izvestia* 15 June 1980, page 4.

[151] *Chelovek i Zakon*, September 1980.

[152] L. H. Gelb, *New York Times Magazine*, 29 November 1981.

[153] L. H. Gelb, *New York Times Magazine*, 29 November 1981.

Table 7.2 (continued)

14 March 1980	Carter Administration decides to oppose inclusion of any language in the Final Document of the Biological Weapons Convention Review Conference (Geneva, 3–21 March 1980) that would suggest that the treaty was being complied with fully
17 March 1980	First US diplomatic demarche: Soviet First Deputy Foreign Minister asked by the US Ambassador in Moscow for an explanation of the Sverdlovsk anthrax epidemic[154]
18 March 1980	US State Department informs the world press that it has sought an explanation from the Soviet Government[155]
19 March 1980	Soviet Foreign Ministry furnishes the US Embassy in Moscow with the explanation requested in the demarche of 17 March[156]
Later in March	Carter Administration reportedly[157] decides to reject the Soviet explanation and, early in April through its Moscow Embassy, may request meetings with senior Soviet scientists to discuss the contraindicative evidence
About the same time	DIA issues an assessment concluding that the anthrax epidemic had been initiated by an accident disseminating bacteriological-warfare facility product[158]
4 April 1980	CIA reportedly[159] issues an assessment concluding that the Sverdlovsk deaths were due to inhalation, not intestinal or cutaneous, anthrax

(continued)

[154] D. K. Willis, *Christian Science Monitor*, 28 March 1980, page 10; and D. T. Twining, Air Force Magazine, March 1981, page 125.

[155] Reported in every major newspaper on 19 March 1980. On the manner in which the press-release was effected, see P. Pringle, *Sunday Times* (London), 23 March 1980, page 10.

[156] D. K. Willis, *Christian Science Monitor*, 28 March 1980, page 10; and D. T. Twining, *Air Force Magazine*, March 1981, page 125.

[157] W. Beecher, *Boston Globe*, 29 March 1980; and R. Burt, New York Times, 29 March 1980, page 21.

[158] J. La Monica, *Journal of Commerce*, 15 May 1980, page 4; R. Evans and R. Novak, *Washington Post* 28 May 1980, page 19; D. Tonge, *Financial Times* (London), 30 May 1980; G. C. Wilson, *Washington Post* 29 June 1980, page 2; and E. T. Tighe (Director, DIA), statement to the Senate Armed Services Committee, 4 Sept 1980.

[159] J. Anderson, *Washington Post* 19 November 1980, page El5.

Table 7.2 (continued)

28 May 1980	State Department announces that the US Administration will seek talks with Moscow about compliance with the 1972 Biological Weapons Convention[160]
June 1980	Possibility of US-Soviet expert-level talks mediated through a non-governmental organisation under active consideration in both Moscow and Washington
29 June 1980	Second US diplomatic demarche had been effected by now[161] No agreed interagency 'estimate' on the Sverdlovsk incident yet issued by US intelligence community.[162]
Mid July 1980	White House decides against seeking clarification through back-channel scientific contacts
Sept 1980	CIA analysts said to be conceding a ten per cent chance to the probability of the Soviet explanation being essentially correct[163]
Feb 1981	An official at the Soviet Embassy in Washington privately draws the Reagan Administration's attention to the fact of a Russian-speaking American scientist, on an official US-Soviet scientific exchange, having resided with his family in Sverdlovsk at the time of the anthrax epidemic[164]
Feb 1981	Reagan Administration begins a re-assessment of the Sverdlovsk incident
August 1981	Third bilateral US-Soviet demarche on the incident
Autumn 1981	New 'second hand witnesses' reportedly[165] confirm CIA analysts in their previous assessment of the incident

[160] R. Burt, *New York Times* 29 May 1980, page 10.

[161] G. C. Wilson, *Washington Post*, 29 June 1980, page 2.

[162] *Soviet Biological Warfare Activities*: Report of the Subcommittee on Oversight, House Permanent Select Committee on Intelligence, 30 June 1980.

[163] N. Wade, *Science* 26 September 1980, page 1501.

[164] L. H. Gelb, *New York Times Magazine*, 29 November 1981.

[165] L. H. Gelb, *New York Times Magazine*, 29 November 1981.

7.5 Extract – Chemical Warfare Arms Control from a European Standpoint[166]

Editors' note. In 1982, Julian was asked to contribute to US Congressional Hearings on chemical warfare. In this extract of his statement, he details the complexities of contemporary discussions on the possibilities of developing a global ban on chemical weapons, focussing particularly on different views about the desirability and acceptability of international on-site inspections in building confidence in such a ban. It showcases one way in which Julian was able to contribute to national policy-shaping and policy-making, and demonstrates his ability to build strong positions based on careful logic, accurate detail and relevant contextual information.

Mr Chairman: In the letter in which you kindly invited my testimony for your hearings on chemical warfare, you asked in particular for my assessment of how Europeans view the two elements of the US chemical warfare program: binary modernization and arms-control policy. On the binary-modernization element, I submit for your consideration a statement that I originally prepared at the invitation of the Chairman of the Senate Committee on Appropriations for his hearings on chemical warfare during 5–6 May 1982. A copy is attached at Annex A. I do not believe there is anything I can usefully add that might possibly aid your consideration of this most important matter. With regard to the arms-control element, I should like to preface the more detailed comments which follow with two introductory remarks:

(a) Perhaps in consequence of Europe having provided the only battleground on which technologically advanced belligerents have used chemical warfare against one another on a large scale, the pursuit of CW arms control has long been a serious and active component of the foreign policies of European governments. Many West-Europeans have therefore been heartened by President Reagan's reaffirmation of US commitment to CW arms-control goals. However, concern continues to be expressed in Europe that active pursuit of these goals is in fact being accorded a lower priority both than the President's reaffirmation suggested and than the potential benefits, including those to the security of the North Atlantic Alliance, would seem to warrant. I have heard similar concern expressed in private by East-European officials about Soviet CW arms-control priorities.

(b) Among the European NATO countries, a further factor sensitizing concern about CW arms control is the likelihood, in the years immediately ahead, of having to confront decisions about the forward deployment of US binary munitions.

166 Editors' note. This extract was taken from: Submission to the Subcommittee on International Security and Scientific Affairs, Committee on Foreign Affairs, US House of Representatives, in lieu of testimony at the hearing on chemical warfare on 13 July 1982 – J. P. Perry Robinson, Senior Fellow, Science Policy Research Unit, University of Sussex, U.K. It is listed in Chap. 9 as item 149. Unless otherwise indicated, the remaining footnotes for this paper are Julian's originals. Apart from correcting obvious typos and renumbering headings and tables in keeping with the requirements of this book, we have retained Julian's formatting and spelling throughout the paper and its notes.

> Alliance harmony on this matter will not be achieved if European opinion is allowed to believe that CW arms-control possibilities have not been thoroughly explored. It is significant that, during the period immediately after President Reagan's certification to the Congress of the need for US production of binary munitions, the governments of both Britain and West Germany put forward constructive new proposals regarding CW arms-control at the 40-nation Committee on Disarmament in Geneva.[167] Europeans were impressed by the progress made during the bilateral US-Soviet CW negotiations that President Ford had initiated in 1976, and rightly appreciative of it. But the view is now taking hold that, because these negotiations have been in abeyance for the past two years, the bedrock of US-Soviet agreement registered during them may soon be lost.

All of this has quickened European interest in the new CW arms-control moves by the Soviet Government and the manner in which Washington responds. It is to this subject that I would like to devote most of my statement.

7.5.1 The June 1982 Gromyko Proposals

On 15th June, at the special session of the United Nations General Assembly then convened in New York, Soviet Foreign Minister Andrei Gromyko announced that the USSR was submitting a "draft document" on CW arms control. He said that the USSR believed the document could be the basis for a "breakthrough in reaching international agreement". Nine pages long, the document[168] was entitled 'Basic Provisions of a Convention on the Prohibition of the Development, Production and Stockpiling of Chemical Weapons and on Their Destruction'. Mr Gromyko made two other remarks when introducing it. The first was to "strongly reject absurd tales implicating the USSR in the use of those weapons". The second was to note that the "draft takes into account other State's wishes, including those that deal with verification". The import of this second remark was spelt out by British Prime Minister Margaret Thatcher when commenting on Mr Gromyko's document during her own speech[169] to the Assembly on 23rd June: "the Soviet Union now seems ready to accept the need for systematic on-site inspection in respect of a chemical weapons treaty".

Prime Minister Thatcher went on to say "We need to redouble our efforts to bridge the gaps that still remain", thereby implying an assessment by her government that

[167] United Kingdom, 'Working paper on Verification and the Monitoring of Compliance in a Chemical Weapons Convention', CD/244 of 18 Feb 82; and Federal Republic of Germany, 'Working paper on Principles and Rules for Verifying Compliance with a Chemical Weapons Convention', CD/265 of 24 March 1982.

[168] Pre-circulated in an English-language version, it was later published in UN doc A/S-12/AC.1/12 of 17 June 1982.

[169] On the genesis of Mrs Thatcher's speech, see Hugo Young, Sarah Hogg and Jon Connell, 'Thatcher's mandarins', *Sunday Times* (London) 11 July 1982, p 17.

the Gromyko document was an initiative to be taken seriously. On the face of it, the document does indeed have major significance. Soviet unwillingness to concede the necessity of systematic on-site inspection for verifying a chemical weapons prohibition has long and widely been seen as the principal obstacle to international agreement on the subject. And if that necessity is now to be conceded for chemical weapons, it can hardly be denied for verification of agreements in areas less peripheral to national security. In short, Mr. Gromyko's advertisement of "breakthrough" seems no hyperbole.

The Gromyko document should therefore encourage those who have been dispirited by the overall outcome of the UN General Assembly special session, which was the second such session in four years given over to the disarmament question. Nor was the Gromyko document the only major proposal then on chemical weapons. In the context of Mr Gromyko's repudiation of the Yellow Rain reports, a particular importance attaches to the detailed memorandum submitted by Belgium proposing for intergovernmental negotiation a novel scheme whereby the international community might monitor the prohibition of the use in combat of chemical, biological and toxin weapons.[170] The manner in which the Soviet Government responds to initiatives such as this Belgian one will colour the future of the Gromyko proposal, for over the past year the US Administration has left the world in no doubt that it is linking the future of arms control in the CBW area with clarificatory actions by Moscow regarding Soviet compliance with the existing agreements. So far, Soviet willingness to discuss the outstanding compliance issues appears to have been expressed only within nongovernmental fora. Attached, at Annex B, is a copy of the report from one such meeting, the March 1982 workshop of the Pugwash Chemical Warfare Study Group (whose American participants included CW specialists from the Organization of the Joint Chiefs of Staff and the Arms Control and Disarmament Agency); see, in particular, the workshop paper 'Problems of CBW Compliance' appended to the joint paper reviewing the workshop by British and East German participants.[171] The compliance paper has much in common with the Belgian proposal.

I shall not presume to review here in detail the contents of the Gromyko document, or offer a full appraisal of its significance. That would in any case require close consideration of the military and political context into which the document has been introduced. It would also require an account of the prior history of the CW arms control negotiations. These obscure and complicated talks have been in progress for more than a decade in both multilateral and bilateral fora, and are not readily summarized, though elsewhere I have published a commentary on their progress during the period 1968–1979 which sets out the principal points at issue[172]; in case it may be of interest, a copy is attached at Annex C. I would instead like to comment on the Gromyko document in relation to two seminal prior documents.

[170] UN doc A/S-12/AC.1/18 of 16 June 1982.

[171] Published, for limited distribution, in Pugwash Newsletter (London: Council of the Pugwash Conferences on Science and World Affairs) 19: 156–66, April 1982.

[172] *Arms Control* (London: Frank Cass) 1: 30–52, May 1980.

7.5.2 Comparison with the 1980 U.S.-Soviet Bilateral Report

The first of these prior documents is the 'USSR-United States Joint Report on the 'Progress in the Bilateral Negotiations on the Prohibition of Chemical Weapons' transmitted by letter dated 7th July 1980 from the US and Soviet representatives to the Chairman of the Committee on Disarmament in Geneva.[173] This was the last such joint report to have emerged from the bilateral US-Soviet CW working group before its present abeyance, and sets out many basic provisions of a Chemical Weapons Convention, some of them almost in treaty language. Its importance of course lies in that it represents all those matters of bilateral agreement which the two sides had reached by that time and were both willing to make public.

In structure, content and language, the Gromyko document appears to have taken the 1980 bilateral report as its starting point.[174] The substance of the Gromyko proposals is therefore concentrated within the points of substantive difference between the two documents. There are about 40 such differences. They may be read in two ways. Either they are statements of Soviet position on the particular matters to which they refer that had not been accepted by the US negotiators. Or they are statements of new Soviet position. Probably there are instances of both in the document. In no significant sense do any of the differences appear to represent a reneging by the USSR on the publicly registered agreements.

The detailed proceedings of the bilateral negotiations are not matters of public record, so it is not possible for me to say in all cases which of the differences are novel Soviet positions and which are simply opportunistic restatements of unaccepted old positions. However, the two most striking differences unquestionably fall within the former category, and may be taken as the essence of Mr Gromyko's proposals – the basis for his claim of "breakthrough". They are paragraphs 3 and 4 of the provisions entitled "Fact-finding procedure regarding compliance with the Convention. On-site inspections":

3. Within the period of destruction or diversion for permitted purposes of the stocks of chemical weapons a possibility of carrying out systematic international on-site inspections (on the basis of an agreed quota, for example) of the destruction of stocks at a converted or specialized facility (facilities) should be provided for.
4. The Convention should provide for a possibility of carrying out international on-site inspections (on the basis of an agreed quota, for example) of the production of super-toxic lethal chemicals for permitted purposes at a specialized facility.

The 1980 bilateral report had recorded agreement on use of "on-site investigations" for verification purposes only as part of a complaints procedure. There could

---

[173] Committee on Disarmament doc CD/112 of 7 July 1980.

[174] In some places the Gromyko document also imports language and concepts from the analogous reports of the multilateral CW talks, in particular from the draft 'elements' of a CW Convention presented in the report by the Chairman of the Ad Hoc Working Group on Chemical Weapons of the Geneva Committee on Disarmament at the close of last year's session (see CD/220 of 17 August 1981).

be resort to such investigations (not "inspections") only after suspicions suggesting its necessity had been declared and substantiated according to a prescribed challenge procedure, and even then it would be permissible for a state-party giving "appropriate explanations" to refuse a site visit. In the parlance of the bilateral and multilateral CW negotiations, the term "systematic" – now used in relation to international on-site inspection for the first time in a public Soviet arms-control proposal – had come to mean inspections of a routine, not challenge or ad hoc, character, acceptance of which would be mandatory (subject to provisions precluding undue encroachment upon national sovereignty). It remains to be seen whether this is in fact the sense in which the Gromyko document uses the term; a matter to which I shall return.

7.5.3 Comparison with the 1979 US Position Paper

The other prior document with which it is useful to compare the Gromyko document is the 'informal paper' distributed within the Geneva Committee on Disarmament by the US delegation on 19th July 1979. It sets out US views on the questions about what should or should not go into a Chemical Weapons Convention that had been circulated nine days previously by the Dutch delegation – an informal questionnaire that was later to appear, unlike the US responses to it, as a formal Committee document.[175] It was generally understood that the Joint Chiefs of Staff had played a key role in formulating the US responses. A copy is attached at Annex D.

Like the Gromyko document although in much less detail, the US informal paper amounts to a set of the basic provisions whose incorporation in a draft Chemical Weapons Convention could then be expected, subject to detailed negotiation, to render the Convention acceptable. The provisions are summarized in the table below [Table 7.3]. In the right-hand column of the table are indications of the extent to which the US basic provisions form part of the Gromyko document. It will be seen at once that there are only two areas in which there are obviously major gaps between the declared positions: (a) provision for confidence-building measures; and (b) provision for the systematic on-site inspection of facilities formerly used for production of chemical weapons. One might, therefore suppose that the Gromyko document has now brought the two sides close together, and that agreement between the USA and the USSR on a Chemical Weapons Convention could be concluded rapidly.

7.5.4 The Gaps Between the Two Sides

There are problems with such a supposition, however. One is that in February 1981 the US representative informed the Geneva Committee on Disarmament that the new Administration in Washington was reviewing US arms-control policy on chemical

[175] CD/41 of 25 July 1979.

Table 7.3 Extent to which the Gromyko document meets requirements of a CW convention stated by the USA in 1979[176]

Basic treaty elements required by the USA, as set out in the US informal paper of July 1979	Is the element a 'basic provision' of the Gromyko document of June 1982?
Convention to provide a complete, effective, and verifiable prohibition of chemical weapons	
Chemicals to be covered by the convention:	
– Categories differentiated by specified toxicity criteria	Yes
— Supertoxic lethal chemicals (STLCs)	Yes
— Other lethal chemicals	Yes
— Other highly toxic chemicals	Yes, as 'other harmful chemicals'
– Including:	
— Incapacitants	Apparently
— Some irritants but not others*	Apparently
– Excluding:	
— Antiplant chemicals	Unclear
– Precursors	Yes
Activities to be expressly outlawed under the convention:	
– Development	Yes
– Production	Yes
– Stockpiling	Yes
– Acquisition	Yes
– Retention	Yes
– Transfer	Yes
Activities to be permitted under the convention involving otherwise proscribed chemicals:	
– Activities necessary for purposes related to protection against chemical attack	Yes
– Limited production of STLCs, subject to reporting requirements	Yes, but only for 'permitted purposes'
– Limited production under appropriate controls of chemicals primarily useful for chemical weapons purposes	Yes, but only for 'permitted purposes'
Actions to be required under the convention:	
– Appropriate national overseeing of implementation	Yes
– Consultative Committee of States-Parties to be established	Yes

(continued)

[176] Editors' note. This table was not numbered in the original.

Table 7.3 (continued)

Basic treaty elements required by the USA, as set out in the US informal paper of July 1979	Is the element a 'basic provision' of the Gromyko document of June 1982?
– Consultative Committee permanent secretariat to be established	Allusion only
– Stockpiles to be declared by each state-party	Yes
– Weapon production/filling facilities to be declared by each state-party	Yes**
– Stockpiles to be destroyed within ten years	Yes
– Weapon production/filling facilities to be mothballed and eventually destroyed	Yes**
– And perhaps also:	
— Reporting by each state-party on listed commercial chemicals that might be diverted to chemical weapons purposes	Yes
Verification measures to be provided for by the convention:	
– International as well as national measures:	Yes
— International measures to be the responsibility of the consultative committee	Apparently**
– Measures for on-site verification by challenge:	Yes
— Challenges may be made bilaterally or through an international mechanism	Yes
– Measures for systematic international on-site verification	Yes
— Applicable at least to:	
—– Destruction of stockpiles	Yes
—– Operations at facilities for permitted production of chemicals primarily useful for chemical weapons purposes	Yes**, for STLCs
—– Mothballing and eventual destruction of weapon production/filling facilities	No
— But not applicable to:	Yes
—– Civilian chemical production	Yes, by implication
– More stringent measures adopted on a regional or some other basis should not be impeded	Unclear
Confidence-building measures:	Term not used; but document notes a particular measure*** that could serve as 'a display of good will'

(continued)

Table 7.3 (continued)

Basic treaty elements required by the USA, as set out in the US informal paper of July 1979	Is the element a 'basic provision' of the Gromyko document of June 1982?
– Should be included in the convention	
– Examples of useful confidence-building measures:	
— General public statements about possession or non-possession of chemical weapons made prior to entry of the Convention into force	No; such statements would be a formal requirement once treaty is in force
— Technical exchange visits	No
— Information exchanges in the field of protection	No

*But the actual language used in the Gromyko document may reflect a difference in underlying concept
**Commencement of stockpile-destruction ahead of the time required under the convention
***Term not used; but document notes a particular measure that could serve as 'a display of good will'

weapons,[177] a task which is apparently still in progress. A question-mark therefore hangs over the status of the 1979 position paper, in any case an 'informal' document, even though it might be supposed that a considered judgement of the Joint Chiefs of Staff – if that was indeed what the July 1979 paper represented – ought in principle to be insensitive to change in the nation's political leadership. By much the same token, a question-mark may also be hanging over the status of the US agreement formally recorded in the 1980 bilateral report.

A second obvious problem is that, with the bilateral US-Soviet CW working group in abeyance, there is no suitable forum within which to negotiate either the gaps or the details. It is true that the Geneva Committee on Disarmament is now in a treaty-elaborating, if not treaty-negotiating, mode on chemical weapons, and will certainly operate to the best of its ability to facilitate agreement between its Soviet and US members. But the delicate bargaining that remains to be undertaken by the two sides will hardly proceed rapidly while there are 38 other delegations (maybe even more next year) also participating. Many of these other delegations will, on past experience, make extensive and not always very constructive use of the strictly limited amount of working time that the Committee can devote to chemical weapons, a subject which is only one among many on its agenda. From this standpoint alone, the case for resuming the bilateral negotiations appears compelling.

Then there are the problems of a substantive rather than procedural nature. The 'yes' entries in the right-hand column of the table above [Table 7.3] may indeed far

[177] Statement of Ambassador Flowerree on 12 Feb 81, CD/PV.105 at page 34. See also his statement of 13 August 1981, CD/PV.146 at page 10. On 9th February 1982, the Director of the US Arms Control and Disarmament Agency, Dr. Eugene Rostow, told the Committee that it was now the US intention "to concentrate its efforts towards the elaboration of a convention banning chemical weapons in the Committee" – i.e. not to proceed with the bilateral US-Soviet CW talks; see CD/PV.152 at page 13.

outnumber the 'no' entries, suggesting many more bridges than gaps between the two sides. But that may not help much if the remaining gaps are wide.

With regard to the apparent rejection of confidence-building measures in the Gromyko document, it looks from one angle as though this may indeed portend major difficulties ahead. Now that the Yellow Rain reports have generated so much more mistrust (mutual mistrust, it seems; for Soviet organs are proclaiming the reports as fabrications aimed at gaining domestic Western acceptance of the binary program), special remedial measures are clearly going to be necessary if there is to be any chance of the United States, and its major Alliance partners, accepting a further CBW arms-control agreement, especially at the ratification stage. However, from another angle this particular gap looks much smaller. Although the Gromyko document may not speak explicitly of "confidence-building measures" – a term which the USSR in any case rarely uses outside the context of the Helsinki Accord – it nonetheless refers to one particular measure as a "display of good will"; and if Moscow can countenance one such display, others too may prove negotiable, not least in view of the extensive provision which the Gromyko document makes for the reporting of various types of data. From a third angle, it may be that the gap looms larger in European eyes than it does in American. It has been the smaller participants in the CW talks thus far that have been pressing for confidence-building measures, probably because such measures could satisfy a need felt much less strongly by the superpowers: where capacious 'national technical means' of verification and other such assets are unavailable, there can be less confidence in the nonexistence of stockpiles, facilities &c that ought to have been declared under the Convention. From whatever angle the gap is viewed, however, sustained US-Soviet dialogue will remain a prerequisite for bridging it.

The other major gap, namely use of systematic international on-site inspections for verifying destruction not only of stockpiles but also of production capacity, lies right at the heart of the CW negotiations. Once again its magnitude changes according to the angle of view. It looks relatively small if four things are assumed: (1) that the Gromyko document does indeed represent a genuine Soviet negotiating position, and is not – as the more impetuous protagonists of binary production can be expected to contend – merely a deception aimed at luring the West away from rapid modernization of its chemical-warfare posture; (2) that it represents the considered view of the Soviet Government, and not just some faction of it; (3) that such an initial statement of position on systematic inspection betokens a fall-back position encompassing wider application of the technique; and (4) that "systematic" means what the negotiators have hitherto taken it to mean. The less acceptable these assumptions appear, the larger may the gap come to seem, and the more problematic some of the other, related, provisions of the Gromyko document. Among the latter may be noted the absence of the word "systematic" [in the above quote] … and the following sentence in an otherwise innocuous section entitled "Promotion of development goals": "And a possibility of intrusion in the areas of activities unrelated to the purposes of the Convention should be precluded". Yet in this matter, as in the confidence-building matter, renewed US-Soviet dialogue will provide the only effective way of probing Soviet intentions.

7.5.5 A Conclusion

A license for states-parties to conduct systematic international on-site inspections is perhaps the most effective measure that can be conceived of for building confidence in the regime to be established by a Chemical Weapons Convention and in compliance with it. The Federal German Government expressed the point as follows to the UN special session ten days after submission of the Gromyko document:[178]

> It should be borne in mind that regular checks – unlike on-challenge investigations which take place when there are specific suspicions and allegations – are a businesslike procedure conducive to cooperation and understanding.

Herein, it seems to me, lies the real significance of the Gromyko document, conferring major importance and value to the Soviet initiative.

For that probably large, if so far inconspicuous, community in the West which may eventually prove to be harbouring serious misgivings about having to submit to systematic on-site inspection, it may be instructive to quote further from the German statement:

> In this field [of verifying nonproduction of chemical weapons] the Federal Republic of Germany disposes of more than 25 years of fruitful experience showing, inter alia, that effective on-site inspection can be carried out without detriment to the legitimate interest in safeguarding commercial and scientific secrets.

In contrast to governments of the leading European NATO countries, the US Government has not yet commented publicly on the Gromyko document, either directly or by allusion. Perhaps it will do so during these hearings. If the comment of the present Administration's staunchest ally in Europe, British Prime Minister Margaret Thatcher, is a proper guide to European attitudes on the matter, intra Alliance discord on chemical-warfare questions will surely increase if the response from Washington is not both positive and constructive.[179]

---

[178] UN doc A/S-12/AC.1/37 of 25 June 1982, at page 3.

[179] Editors' note. The original included additional material in these four annexes: A Statement before the Senate Appropriations Committee, 5th May 1982; B Report of the Ninth Pugwash Workshop on Chemical Warfare, Geneva, 12–14 March 1982; C 'The negotiations on chemical-warfare arms control', May 1980; D US delegation to the Committee on Disarmament, 'US views on Dutch questions', 19th July 1979. These are available from the Sussex end of the Harvard Sussex Program.

7.6 Biological Weapons[180,181]

Editors' note. In 1978, Julian and Mary Kaldor co-founded the Armament Disarmament Information Unit (ADIU) at the Science Policy Research Unit (SPRU), University of Sussex, establishing it as an interdisciplinary research group renowned for its pioneering work on weapons systems and regulations. This ADIU fact sheet provides a succinct overview of biological weapons, useful for both general and specialist audiences.

A biological weapon has two components: a biological-warfare (BW) agent, such as a preparation of disease-causing germs, and delivery-means whereby the agent is conveyed to its target in an effective form. The label 'biological weapon' has been applied to devices dispensing all the following categories of agent:

a. Disease-causing microbes (viruses, bacteria and other germs)
b. Toxic substances generated by microbes
c. Toxic substances generated by nonmicrobial organisms, such as poisonous plants and venomous animals
d. Synthetic chemicals identical in molecular structure to (b) or (c)
e. Close chemical relatives of (d)
f. Chemical herbicides and defoliants.

Although weapons dispensing the last of these categories are rarely designated 'biological' nowadays, past writings which called them that remain a source of confusion. Current usage is most properly governed by the 1972 Biological Weapons Convention, a treaty outlawing the development, production and stockpiling of germ and "toxin" weapons that now binds more than a hundred countries, including all of NATO and the WTO. The 1972 Convention applies to categories (a) to (e) inclusive, even though weapons based on, for example, category (d) – a category which includes hydrogen cyanide – might more logically be considered as chemical weapons. "Toxin" in the sense of the 1972 Convention denotes everything in categories (b), (c), (d), and (e).

It is probably the case that all microbes known to cause human disease have been studied in BW laboratories. Toxins, in contrast, are far too numerous and, for scientists, still such relatively uncharted terrain, for this to have been possible. It is the toxin categories, then, that present the greater risk of 'technological surprise' – of upsetting current conceptions of the military value of biological weapons.

---

[180] Editors' note. This was originally published as ADIU Factsheet No 6, 15 March 1984. 4 pp. It is listed in Chap. 9 as item 172. Unless otherwise indicated, the remaining footnotes for this paper are Julian's originals. Apart from correcting obvious typos and renumbering headings and tables in keeping with the requirements of this book, we have retained Julian's formatting and spelling throughout the paper and its notes.

[181] For further reading, see ADIU Factsheets no. 5 (Chemical/Biological Warfare) and no. 7 (Chemical Weapons).

Table 7.4 Different casualty agents: effective doses per kilogram (from US defense dept data)[183]

Modern assault-rifle bullets	Hundreds
Steel, when fragmented by high-explosive	Tens of thousands
Nerve gas	Millions
Shellfish poison	Millions
Botulism toxin, 1943 weapons grade	Tens of millions
Anthrax spore slurry, 1945 weapons grade	Hundreds of millions
VEE virus powder, 1964 weapons grade	Hundreds of thousands of millions
Tularemia bacteria powder, 1955 weapons grade	Millions of millions
Botulism toxin, pure	Millions of millions

The toxin agents that are known to have been studied in past biological-weapons development programmes have all been antipersonnel agents. Germ agents have been developed for weapons designed to attack either people or animals (or both) or plants. Anticrop biological weapons – such as those which the USA first deployed overseas in the early 1950s for targetting on Ukrainian wheatfields[182] – have at times been considered more important than antipersonnel ones. The latter have been developed in both 'lethal' and 'incapacitating' forms according to whether the disabling diseases which they initiate are of high or low mortality. Incapacitating BW agents have probably received the greater attention, largely because experimentation on human subjects has been the more practicable.

In the biological-weapons development programmes of the past about which there is now public information, the antipersonnel BW agents that ultimately came to be favoured were ones that cause diseases which do not subsequently spread of their own accord. Epidemics were judged to be too poorly predictable for transmissible-disease agents to have significant military utility: anthrax was in, but plague was out. For anticrop biological weapons, on the other hand, the reverse was true.

Certain biological weapons could conceivably kill as many people as high-yield nuclear weapons. These are ones designed to dispense their BW agent as a cloud of particles small enough to remain airborne for very long periods of time. Scientists use the term 'aerosol' to describe such suspensions. An aerosol of BW agent could travel with the wind to pollute the air over wide expanses of ground. With some BW agents, inhalation of a few particles only might constitute an effective dosage, so that in principle enormous areas – even thousands of square miles – can be targetted by a single biological weapon (Table 7.4).

[182] Dorothy L Miller, History of Air Force Participation in the Biological Warfare Program, 1944–1951 and 1951–54 (US Air Force Historical Study no 194, Sept 1952, and no 313, Jan 1957, both declassified in 1978).

[183] Editors' note. This table was not numbered in the original.

There are several ways of generating BW aerosols. One way is to discharge a tank containing a slurry of BW agent into the slipstream of an aircraft flying upwind of the target. Or the spraytank might be fitted with dried BW agent pre-sized into particles of the requisite fineness. Or special aerosol generators resembling household 'aerosol cans' might be built into aircraft bombs, artillery shell or rocket warheads. The simplest devices have used just a charge of high explosive to shatter the agent payload into aerosol form.

Other delivery means, but ones of much less destructive potential, include devices for contaminating water supplies or food with BW agent. Or biting insects may be infected with the agent and released over the target, a technique sometimes called 'entomological warfare'. Or inanimate self-dispersing carrier materials, such as feathers, may be dusted with BW agent and used in a similar fashion.

Biological weapons of all the foregoing types have been designed, and some of them stockpiled. The pioneering work appears to have been done in Japan during 1934–45 and in a tripartite British-American-Canadian programme that began during World War II, ending – essentially as a US programme – in 1969. Other countries, too, have conducted such programmes, but virtually nothing is known in public about them. The US Defense Secretary has recently asserted that the USSR is still developing biological weapons.[184] On the evidence publicly available, however, this charge has to be regarded as non-proven. That Soviet efforts in this area might have at least as long a history as those of other countries is suggested by the frequency with which biological-warfare charges, typically of sabotage, appeared in indictments during Soviet state trials in the 1930s.[185]

Most germs and toxins lack the properties which scientists have shown to be essential for pathogenic air contamination on a large scale. Some of the exceptions are included in the lists below. Few species of germ can survive the stresses to which they would be subject upon dissemination from a weapon as aerosol, or upon subsequent exposure to other air pollutants, to sunlight or to a humidity differing from that to which they were naturally (or artificially) adapted. Some resistance to some of these factors can be bred into some microbes, and new techniques such as DNA hybridization might somewhat extend the possibilities. But the chances of ever being able to tailor the biological-decay rate of disseminated BW agent so as to match military requirements, especially that of predictability and hence controllability, must probably be judged very slight. This was the principal reason, it may confidently be assumed, why military staffs were willing to accept the restrictions imposed by the 1972 Biological Weapons Convention (Table 7.5).

---

[184] Annual Report to the Congress, Fiscal Year 1985, 2 Feb 1984.

[185] Robert Conquest, *The Great Terror.* Oxford University Press (1968, rev 1971).

Table 7.5 Some historical examples of biological weapons[186,187,188,189]

90-pound air-droppable fragmentation bomb incorporating a slurry of anthrax spores to contaminate the shrapnel (Japan, 1941)
4-pound air-droppable clusterable bomb filled with a slurry of anthrax spores and an explosive burster charge (UK, 1942)
Air-droppable cattle-feed cakes contaminated with anthrax spores (UK, 1942).
Aircraft cannon shell charged with anticrop fungal spores (USA, 1952)
Long-range unmanned balloons adapted to release clusters of containers filled with feathers dusted with anticrop fungal spores (USA, 1955)
Aircraft cluster-munitions and battlefield-support missile-warheads adapted to release numerous spheres containing mosquitoes carrying Yellow Fever virus (USA, 1959)
Unmanned drone aircraft (RPV) adapted to spray a slurry of tularemia bacteria (USA, 1959)
Disseminator shaped like a small hip-flask using compressed gas to discharge a powder of freeze-dried germs or toxins (USA, 1962)
Ton-sized spraytanks for high-performance aircraft adapted to discharge micro-pulverized germ or toxin agents (USA, 1968)

So, although it is in their large-scale aerosol threat that biological weapons are most to be feared, the forms of biological weapon perhaps least unlikely to be used are ones designed for strictly localized small-scale effects during clandestine operations. Of the six BW munitions that were being procured by the Army and Air Force during the last five years of the now defunct US biological-weapons programme, four were in this category (Table 7.6).[190]

[186] Editors' note. This table was not numbered in the original.

[187] Dorothy L. Miller, History of Air Force Participation in the Biological Warfare Program, 1944–1951 and 1951–54 (US Air Force Historical Study no 194, Sept 1952, and no 313, Jan 1957, both declassified in 1978).

[188] British Intelligence Objectives Sub-Committee, Report on Scientific Intelligence Survey in Japan, volume V (1946).

[189] R. Harris and J. Paxman, *A Higher Form of Killing.* (1982).

[190] US Defense Department, reports of ammunition procurement programmes during FY 1964 to FY 1972 inclusive (annual returns of the four armed services on form DD 1416).

Table 7.6 Some BW agents judged especially important by American authorities[191]

Lethal antipersonnel	Incapacitating	Anticrop
Produced and/or standardized for biological weapons[192,193,194]		
Tularemia bacteria	Brucellosis bacteria	Wheat stem-rust spores
Yellow fever virus	Q fever rickettsiae	Rye stem-rust spores
Anthrax bacterial spores	Venezuelan equine encephalitis virus	Rice blast spores
Botulism toxin	Staphylococcal enterotoxin	Two others
Shellfish poison (saxitoxin)		
Others for which stockpiles of BW-defence vaccine are maintained[195]		
Smallpox virus	Rift Valley Fever virus	–
Others for which such vaccines/antidotes are in the 1982–92 development schedules[196]		
Trichothecene mycotoxins	Chikungunya virus	–
Lassa fever virus		
Haemorrhagic fever viruses (Argentinian, Korean &c)		
Tick-borne encephalitis virus		

7.7 The Yellow Rain Affair: Lessons from a Discredited Allegation[197]

Editors' note. This article was one of several that Julian co-authored with Matthew Meselson,[198] in which they documented their own and others' work disproving US allegations of chemical weapons use in Southeast Asia in so-called "Yellow Rain" incidents.[199]

---

[191] Editors' note. This table was not numbered in the original.

[192] Dorothy L Miller, History of Air Force Participation in the Biological Warfare Program, 1944–1951 and 1951–54 (US Air Force Historical Study no 194, Sept 1952, and no 313, Jan 1957, both declassified in 1978).

[193] US Dept of the Army, US Army Activity in the US Biological Warfare Programs (1977).

[194] US Army Medical Research and Development Command briefing documents, 1982.

[195] US Army Medical Research and Development Command briefing documents, 1982.

[196] US Army Medical Research and Development Command briefing documents, 1982.

[197] Editors' note. Julian co-authored this with Matthew Meselson. The original reference is: Matthew Meselson and Julian Perry Robinson (2008), "The Yellow Rain Affair Lessons from a Discredited Allegation", in Anne L Clunan, Peter R. Lavoy and Susan B. Martin (eds), *Terrorism, War, or Disease: Unraveling the Use of Biological Weapons*, Stanford California: Stanford University Press. It is listed in Chap. 9 as item 454. Unless otherwise indicated, the remaining footnotes for this paper are Julian's originals. Apart from correcting obvious typos and renumbering headings and tables in keeping with the requirements of this book, we have retained Julian's formatting and spelling throughout the paper. Full references can be found in the original.

[198] As detailed in Chaps. 2 and 9, Jeanne Guillemin was also pivotal to this research.

[199] Editors' note. See Item 219 in Chap. 9 for details of their other publications on this.

By carefully accessing, comparing, and analysing data from diverse sources, they were able to demonstrate that samples of the alleged CW agent actually comprised bee faeces.

U.S. Secretary of State Alexander Haig, in a speech in West Berlin in September 1981 and in a detailed report to the Congress the following March, charged Soviet-backed Laotian and Vietnamese forces with waging toxin warfare against Hmong resistance fighters and their villages in Laos and against Khmer Rouge soldiers and villages in Cambodia. The charges were repeated with additional details in a further report to the Congress and to the member states of the United Nations in November 1982 by Haig's successor, Secretary of State George Shultz.

The investigation on which the allegation was based, however, failed to employ reliable methods of witness interrogation or of forensic laboratory investigation; it was further marred by the dismissal and withholding of contrary evidence and a lack of independent review. When the evidence for toxin attacks or any other form of chemical/biological warfare (CBW) was subjected to more careful examination, it could not be confirmed or was discredited. In what became known as the "Yellow Rain" affair, these charges – that toxic substances called trichothecenes were used in CBW – were initially pressed vigorously by the U.S. government and, even when the allegations proved unsustainable, they were not withdrawn.

This chapter reviews all of the evidence adduced at the time that is now public, including a large body of material declassified and provided to us and others starting in 1986 under the Freedom of Information Act. The evidence supports explanations of the Yellow Rain events that involve no CBW weapons at all, beyond the use of riot-control agents. Focusing on the evidence rather than the politics of the affair, the chapter seeks lessons that could increase the accuracy and credibility of possible future assessments of alleged or suspected CBW. There is first a historical introduction and then each category of evidence is described and assessed in turn. In the concluding section, the principal lessons drawn from the Yellow Rain episode for investigations of CBW allegations concern the taking and confirmation of testimony from alleged witnesses, the conduct of forensic chemical analysis, and the importance of wide consultation and critical review.

7.7.1 The U.S. Allegations

The Haig and Shultz reports cited three main types of evidence in support of their allegations. First, Hmong villagers who had fled from Laos to refugee camps in Thailand and Khmer soldiers operating in the Thai-Cambodian border region claimed to have witnessed chemical attacks, including several that were said to have caused hundreds of deaths. Altogether, the two reports list some 400 alleged attacks between the summer of 1975 and June 1982, nearly two-thirds of them in Laos, most of the rest in Cambodia, and a few in Thailand near the Cambodian border. Second, samples of a yellowish material were turned in by people who claimed to have witnessed attacks and who stated that the samples were air-dropped poison; this was the so-called

"Yellow Rain." A third type of evidence comprised chemical analyses reporting the presence of trichothecene mycotoxins, toxic substances made naturally, especially by fungi of the genus Fusarium, in environmental samples from sites of alleged chemical attack and in biomedical samples from individuals allegedly exposed. In addition, the Haig report cited an account given to U.S. Embassy personnel in Thailand by a Lao pilot who had defected and who, according to the report, had been "directly involved in chemical warfare."

The Hmong are an ethnic group inhabiting the highlands of South China, Thailand, Vietnam, and Laos. Starting in 1960, the CIA secretly recruited and supplied an indigenous Lao army, composed mainly of Hmong, to oppose the communist-led Pathet Lao, the North Vietnamese, and other Hmong who were allied with the Pathet Lao. After the U.S. departure from Vietnam in 1975, the Hmong who had been allied with the United States continued to resist Lao government control and relocation from their highland settlements to the lowland plains; in the late 1970s they came under heavy attack by the Pathet Lao and the Vietnamese air force and, in 1978, their Phu Bia Mountain stronghold fell to the Pathet Lao, leaving their villages exposed to continuing harassment and attack by Pathet Lao and Vietnamese ground and air forces. By 1983, tens of thousands of Hmong had fled from Laos across the Mekong into Thailand, where they were interned in refugee camps, the largest of which was at Ban Vinai.

Soldiers of the Khmer Rouge regime of Pol Pot and soldiers of the anti-communist Khmer People's National Liberation Front (KPNLF) also claimed to have experienced chemical attacks. They were fighting each other and both were fighting Vietnamese forces that had invaded Cambodia in December 1978. Most of the attacks alleged to have occurred in Cambodia were said to have taken place in the Thai border region, while most of the alleged attacks in Laos were claimed to have taken place around Phu Bia Mountain, about 80 km from the Thai frontier.

A telegram from the U.S. Embassy in Bangkok to Washington dated September 2, 1978, appears to have been the first to report Hmong accounts of lethal chemical attack. It cited interviews conducted by Embassy personnel with Hmong who had arrived recently from Laos, three of whom claimed they had witnessed Hmong fatalities due to what they said was poison gas delivered by small aircraft, including "rocket-fired smoke" and "blue or yellow spray." The telegram said that for several months the Embassy had heard reports from refugees claiming that poison gas was being used against the Hmong in Laos, but cautioned that the telegram was "based on refugee interviews and should therefore be regarded as raw information rather than confirmed fact."

The earliest Khmer Rouge claims of chemical attacks of which we are aware were made over Radio Phnom Penh in November 1978 and alleged the use of "poisonous gas" artillery munitions by invading Vietnamese forces. A note distributed in April 1980 by the Khmer Rouge diplomatic mission in Geneva claimed there had been some two dozen attacks by Vietnamese aircraft and artillery with "toxic chemicals" that had caused vomiting, dizziness, fever, and at least 455 deaths between July 1979 and March 1980. The alleged agent was described as having a "yellow colour" and as sticking to "leaves of the trees."

As claims of chemical warfare continued to be made, the U.S. State Department dispatched a Foreign Service officer to Thailand to conduct interviews regarding the allegations. In June 1979, he and a U.S. Embassy consular official interviewed, through interpreters, 22 Hmong refugees who told of chemical attacks in Laos, some of which were said to have been associated with hundreds of deaths. Nineteen of the refugees told of a CBW agent that was yellowish; two handed in samples of vegetation with yellow spots about 3 millimeters in diameter.

In October 1979, four U.S. Army medical and toxicological specialists went to refugee camps at Ban Vinai and elsewhere in Thailand. There they completed 38 more interviews with Hmong who claimed they had witnessed or knew of attacks. The alleged agent was said to have been of various colors – red, green, black, white, blue, grey – but most often yellow. The Army interviewers also were given samples from an alleged attack: yellowish spots a few millimeters in diameter on pieces of bark from a village rooftop. Over the next few years, samples of the yellowish alleged agent continued to be handed in to U.S. officials and to officials of other countries, including Australia, Canada, France, Sweden, Thailand, and the UK, by Hmong refugees and Khmer soldiers claiming them to have been collected from sites of chemical attack.

Throughout all the available interviews with Hmong refugees conducted by U.S. and Canadian officials, by UN investigators, and by relief agency and hospital personnel from January 1979 through June 1983, the alleged agent was most often said to have been yellowish and to have fallen like rain. As described in the Shultz report, "Usually the Hmong state that aircraft or helicopters spray a yellow rain-like material on their villages and crops." The U.S. assessment that the yellow-rain-like substance was a chemical-warfare agent was repeated in a 1985 article co-authored by Christopher Green, the CIA official in the Agency's Directorate of Intelligence who chaired the CBW/Toxin Use Working Group, an inter-agency body formed to coordinate U.S. investigations of the CBW allegations. According to the article, "Aerial attacks, usually by spray, dispersed yellow to yellow-brown liquid or semi-solid particles that fell and sometimes sounded like rain when striking thatched rooftops."

7.7.2 Composition of the Alleged Agent

None of the alleged attacks was witnessed by a Western observer. The most tangible evidence bearing on the allegations consisted of the samples of alleged agent turned in by refugees and the laboratory analyses of these and other environmental samples, of blood and urine from alleged victims, and of tissues taken at autopsy from a Khmer Rouge soldier.

Samples of the alleged agent – the yellow spots and powders handed in by Hmong refugees and Khmer soldiers – first became available to U.S. investigators in 1979. But it was three more years before their principal component was correctly identified. In January 1982, a scientist at the UK Chemical Defence Establishment at Porton Down saw under a microscope that the principal component of a Yellow Rain sample was pollen. This was found to be the case for all samples of the yellow alleged agent accumulated at Porton, some 30 in number. Soon afterward, the discovery of pollen in Yellow Rain samples was made independently in Australia, Thailand, and Canada. The Porton finding was communicated to the United States and was confirmed with many additional samples at what was then called the Army Chemical Systems Laboratory (CSL), later named the Chemical Research and Development Command (CRDC), at the Edgewood Area of Aberdeen Proving Ground, Maryland. It was also confirmed at defense establishments in France and Sweden. Altogether, at least 100 samples of yellowish spots and powders from alleged attacks in Laos and Cambodia were examined under the microscope and, without exception, found to consist principally of pollen.

The presence of pollen in the samples of the alleged CBW agent that had been accumulated by the United States was first made public by Emory Sarver of CSL on November 29, 1982, during a State Department press briefing on the Shultz report. The next day, in a second Washington press briefing, an explanation for the presence of pollen in the Yellow Rain samples was given by Gary Crocker, a State Department intelligence officer who chaired the Assessments and Policy Support Sub-group of the CBW/Toxin Use Working Group, and by Sharon Watson, an intelligence analyst at what was then called the U.S. Army Medical Intelligence and Information Agency (MIIA), later the Armed Forces Medical Intelligence Center (AFMIC), at Fort Detrick, Maryland. It was Watson who, in 1981, had suggested that trichothecenes had been used and who was responsible for scientific and intelligence aspects of the subsequent MIIA/ AFMIC program under which biomedical samples collected in Southeast Asia were analyzed for trichothecenes by contractor laboratories.

Crocker is recorded as having said:

> It contains pollen, and not windborne pollen, but pollen that would be commercially collected or is collected, if you will, by insects, the type of thing you would get in a – a honey bee would take from flowers. And they are the particular size. It happens to be the right size to be retained in the body.

This was the first public suggestion that the pollen in Yellow Rain might have something to do with honeybees. Crocker's explanation was then amplified by Watson:

> Well I think I might just clarify a point, and that is the role of pollen in the mixture. The agent, as it comes down, is wet, and at this time the primary exposure appears to be through the skin, and the toxins are dissolved in the solvent, going through the skin very quickly. But as the agent dries, a secondary aerosol effect can be caused by kicking up this pollen-like dust that is of a particle size that will be retained in the bronchii of the lung. Now we've shown in studies with animals that the internasal LD-50 [the dose lethal to 50 percent of an exposed population] for the Trichothecenes is much lower than we would have expected, and that the Trichothecenes, if they come in contact with the mucous membrane, were very rapidly absorbed and are very toxic by this route. So if you could bring the compound into contact with the mucous membranes of the bronchii, then it's a very effective way of getting it across. So there are two different ways that the compound is absorbed. It's [a] very clever, clever mixture.

Even then, the theory announced by these two U.S. intelligence analysts could have been dismissed for a number of reasons, not the least of which was that the samples of Yellow Rain had no such tendency to disperse.

U.S. officials nevertheless continued to assert this explanation for the presence of pollen in the yellow spots and powders that they claimed to be an agent of toxin warfare. In a December 1982 address to the First Committee of the United Nations General Assembly, Kenneth Adelman, then U.S. Deputy Representative to the UN, declared: "We are now, however, able to isolate the components of yellow rain. There is good evidence for the presence of commercially-produced pollen as a carrier and to help ensure the retention of toxins in the human body."

Starting in early 1983, however, the government's theory that the yellow spots and powders handed in by refugees were samples of a chemical warfare agent was being undermined by studies done at the Smithsonian Institution in Washington by Joan Nowicke, an authority on pollen identification. Both the U.S. Army and one of the present authors (MM) had brought her samples of the alleged agent, in the latter case including samples from alleged attacks provided by Australian and Canadian government scientists. Examining the samples under an electron microscope, Nowicke had found by mid-April 1983 that the several plant families represented in the samples are common in Southeast Asia. She also found that no two spots, even on the same leaf or rock, had the same mixture of pollen types. The tropical plant families represented in the pollen and the diversity between spots were attributes not to be expected for a chemical warfare agent made in the Soviet Union and dispersed from exploding munitions or spray tanks.

The implausibility of the government's explanation of pollen in the alleged agent and questions regarding other aspects of its Yellow Rain investigation had caused one of us (MM), together with an MIT colleague with experience in scientific intelligence matters, to convene a workshop of government and university experts in diverse fields of potential relevance, including anthropology, botany, toxicology, and analytical chemistry, to review the allegations of toxin warfare in Southeast Asia. It was held during April 15–16, 1983, at the American Academy of Arts and Sciences in Cambridge, Massachusetts. The 35 participants included Joan Nowicke; David Jarzen, a Canadian pollen expert who had examined samples of the yellowish alleged agent obtained by Canadian investigators; Gary Crocker of the State Department; Emory Sarver of the U.S. Army Chemical Systems Laboratory; Joseph Rosen, a Rutgers University chemist who had reported finding trichothecenes in a sample of the alleged agent obtained by ABC News that was later found by Nowicke to be mainly pollen; Amos Townsend, a retired U.S. Air Force doctor who had interviewed Hmong refugees about the alleged chemical attacks; Peter Ashton, a Harvard botanist who had studied bee pollination of Southeast Asian trees; and both of the present authors. By the end of the workshop, it seemed clear that the findings of Nowicke and of Jarzen meant that the samples had to be of natural origin and that, as Crocker had suggested, bees were somehow involved – but how?

As the workshop ended, Ashton, the Harvard botanist, suggested that we contact Thomas Seeley, a Yale zoology professor who had studied wild honeybees in Thailand. Upon hearing our description of the suspected agent – yellow, pollen-laden spots 2–5 mm in diameter that were said to fall like rain – Seeley concluded that the State Department explanation was "not parsimonious" and that the yellow spots were bee feces.

At the time, entomologists did not know whether tropical Asian honeybees defecated individually or might do so collectively, producing showers that could have been mistaken by witnesses for sprays from aircraft. In order to address the question, Seeley, Pongthep Akratanakul (a Thai honeybee expert), and one of us (MM) conducted a field study in Thailand in March 1984. On one occasion, near Chiang Mai, we observed great numbers of wild honeybees suddenly leaving their nests, flying too high to be easily seen, and producing a shower of feces that lasted for several minutes and covered an area of an acre or more with hundreds of thousands of yellow spots indistinguishable in appearance from those handed in and described as "poison from the sky." While this was the most intense shower we encountered, we also observed lesser showers of wild honeybee feces in other parts of the country.

Reporting our findings at the U.S. Embassy in Bangkok, we were told that the Embassy had requested and Washington had eventually agreed to the creation of a joint Defense-State CBW team to conduct a more systematic field investigation than had so far been done. One of the activities of the CBW Team was a field trip, in April 1984, together with Akratanakul, to corroborate our discovery of honeybee fecal showers. Their telegram to Washington reported showers that they observed, including one in which the team physician and their Hmong interpreter were caught, during which "the bees were seen flying approximately 40 to 50 feet above and passing over at a speed of approximately 20 mph," noting that the defecating bees were "seen only with difficulty." The telegram also reported that, "Dr. Akratanakul stated that he never noticed bee feces deposits in all his years of work until the present yellow rain controversy arose." Both we and the CBW Team spoke on a number of occasions with Hmong refugees who said they had witnessed chemical attacks but who, when shown spots of honeybee feces on leaves, identified them as the alleged agent.

Further tests by U.S., British, Australian, Canadian, French, and Swedish government investigators of the yellowish samples of the alleged agent found that they were indistinguishable from the feces of wild Asian honeybees. This included not only the gross appearance of the samples but also the concentration and species distribution of pollen seen under the electron microscope and the results of various other laboratory tests. One of these tests, on the ABC News sample in which the Rutgers laboratory had reported the presence of trichothecenes, showed the pollen grains to be hollow shells, just like the pollen grains in feces taken directly from honeybees, from which the nutritive protein and lipid has been removed by the bee's digestive processes. There was little room to doubt that what the Hmong had frequently mistaken as poison from the sky and which the Haig and Shultz reports, Crocker, Watson, and the 1985 article co-authored by the Chairman of the CBW/Toxin Use Working Group all asserted was a chemical warfare agent, was, in fact, the feces of Asian honeybees.

The finding that the yellowish materials that Hmong refugees and Khmer soldiers had repeatedly claimed to be a chemical warfare agent were in fact only honeybee feces showed that U.S. government investigators had, for years, given inadequate attention to possible natural explanations of the Yellow Rain. Moreover, it raised grave questions about the reliability of the interviews. If the Yellow Rain that was described and handed in by the alleged witnesses was honeybee feces, how could there have been some other substance delivered from the air simultaneously in sufficient quantity to cause the claimed effects that was neither observed nor sampled?

It could be argued that, among the minority of refugee accounts that are recorded as having described other kinds or colors of agent – white, red, black, or green – some describe actual chemical attacks. Yet the casualties described even in these accounts, if authentic, could be ascribed to natural disease or to the effects of blast or of inhalation or contact with materials released from rocket or artillery marker rounds, explosive munitions, riot control munitions, or phosphorus bombs and projectiles, all of which are known to have been in use at the time in Laos and Cambodia. Moreover, the widespread and long- standing misidentification by alleged witnesses of honeybee feces as a chemical warfare agent casts serious doubt on the entire body of interview evidence.

Could some of the Hmong and Cambodians have misinterpreted a natural phenomenon as a weapon of war? Unknown at the time of the first U.S. allegations of chemical warfare in Southeast Asia, a similar misidentification of honeybee feces as poison from the sky had occurred in China in 1976. As described in an article published in 1977 in the Chinese science journal Kexue Tongbau, drops of a viscous sticky liquid fell from the sky like rain in many regions of Jiangsu Province in August and September 1976. The showers usually lasted for a few minutes and generally extended over areas of up to an acre and sometimes considerably more. Upon drying, the drops formed yellowish spots a few millimeters in diameter. Local people called it "Hwang Yu" (Yellow Rain) and some of them said it was poison sprayed from aircraft and that they were afraid to consume food or water that might be contaminated with it. Investigating scientists from Nanjing University, however, found no clinical evidence of toxicity, nor is there any evidence that honeybee feces are toxic. Upon examining the yellowish material under a microscope, the Chinese scientists saw that it contained concentrated pollen and that the types of pollen were those produced by local plants. They concluded that the phenomenon resulted from mass flights of defecating honeybees. The Chinese yellow rain first became known to Western investigators of Yellow Rain only in 1983, when the 1977 article was brought to the attention of scientists and officials at Porton by a British academic.

What if the villagers of Jiangsu, like the Hmong in Laos who fled to Thailand and like the soldiers and villagers in Cambodia, had been under frequent attack by aircraft and artillery delivering explosives, napalm, phosphorus, riot-control agent, and smoke markers? And what if, instead of university scientists and village officials telling them that there was nothing to fear, they had been interviewed by apparently authoritative Westerners asking them to describe the poison attacks? Under these conditions, would not the mistaken beliefs of the people in Jiangsu have been repeated and greatly amplified? If so, is it possible that Hmong and Cambodian "witnesses" had mistakenly associated naturally occurring disease and battle injuries with showers of honeybee feces; that others were repeating unfounded rumor, even if sincerely believed, sometimes saying they had seen things they had only heard about; and that some said what they had been told to say by others, or what they thought would conform to the expectations of their questioners? These and other questions raised by the bee-feces findings were never seriously addressed by MIIA/AFMIC, by the CBW/Toxin Use Working Group, or in any of the commentaries holding that there had been chemical attacks.

7.7.3 Interview Evidence and the Defense-State CBW Team

Until 1983, nearly all interviews with alleged witnesses had been conducted in ways that fell far short of standards that would allow fact to be distinguished from rumor, misperception, miscommunication, or fabrication. Testifying to a committee of the Congress on December 12, 1979, Army dermatologist Colonel Charles W. Lewis, leader of the Army team that had conducted interviews with Hmong refugees two months earlier, said that, "Our team attempted to be very selective in talking only with Hmongs who had either been direct witnesses of attacks or those who had been exposed to chemical agents themselves." The team's ethnic Hmong interpreters selected only those refugees who said they had been eyewitnesses to chemical attacks, thereby excluding those who might have disputed the claims of chemical warfare. Colonel Lewis said that, "To achieve conformity in our results, we used a prepared questionnaire that we took with us." No attempt was made to avoid leading questions: the first two questions in the questionnaire were: "How was your health prior to this incident?" and "Where were you when it happened?"

Not until the arrival of the Joint Defense-State CBW Team in 1983 were appropriate efforts made to check the reliability of the interview claims of chemical attack. The team consisted of an Army Chemical Corps officer, a military physician, a Foreign Service officer, a secretary, and two interpreters, one who spoke Hmong and one who spoke Khmer. It operated in Thailand for three years, from November 1983 until October 1986. The methods the team employed to test the truth of earlier allegations included standard investigatory procedures for ascertaining the accuracy of witness testimony. The team used double-checks, seeking out and re-interviewing refugees whose previous interviews had portrayed them as victims or witnesses of attacks. Using cross-checks, the team questioned other refugees who had also been present at the same time and place or approximate locale as the interviewee who alleged chemical attacks, in an effort to determine whether their accounts supported what others had said. The CBW team's interviews took place at Ban Vinai refugee camp in Thailand and elsewhere in Thailand and the Thai-Cambodian border region.

Characteristic of the careful approach taken by the CBW Team was a report telegraphed back to Washington in January 1984. The team re-interviewed a Hmong man who had said in an earlier interview that he had seen "with his own eyes" chemical attacks on a village in Laos, had witnessed six deaths, and had suffered chest pains, chills, and dizziness. The CBW Team asked him how, while he was a resistance fighter in the distant jungle, he could know so much about attacks on the village. He then changed his account to say that he "did not personally see the attacks but rather had received accounts of the attacks from others." In another example, the team was able to locate a Hmong woman previously interviewed by a British nurse. The woman reportedly said in the earlier interview that she had experienced a chemical attack and had become very ill from it. When interviewed by the CBW Team, however, the woman gave a different account: she confirmed that airplanes flew over her village every two or three days at high altitude but she denied ever having experienced chemical warfare. The team also interviewed a Hmong man who, when in a group with other Hmong, "told of gassing attacks but denied its reality when the interviewer talked to him alone outside."

The CBW Team also questioned people who had not been contacted before. In May 1984 they interviewed a former resistance leader who had recently led a group of 93 other Hmong out of Laos to Thailand. He had commanded about 40 men and their families living since 1975 on the slope of Phu Bia Mountain, where the Haig and Shultz reports said that most of the attacks in Laos had occurred. However, as the CBW Team reported in a telegram to Washington in 1984, "he denied ever having experienced a CBW attack nor ever having seen any evidence of CBW use." When questioned why other Hmong refugees gave accounts of CBW attacks and he, a resident of Phu Bia for eight years, had seen nothing, he explained that as an educated man, he related what he saw and not what he felt, whereas other Hmong would relate what they heard and felt to be true. A few days later, the team telegraphed to Washington their view that, "The question is not whether Hmong refugees lie but whether Hmong refugees are accurate reporters of reality. Generally, we have not found them to be so and believe that their stories must be supported by external and, if possible, objective means." In the same telegram, the CBW Team reported that "The present Ban Vinai camp leader has told every member of the team that he doesn't know whether the gassing stories result from chemical warfare or from bee feces but that he would like the team to find out."

The CBW Team identified a further reason to question the reliability of interviews conducted by previous investigators. Many of the interviews conducted in 1979 by State Department personnel and by U.S. Army medical and toxin specialists specified the exact day of alleged attacks. The problem with such precision is seen in a telegram that the CBW Team sent to Washington in April 1984. The telegram described an interview the team had conducted with three family groups from the same village in Laos whose houses were ten meters apart. The family members could not agree among themselves on the date of an alleged attack within a four-month period, nor on the location of a Yellow Rain incident to which they had testified earlier. The team recognized the obstacles to resolving such conflicting details: "The Hmong culture does not compartmentalize units of time as tightly as we who have broken our lives into seconds, minutes, hours and days. Their time blocks are by seasons and as a result any effort to confirm a specific date of a given incident is usually frustrated." Earlier interviewers and interpreters appear thus to have led the Hmong to stipulate Western calendar dates to a misleading degree of false precision, and therefore may have elicited other erroneous testimony as well. Moreover, the unlikelihood that Hmong refugees could accurately specify the exact dates of chemical attacks calls into question the extraordinary assertion in the Haig report that intelligence information "indicates that military activity took place at the time and place of every incident reported to involve lethal chemical agents."

After working in Thailand for two years, in November 1985 the CBW Team telegraphed Washington their "conclusions to date." Their Sixth Quarterly Report stated that "Information regarding the use of 'yellow rain' against the Hmong in Laos remains too incomplete or implausible on which to base conclusions". Despite its long and careful investigation, the negative conclusions of the CBW Team were never publicly acknowledged by the U.S. government and became known to us only from declassified telegrams obtained under the Freedom of Information Act.

7.7.4 Alleged Symptoms

The finding that many of the alleged witnesses had mistaken honeybee feces for an agent of chemical warfare inevitably raises questions about the reliability of their claimed association of specific medical symptoms with chemical attacks. The Haig report presents tables comparing symptoms claimed by alleged victims of chemical attack in Laos and in Cambodia with the symptoms its authors believed would result from trichothecene poisoning and with symptoms characteristic of six well-known chemical warfare agents. The report claims a significant match of symptoms with those expected of trichothecenes. No attempt was made to compare the listed symptoms with those that would be expected to result from natural afflictions, smoke inhalation, blast damage, or other battle injuries. The symptoms listed in the Haig report as most often reported for alleged chemical attacks in Laos – vomiting, diarrhea, hemorrhage, breathing difficulty, and skin irritation – are not unusual given the level of natural illness, dislocation, and battle casualties experienced by the Hmong. Moreover, instead of the common symptomology suggestive of trichothecene exposure claimed in the Haig report, there was wide variation between individuals in the particular constellation of symptoms they described, as may be seen in the State Department and Army interviews of 1979 and as is underscored in the Shultz report, which notes that "different symptoms are ascribed to men, women, children and animals."

Symptoms ascribed to alleged chemical attack were repeatedly judged by experienced physicians on the scene to be of altogether different origin. In March 1981, for example, the U.S. Embassy's Regional Medical Officer examined purported Cambodian and Hmong victims, but he found no clear evidence to support CBW allegations. The medical investigations of the U.S. Defense-State CBW Team in Thailand did not reveal any clear evidence of exposure to chemical attack: during the period December 1983–June 1984, it investigated 36 suspected chemical warfare attacks, but not one yielded confirmatory medical evidence. For example, on several occasions the team investigated claims that Khmer Rouge soldiers had been gassed in combat. In a typical instance, extensive medical workup on an alleged victim who had complained of dizziness, nausea, and vomiting blood yielded the conclusion that his "symptoms, which he reported to be due to 'toxic gas exposure' could be easily explained by the clinical course resulting from the blast effects alone." In another case, the team's chemical and medical officers examined five patients in a Bangkok hospital who complained of dizziness, vomiting, chest pain, and temporary unconsciousness, which they attributed to chemical attacks. After examining them, however, the team "conclude[d] that their symptoms were as a result of battle fatigue, smoke inhalation, heat stress, or a combination of these effects." A few days later, the CBW Team and a Royal Thai Navy physician examined four other patients complaining of the same symptoms. The CBW Team agreed with the Thai physician's diagnosis that "all four victims suffered smoke inhalation and there is insufficient clinical evidence to support a finding of chemical warfare."

Regarding claims that KPNLF soldiers fighting the Vietnamese in the Thai-Cambodian border region had come under lethal chemical attack, General Sak Sutsakhan, Vice President of the KPNLF and Chief of its General Staff in April 1984, "denied that Vietnamese had used toxic chemicals so far" against his forces. He said, "The blast of big shells can damage your skin and make a lot of blood vessels explode, so you vomit blood. This must have happened to some of our soldiers. It is also possible that tear gas has been used." The Bangkok Post article quoting him also noted that "General Sutsakhan's assessment on toxins was confirmed by other KPNLF sources and medical teams at the border."

From the first allegations of chemical warfare in Laos and Cambodia in 1976 to the conclusion of the work of the CBW Team in 1986, we know of no case in which diagnostic examination or autopsy provided clear evidence of exposure to biological or chemical warfare agents other than riot control agent. With only the claims of alleged victims, with little or no attention to the problems caused by leading questions, hearsay, and possible fabrication, and no adequate consideration given to the likely effects of natural illness and the effects of battle, the claims of medical evidence for chemical warfare are not sustainable.

AFMIC argued that CBW attacks had largely ceased by the time in 1983 that the CBW Team implemented its more trustworthy methods of investigation, but that earlier reports were nonetheless authentic. However, a presumption that earlier reports were authentic but the later ones were not would imply a severe lack of reliability in the Hmong and Cambodian testimony generally, nor would it account for testimony like that of the Hmong resistance leader or the Ban Vinai camp leader described above. It also fails to take into account the fact that all of the symptoms claimed to have resulted from chemical attack could be, and in many cases were, judged by U.S. and Thai military medical examiners to be the result of natural illness and battle casualties. It is more likely that, aside from the use of riot-control agents, there never was any good evidence for chemical attacks, and that the conclusions reached starting in 1983 differ only because reliable methods of investigation were then instituted, both by the CBW Team and also, as discussed below, by laboratories seeking to corroborate the reports of trichothecenes in samples from alleged attacks.

7.7.5 The Trichothecene Analyses

The concentrations of trichothecenes said by the United States to have been found in samples from sites of alleged toxin attack in Laos and Cambodia are very low, ranging from a high of 143 parts per million of T-2 in a sample of the pollen-laden alleged agent, down to concentrations of only a few parts per billion in some samples of blood and urine. Formidable problems face the chemist seeking to detect a substance present in such tiny amounts in the sample to be analyzed. The problem is greatly compounded when, as in the environmental and biomedical materials from alleged attack sites, the sample is a complex mixture of substances, some of which may mimic the one being sought, thereby giving a false positive. False positives can also result from laboratory contamination, as may occur when the substance being tested for at trace levels is present in the laboratory in more substantial amounts, or when some of the substance present in one analytical run is carried over into a subsequent run. These difficulties can be avoided only by taking the most stringent precautions. The avoidance of false positives in analyses of environmental or biomedical samples for trace amounts of trichothecenes requires highly discriminating analytical methods and the parallel analysis of matched control samples. It is particularly risky to conduct trace analysis in a laboratory that has handled substantial amounts of the substance for which trace analysis is being attempted. The risk is that minute amounts adhering to containers, equipment, or other surfaces may inadvertently and unpredictably enter at various steps in the preparation of a sample or in its analysis. Indeed, both CSL and Porton found that reliable analysis of trichothecenes required the use of new glassware for each analysis. Trace analysis should never be undertaken in a laboratory that has previously handled anything but trace amounts of the substance or any substance that could mimic it. An example of a "clean" laboratory, so far as trichothecenes are concerned, is the U.S. Army Chemical Systems Laboratory in Maryland. Another is that of the UK Chemical Defence Establishment at Porton Down. A further essential precaution is to analyze blanks – artificial samples not containing the substance of interest but prepared for analysis just as one would prepare an unknown sample – preferably before and after each analysis of an unknown. Finally, if high confidence is required, no report should be accepted as correct unless it is confirmed in an independent laboratory.

Essentially all of the analyses for trichothecenes on which the United States relied in making its charges of trichothecene warfare were done, starting in 1981, in a contractor laboratory at the University of Minnesota. The Haig and Shultz reports as well as a 1983 publication from the Minnesota laboratory state that trichothecenes were found only in biomedical and environmental samples from alleged attacks and not in controls. The publication reported analyses of some 80 blood and urine samples from alleged victims, nearly 40 percent of them positive for trichothecenes. Important questions, however, were left unanswered. The number of controls, the manner and sequence in which they were analyzed, and the specific results were not adequately reported and, for most samples, not reported at all. The lack of satisfactory information regarding control analyses is all the more problematic because the Minnesota laboratory had previously done a great deal of work with substantial amounts of trichothecenes, having published more than twenty papers about such work during the four years before undertaking trace analysis of samples from the alleged Yellow Rain attacks. The specific trichothecenes with which that earlier work was done included the same ones which the Minnesota laboratory subsequently reported in environmental and biomedical samples from alleged attacks, principally T-2 and its catabolite, the trichothecene HT-2.

The possibility that the reports on which the United States relied in making its accusations had been compromised by false positives did not cause much concern until it became apparent that government laboratories in the United States (CSL) and the UK (Porton Down) were failing to confirm the presence of trichothecenes in any of the numerous environmental and biomedical samples from alleged attacks they analyzed, not even in samples that the Minnesota laboratory had reported positive. One of these, for example, was a yellow pollen – containing material alleged to have been sprayed by a twin-engined plane on the afternoon of March 13, 1981, near Phu Bia, in Laos. Known as sample C-168-81, it had been sent seven months later to the Minnesota laboratory which reported it to contain 143 parts per million of T-2 and 27 parts per million of DAS (another trichothecene), the highest trichothecene levels the Minnesota laboratory reported for any sample from an alleged attack. Even though CSL and Porton should have been able to detect even a few parts per million of T-2 or DAS in their portions of this sample, neither could confirm the Minnesota finding.

According to information provided by the Army to the U.S. Senate Committee on Foreign Relations in 1992, CSL had analyzed a total of 251 environmental samples from alleged attacks and had found no trichothecenes in any of them. Neither were trichothecenes found in any of the approximately 50 environmental and 20 biomedical samples from alleged attacks that were analyzed at Porton during the period March 1982 to September 1984, nor in the approximately 50 samples analyzed at the Swedish Research Institute of National Defense, nor in samples analyzed at Centre d'Etudes du Bouchet, the French CBW defense research center.

Defending the Yellow Rain allegations, AFMIC argued that the trichothecenes had decomposed during the time between the Minnesota analyses and those done at CSL and Porton. But pure T-2 and DAS are stable; there appears to be no experimental evidence showing that they would decompose in environmental samples under the conditions in which they were kept at CSL and Porton. A problem of trichothecene instability does arise, but it does so in a manner that raises additional doubts about the reliability of reports of trichothecenes in blood and urine samples. Because of the long trek through Laos from most of the alleged attack sites to refugee camps in Thailand, the biomedical samples were taken from alleged victims weeks after their alleged exposure. A 1988 report from the U.S. Army Medical Research Institute of Infectious Diseases at Fort Detrick describes experiments measuring the amounts of T-2 and its catabolite HT-2 in blood and urine taken at various times after intravenous injection of monkeys with T-2. The report concludes that disappearance of HT-2 "was so rapid that its detection in serum as an indicator of T-2 intoxication would be useful only within the first 6 hr [six hours] following exposure." Measurements of the amount of

T-2 in blood 28 days after its administration to monkeys and studies of trichothecene metabolism at Porton pointed to a similar problem.

There is yet further reason to question the claim that trichothecenes were associated with alleged attacks. An April 1987 report prepared by a member of the Collection and Requirements Division at AFMIC tabulates the findings communicated to AFMIC by the Minnesota laboratory for T-2 and HT-2 in blood and urine samples collected from 1981 through 1985 from 146 individuals identified as controls. Of these controls, the report lists 24 as positive and therefore concludes that "the blood and urine samples from CW victims in Southeast Asia cannot be used as evidence that mycotoxins were used as CW agents."

Thus, the claim of an association between alleged attacks and the presence of trichothecenes in environmental and biomedical samples is undermined by the lack of reliable information about the performance and outcome of control analyses; by the substantial risk of laboratory contamination; by the lack of later confirmation by CSL, Porton and other defense laboratories; and by the problem of trichothecene decomposition in vivo. A thorough, independent, and objective investigation, including a forensic audit of the original records, might be able to determine whether the conclusion of the above report is correct. If it is, the most senior officials of the U.S. government had been gravely misinformed.

The Shultz report charged that T-2 had also been detected on the surface of a Soviet gas mask obtained near Kabul, Afghanistan, in 1981 and concluded that trichothecene weapons had been used there since at least 1980. Little more has appeared in public about this allegation. While it is now known that the USSR, in violation of the 1972 Biological and Toxin Weapons Convention, had a massive bioweapons development and production program, there is no evidence even from recent inquiries that the Soviet program extended to trichothecenes. The State-Defense CBW Team spent two weeks in Pakistan in October 1985 to conduct an exploratory investigation, but found itself "unable to provide a scientifically sustainable conclusion regarding agent usage."

7.7.6 Defector and Prisoner Debriefings and the Absence of Recovered Munitions

The Haig report also claimed that "One of the most complete descriptions of chemical warfare activities in the 1976–78 period came from a Lao pilot who was directly involved in chemical warfare" and who defected to Thailand where he was interrogated by U.S. Embassy personnel. According to the Haig report, the pilot described numerous missions on which he said that he fired rockets at Hmong targets, dispensing smoke of various colors which he "surmised" was poison. The report asserts that several Hmong accounts support the testimony of the Lao pilot, and cites one of these in which rockets dispensing red and green smoke fired by aircraft over a seven-day period in June 1979 caused the deaths of ten villagers by gassing and 30 by shrapnel.

However, the portrayal of the Lao pilot's account in the Haig report had already been undercut by a classified telegram to Washington from the U.S. Embassy in Bangkok. It stated that when the Lao pilot was re-interrogated, this time not by Embassy personnel but by a U.S. Air Force pilot, an entirely different conclusion was reached. The U.S. pilot, an Air Force major, reported in the telegram his conclusion that "It seems very unlikely that the rockets fired contained toxic [word illegible in available copy] CW agents. A rocket delivered gas attack would require at least some sort of rudimentary knowledge of prevailing winds along with a few basic safety procedures. Although there have been no Hmong reports to verify the contention, the RO [reporting officer?] suspects that, based on the Lao pilot's description of the aerial detonation of the rocket, the weapon in question may have been a flechette round," that is, a munition discharging many small dart-like projectiles.

It might nevertheless be asked if other pilots flying other missions had engaged in chemical attacks like those alleged by the Lao pilot. If so, many spent and dud chemical munitions would have reached the ground, for more than half of the 60 interviews summarized in the Haig report cite rockets as the means of dissemination. In order to achieve the target effects claimed, some of the attacks would have had to involve firing very large numbers of rockets. Nevertheless, not a single authenticated chemical munition was ever recovered, neither from Cambodia nor from Laos, where Hmong refugees had collected numerous samples of what they claimed to be the alleged agent itself.

Debriefings of Vietnamese soldiers who had fought in Laos and Cambodia during the same years in which the Haig and Shultz reports claimed that there had been hundreds of chemical attacks also failed to obtain any evidence for chemical warfare. A CIA case officer who served in Southeast Asia during 1981–83, now retired, recently wrote in an article cleared by the CIA for publication that "none of the hundreds of Vietnamese soldiers, both defectors and prisoners, that we debriefed over the course of my two-year tour ever provided us a shred of information about anything remotely resembling 'yellow rain'. We talked to Vietnamese soldiers – officers, NCOs, and privates – who had fought in battles where chemical weapons had supposedly been used. They all said they had no knowledge of Vietnamese possession or use of chemical or biological weapons." He and his fellow working-level case officers in the CIA Directorate of Operations, who were specifically tasked to collect information on the alleged use of CBW weapons in Southeast Asia, concluded from their investigations that there was "absolutely no credible evidence of the existence or use of such weapons" but that "our Directorate of Intelligence (DI) analysts and senior administration policymakers did not like the answer we gave them."

7.7.7 *Conclusion*

The U.S. accusations appear to have been based on no credible evidence: without confirmation of a single alleged witness report, without confirmation of an association between trichothecenes and any alleged attacks, without any sample of the agent itself, without any recovered rocket or other munition, without any otherwise inexplicable claimed symptoms, and without any credible defector or prisoner testimony in all these years, counterfactual analysis leads to the conclusion that, except for riot-control agent, CBW weapons were not used in Laos or Cambodia.

The lessons to be learned from the Yellow Rain episode are straightforward: reliable procedures must be used to acquire and evaluate interview evidence, including the use of corroborative cross-checks and double-checks and careful avoidance of leading questions. Chemical identification of trace components must adhere to appropriate standards for forensic analysis, and results must be corroborated by an independent laboratory. Hypotheses must be subjected to wide consultation and objective criticism. Failure to apply these lessons, whether through incompetence or because of political exigencies and pressures, imperils the credibility of subsequent investigations of situations in which CBW weapons may actually have been used.

7.8 Some Consequences of the General Purpose Criterion Having Been Used to Define the Scope of the Chemical Weapons Convention[200]

Editors' note. Both the Biological Weapons Convention and Chemical Weapons Convention root their prohibitions in a general purpose criterion. This means that, rather than limiting proscriptions to a set of material components such as particular agents, precursors, or delivery systems, the treaties outlaw all action intended for any purpose that violates treaty strictures. In this paper, Julian sets out the history of negotiations that led to the Chemical Weapons Convention's general purpose criterion, and explains the central importance of the criterion to preventing misuse of dual-use materials as well as to ensuring that the treaty covers all chemical weapons (including ones that have not yet been invented). The paper also notes the role of the non-governmental groups SIPRI (the Stockholm International Peace Research Institute) and the Pugwash movement in the debate.

---

[200] Editors' note. The original reference for this is J. P. Perry Robinson. 'Some consequences of the General Purpose Criterion having been used to define the scope of the Chemical Weapons Convention.' Working Paper for Pugwash Meeting no.207, 3rd Workshop of the Pugwash Study Group on Implementation of the Chemical and Biological Weapons Conventions, 19–21 May 1995, Noordwijk, The Netherlands. Rev.1 dated 21 June 1995. It is listed in Chap. 9 as Item 308 Rev.1. Unless otherwise indicated, the remaining footnotes for this paper are Julian's originals. Apart from correcting obvious typos and renumbering headings and tables in keeping with the requirements of this book, we have retained Julian's formatting and spelling throughout the paper and its notes.

Julian shows that while the Convention's general purpose criterion provides a comprehensive, future-proof approach, it can be difficult to implement such broad and open-ended provisions, and notes that because of this, the Chemical Weapons Convention includes schedules of chemicals in separate annexes to guide the operation of the Convention's verification regime. Nevertheless, Julian points out that unduly focusing on the schedules, and overlooking the full extent of the general purpose criterion, risks weakening the Convention.

Abstract

An annex to the Chemical Weapons Convention [CWC] contains schedules of chemicals that exemplify the chemical weapons banned by the treaty. The Convention defines the weapons it bans in terms of a general purpose criterion, meaning that it is the purpose to which a toxic chemical is to be put, not its chemical identity, which determines whether the strictures of the treaty apply to it. A history of the negotiation of this approach is presented. The approach allows the treaty not to preclude peaceful applications of toxic chemicals. It also extends the treaty to chemical weapons that are still secret or not yet discovered. But such open-endedness creates practical difficulties for administrators and for industry, so much so that the treaty is liable to be implemented as though it were the annexed schedules that actually defined the banned weapons. The authority of the general purpose criterion is thereby placed in jeopardy, and with it the heart of the Convention. It is to national implementing legislation and to the nascent CWC National Authorities that the search for safeguards may best be directed.

The objective of the Chemical Weapons Convention [CWC] is set out in the Preamble. There, States Parties record their determination, *for the sake of all mankind, to exclude completely the possibility of the use of chemical weapons, through the implementation of the provisions of this Convention, thereby complementing the obligations assumed under the Geneva Protocol of 1925.* These words also record the basic bargain underpinning the treaty. As quid pro quo for furtherance of the objective thus enunciated, each State Party undertakes to accept the burden which the demands and strictures of the Convention will place upon it. The burden may be substantial for some States Parties, whether it be measured in terms of administrative resources committed, intrusion of the verification system into private business or national security affairs, reduction of military capabilities, or restriction of commerce. Whether the bargain is worth striking and whether it will remain worth having – whether, in other words, the international regime to be established by the Convention will actually come into existence and thereafter prove stable – will depend not least on the meaning attached to "chemical weapons".

The negotiators came to use the twin devices of the General Purpose Criterion and the Schedules to impart an acceptable scope to what the Convention was to mean by "chemical weapons" and how it was to deal with them. The present paper identifies a currently topical implication of this approach.

The implication may be summarized as follows. Implementers of the Convention have now taken over from the negotiators. At the national level, they are devising the detailed rules and procedures necessary for domesticating the treaty's international obligations, while at the hardly less important international level, they are struggling to build an effective oversight organization, the [Organisation for the Prohibition of Chemical Weapons] OPCW. At both levels the onus is upon them to preserve the attractiveness of the basic bargain. They must, in other words, protect the scope of the Convention while contending with the complexities of putting it into practice. Compromise here could be dangerous. If the subtlety of the negotiators' definition were to be lost – if, for example, the scope of the treaty were somehow to become locked into the technology currently depicted in the Schedules – the basic bargain could soon become outdated, no longer attractive. Implementing the treaty might then destroy it.

It is useful to begin by recalling how it was that the scope of the Convention came to be set the way it has. The brief summary that follows rests on the increasingly unreliable memory of the author. Other people may remember things differently.

There is a sense in which virtually all weapons are "chemical". Explosives, propellants, toxicants, flame agents, incendiary and smoke-generating compositions, in fact most substances that have formed the payload of munitions, are materials that can be characterized in chemical terms. They are, in a word, chemicals; and weapons into which they are loaded are, in this sense, "chemical weapons". The negotiators of the CWC were not, however, negotiating general and complete disarmament; they were, as the Preamble says, seeking to complement the Geneva Protocol. So it was in the far narrower sense of a chemical weapon being an instrument of the chemical warfare outlawed by the Protocol that they sought consensus on an international prohibition.

When the talks on CBW disarmament began in Geneva in 1968, two major powers, Japan and the USA, were not yet parties to the Geneva Protocol, and the United States was actually employing toxic chemicals as weapons in the Vietnam War. Nevertheless, it was clear to the negotiators that the concept of toxic chemical as used in the Geneva Protocol (*... asphyxiants, toxiques ou similaires ...*) provided the only sound rationale for determining which varieties of chemical weapon should and should not fall within the scope of the chemical-weapons negotiation. Explosives, napalm and white-phosphorus weapons do not work through toxicity, so they did not come to be treated as chemical weapons in the special sense of the projected Chemical Weapons Convention; nerve- and mustard-gas weapons do, and therefore did. A fuzziness around the edges of the concept of toxic chemical was preserved for many years in order to accommodate the United States. Only in the final months of the negotiation – almost two decades after US forces had stopped using temporary-incapacitant and chemical-herbicide weapons (such as CS and Agent Orange) in Vietnam – was consensus secured on a formulation of the concept. This was the language now to be found in CWC Article II.2:

> "Toxic chemical" means: Any chemical which through its chemical action on life processes can cause death, temporary incapacitation or permanent harm to humans or animals. This includes all such chemicals, regardless of their origin or of their method of production, and regardless of whether they are produced in facilities, in munitions or elsewhere.

Having opted to follow the precedent of the Geneva Protocol, the negotiators then found themselves having to cope with two awkward facts. One was that some chemical based weapons, although not intended to work through toxicity, can nevertheless have toxic side effects. British forces fighting in South Africa at the turn of the century, for example, were accused of resorting to poison-gas warfare when they used artillery munitions charged with Lyddite. This substance was one of the new "high" explosives (in fact a formulation of trinitrophenol), the detonation of which yielded toxic nitrous fumes. The other awkward fact was that almost any chemical can be harmful, even toxic, in sufficient quantity, including a great range of industrial and other civil-use chemicals that could not possibly be outlawed. So the concept of toxic chemical, essential for defining the scope of the negotiation and then the CWC itself, was not in itself sufficient. The negotiators had to bring in additional elements.

Their initial idea was that the prohibitions and limitations of the projected treaty could be focussed on "chemical warfare agents" or, as the Nixon-Brezhnev communiqué of July 1974 put it, "means of chemical warfare". This was a concept that embraced both toxicity and the purpose for which the toxicity was to be exploited, namely an agency of warfare. What might be understood by it had already been debated in the United Nations, for in December 1969 the General Assembly had adopted a resolution initiated by Sweden declaring as contrary to international law *the use in international armed conflicts of .. [a]ny chemical agents of warfare – chemical substances, whether gaseous, liquid or solid – which might be employed because of their direct toxic effects on man, animals or plants.* The memorandum prepared in support of this definition by a Swedish Foreign-Ministry lawyer (now Director General of the IAEA[201]) received wide distribution and acclaim.

The fact that the Swedish resolution extended to chemical herbicides of course made it unacceptable to the United States and its co-protagonists in Vietnam. Western countries either abstained or, in the cases of Australia, Portugal and the USA, voted against the resolution. It was not this, however, which ultimately drove the concept of chemical-warfare agent out of the CWC talks in Geneva. Far from it: in variant forms the concept figured in, for example, the Japanese draft convention of April 1974, in the "phased approach" proposed by Canada in July 1974 and again in July 1976, in the British draft convention of August 1976, and in the initial communiqués from the USA-USSR bilateral talks that commenced in March 1977.

By then, however, the notion long nurtured by [the Stockholm International Peace Research Institute] SIPRI and the Pugwash CW Study Group was starting to grip negotiators that, if international on-site inspection of possible chemical-weapons locations were to remain unacceptable to some participating delegations, then "non-production verification" might provide an alternative. The notion took hold, and civil chemical industry thereupon moved centre-stage.

But it was soon discovered that, while chemical companies might be prepared to accept international inspection in the interests of chemical disarmament, they were highly unwilling to chance the outside world thereby supposing that their facilities were somehow associated with "chemical warfare agents". Nor, especially with the advent of binary munition technology, was it possible to specify objective technical criteria that could be used to distinguish proscribed chemical-warfare agents from

[201] Editors' note. Hans Blix was Director General of the IAEA when Julian wrote this paper.

legitimate industrial chemicals. The thiodiglycol that can be made into mustard gas is identical to the thiodiglycol that is used in the dyeing of textiles; the CS that is manufactured for controlling riots and suchlike police purposes is identical to the CS that was loaded into 750-pound bombs for use in the Vietnam War. Possible applications of quantitative toxicity and other such criteria caused the concept of "chemical warfare agent" to linger on for a while, but by this time the many advantages of the competing approach that used the so-called "general purpose criterion" were becoming more widely appreciated. In the bilateral USA-USSR communiqué of May 1978 the concept was supplanted altogether by the General Purpose Criterion; at the multilateral level it disappeared by the time of the Turbanski rolling text of August 1985.[202]

The new approach, as it came to be developed, followed the precedent which had been set by the 1972 Biological Weapons Convention. This treaty expressly applies to biological agents and toxins *that have no justification for prophylactic, protective or other peaceful purposes.* In other words, any supply of microbes that could not be explained in terms of a peaceful purpose was liable to be treated by the Convention as a biological weapon. The beauty of so all-embracing an approach was that it would capture biologicals whose existence was not yet known but which, if they were to be discovered, might then be made into horrible new weapons: some genetically modified organism of huge virulence, for example, or a new haemorrhagic fever virus which might, like Ebola, suddenly emerge from a tropical wilderness. This was a form of technological surprise against which it was no less important that the projected Chemical Weapons Convention should also be guarded. The negotiators duly adopted an analogous formulation which, in its final form in CWC Article II.1 (a), states that the "chemical weapons" of the Chemical Weapons Convention include all *toxic chemicals and their precursors, except where intended for purposes not prohibited under this Convention, as long as the types and quantities are consistent with such purposes.* So, in determining whether an activity involving a toxic chemical or a chemical from which a toxic chemical can be made is actually banned – an activity such as "development", "production", "transfer" or "other acquisition" – the criterion to be applied is the purpose of the activity. Unlike the biological treaty, the Chemical Weapons Convention actually lists the different purposes that are not prohibited. They include *inter alia* industrial, agricultural, research, medical, protective and law enforcement purposes.

Herein lay the second great advantage of the General Purpose Criterion. As well as embracing toxic chemicals not yet discovered, it also provided a mechanism whereby dual-use chemicals could be accommodated. For what the Criterion does is to ensure that the prohibitions laid down by the Convention are not directed at chemicals *per se.* They are instead directed at the purposes for which chemicals are intended. Thus holdings of phosgene, the leading killer gas of the First World War, are not proscribed if they are intended for production of, say, polyurethane plastics, but they are proscribed if they have no such nonprohibited purpose. Sarin nerve-gas

---

[202] Editors' note. In international negotiations, rolling texts comprise draft language anticipating areas of final agreement, and form the basis of discussions aimed at arriving at a final agreement.

is not itself necessarily a chemical weapon within the meaning of the Convention; it becomes such a weapon if it is not being held for, say, neurophysiological research or for the testing of protective equipment. CS tear-gas is a chemical weapon if it is held in forms and quantities inconsistent with law enforcement or other such nonprohibited purposes.

Not only, then, does this device of the General Purpose Criterion protect the Chemical Weapons Convention against technical change; it also protects legitimate uses of all toxic chemicals and their precursors. Whole sectors of chemical and biotechnological industry and enterprise are thereby shielded from the strictures which their dual-use character might otherwise invoke. The General Purpose Criterion is thus the heart of the treaty. It is also, very probably, the least understood part of the treaty, and vulnerable, therefore, to heedless implementation policies.

What, then, of the Schedules? The role which the negotiators of the Chemical Weapons Convention assigned to them is set out in parentheses at the ends of Articles II.2 and II.3: (*For the purpose of implementing this Convention, toxic chemicals [and] precursors which have been identified for the application of verification measures are listed in Schedules contained in the Annex on Chemicals*). The Schedules in effect define the scope of the "routine verification" arrangements of the Convention.

These arrangements rest on a division of labour between, on the one hand, individual States Parties, which are required to collect and declare Schedule-related data about themselves, and, on the other hand, the OPCW Technical Secretariat, which is required to receive those data from State Parties and validate them by means of routine on-site inspection. For "effective liaison" with the OPCW, States Parties are required under Article VII.4 to *designate or establish a National Authority*.

The division of labour between the National Authorities and the Technical Secretariat will in fact extend beyond routine verification. The Convention's arrangements for verification-by-challenge will add to the tasks that States Parties are likely to assign to their National Authorities. In addition, since the verification measures set out in the Convention cover only a fraction of the treaty's main obligations, responsibility for overseeing compliance with the remaining fraction falls by default to individual States Parties. Table 7.7 … indicates something of the relative magnitude of these two fractions. This further aspect of the division of labour is by no means a notional one, for each State Party is required, under Article VII.I, to *adopt the necessary measures to implement its obligations under this Convention* and, under Article VII.4, to set up its National Authority *[i]n order to fulfil its obligations under this Convention*. Article VI.2 further emphasizes this responsibility in regard to the particular prohibitions contained in Article I.1 (a) and (b): *Each State Party shall adopt the necessary measures to ensure that toxic chemicals and their precursors are only developed, produced, otherwise acquired, retained, transferred, or used within its territory or in any other place under its jurisdiction or control for purposes not prohibited under this Convention*. … Although this language is commonly portrayed by some signatory states as endorsing the work of the Australia Group and its associated national monitoring and data-collection efforts, the language is certainly not limited to exports.

Table 7.7 The division of labour between the OPCW and states parties in verifying compliance with the main prohibitions laid down in the chemical weapons convention[203]

Negative obligations under the CWC		*Routine verification* of compliance by the OPCW Technical Secretariat*	
CWC element	*Obligation*	*CWC element*	*Verification measure*
Art I.1(a)	Never to develop CW	None	[left to States Parties]
Art I.1(a)	Never to produce CW	Art III.1(c) Art V.3 VerAx V	At declared CW production facilities, on-site inspection and monitoring with on-site instruments
		Art VI.3 VerAx VI-IX	At declared industrial facilities producing more than threshold quantities of specified chemicals, on-site inspection [Possibility of undeclared production left to SPs]
Art I.1(a)	Never otherwise to acquire CW	None	[left to States Parties]
Art I.1(a)	Never to stockpile or retain CW	Art III.1(a) Art IV.3 VerAx IV	At declared CW storage facilities, on-site inspection and monitoring with on-site instruments [Possibility of undeclared stockpiles left to SPs]
Art I.1(a)	Never to transfer CW to anyone, directly or indirectly	None	[left to States Parties]
Art I.1(b)	Never to use CW	Art IX.19 Art X.9 VerAx XI	In cases of alleged use and if so requested by States Parties, on-site inspection
Art I.1(c)	Never to engage in any military preparations to use CW	None	[left to States Parties]
Art I.1(d)	Never to assist, encourage or induce, in any way, anyone to engage in any activity prohibited to a State Party under this Convention	None	[left to States Parties]
Art I.5	Not to use riot control agents as a method of warfare	Art IX.19 Art X.9 VerAx XI	In cases of alleged use and if so requested by States Parties, on-site inspection

*The Convention also provides for verification by challenge, in which virtually any site may be inspected by the OPCW Technical Secretariat if the prescribed procedures for triggering such visits have been followed. But these measures can only be activated by States Parties, not by the OPCW

Note The "chemical weapons" (CW) to which the prohibitions of the Convention apply are defined, in Art II.1, as meaning "the following, together or separately: (a) Toxic chemicals [defined in Art II.2 as "Any chemical which through its chemical action on life processes can cause death, temporary incapacitation or permanent harm to humans or animals"] and their precursors, except where intended for purposes not prohibited under this Convention, as long as the types and quantities are consistent with such purposes; (b) Munitions and devices, specifically designed to cause death or other harm through the toxic properties of those toxic chemicals specified in subparagraph (a), which would be released as a result of the employment of such munitions and devices; (c) Any equipment specifically designed for use directly in connection with the employment of munitions and devices specified in subparagraph (b)"

All of this suggests that the "necessary measures" of Article VII.1 will need to include, not only the penal legislation called for by Article VII.l (a) and associated lawenforcement measures, but also active domestic compliance-monitoring measures as well.

In embryo, such domestic compliance-monitoring systems will anyway exist in the data-collection and data-validation procedures operated by the National Authorities in relation to scheduled chemicals and any scientific or industrial facilities in which such chemicals occur. The people who are now engaged in setting up the National Authorities are presumably wrestling with the question of how much further

[203] Editors' note. This table was 'Table 5.1' in the original.

these embryos will have to be developed. In particular, how deeply must the active monitoring systems extend into the domain of unscheduled chemicals?

Two considerations can at once be brought to bear on that last question. The first is that, from the all-important standpoint of chemical industry, the Chemical Weapons Convention exists to protect chemical firms from unfounded accusations of involvement in chemical-warfare armament. Legitimate producers or consumers of unscheduled dual-use chemicals will rightly expect the same protection from the Convention as will legitimate users of scheduled chemicals. The second and in fact closely related consideration is that if the purview of national active monitoring does not extend to unscheduled chemicals, the country concerned is disregarding the General Purpose Criterion. It would have difficulty, therefore, in persuading others that it was serious about the Convention itself. On both counts, that country would risk losing much of the benefit of joining the Convention in the first place.

This is the point at which a drawback of the General Purpose Criterion is displayed: its extreme administrative awkwardness. If the Criterion were to be applied rigorously, each national compliance-monitoring system would have to treat each and every toxic chemical and every chemical from which a toxic chemical could be made – ten million different substances, maybe? – as a chemical weapon unless possessors could show otherwise, by demonstrating that the possession was for a purpose not prohibited under the Convention. That would require the operators of the active monitoring systems – the Article VII National Authorities, presumably – to be given quite outrageously intrusive powers of access and discovery and, in even modestly industrialized countries, legions of inspectors to roam through factories, research establishments and academic laboratories. Clearly that would be absurd.

As it happens, and for reasons not directly stemming from the General Purpose Criterion, many States Parties will anyway be obliged to extend their domestic monitoring systems some way into the domain of unscheduled toxic chemicals and their precursors. There are three reasons for this. One is that Article III.1 (e) requires each State Party to declare to the OPCW Technical Secretariat the identities *of each chemical it holds for riot control purposes*. Article II.7 of the Convention defines such "riot control agents" as *Any chemical not listed in a Schedule, which can produce rapidly in humans sensory irritation or disabling physical effects which disappear within a short time following termination of exposure*. The second reason, currently affecting 28 CWC signatory states, is the Australia Group [AG]. Of the 54 chemicals on the AG export control list, no less than 20 – identified in Table 7.8 below – do not appear in the Schedules of the Chemical Weapons Convention. The third reason is Article VI.6 of the Convention. This requires that, within 30 days of the Convention entering into force for it, each State Party must provide the OPCW Technical Secretariat with a list of each and every plant site within its jurisdiction where more than 200 tonnes of "unscheduled discrete organic chemicals" (30 tonnes if the chemicals contain phosphorus, sulphur or fluorine) were produced during the previous year.

The monitoring work necessitated for those three particular purposes, although extensive, will not, it has to be said, fulfil what needs to be done if the General Purpose Criterion is to be properly served. Fortunately the further monitoring tasks are rather more limited and less labour-intensive than might otherwise appear. The reason is

Table 7.8 CWC schedules 2 and 3, and the Australia group *chemical weapons precursors export control list*[204]

CWC Schedule 2 chemicals	***CWC Schedule 3 chemicals***	***Unscheduled chemicals on the AG list***
Amiton and corresponding alkylated or protonated salts Perfluoroisobutene 3-Quinuclidinyl benzilate Chemicals containing a phosphorus atom to which is bonded one methyl, ethyl or propyl (normal or iso) group but not further carbon atoms [Exemption: Fonofos] *N,N*-Dialkyl (Me, Et, n-Pr or i-Pr) phosphoramidic dihalides Dialkyl (Me, Et, n-Pr or i-Pr) *N,N*-dialkyl (Me, Et, n-Pr or i-Pr) phosphoramidates *N,N*-Dialkyl (Me, Et, n-Pr or i-Pr) aminoethyl-2-chlorides and corresponding protonated salts *N,N*-Dialkyl (Me, Et, n-Pr or i-Pr) aminoethane-2-ols and corresponding protonated salts [Exemptions: N,N-dimethyl- and N,N-diethyl-aminoethanol and corresponding protonated salts] *N,N*-Dialkyl (Me, Et, n-Pr or i-Pr) aminoethane-2-thiols and corresponding protonated salts ***On AG list as well:*** 17 specific chemicals from the families listed above Thiodiglycol Quinuclidine-3-ol Pinacolyl alcohol Arsenic trichloride Benzilic acid	Phosgene Cyanogen chloride Hydrogen cyanide Chloropicrin Ethyldiethanolamine Methyldiethanolamine ***On AG list as well:*** Phosphorus oxychloride Dimethyl phosphite Phosphorus trichloride Trimethyl phosphite Thionyl chloride Diethyl phosphite Triethyl phosphite Phosphorus pentachloride Triethanolamine Sulfur monochloride Sulfur dichloride	3-Hydroxy-1-methylpiperidine Potassium fluoride 2-Chloroethanol Dimethylamine Dimethylamine hydrochloride Hydrogen fluoride Methyl benzilate 3-Quinuclidone Pinacolone Potassium cyanide Potassium bifluoride Ammonium bifluoride Sodium bifluoride Sodium fluoride Sodium cyanide Phosphorus pentasulphide Diisopropylamine Diethylaminoethanol Sodium sulphide Triethanolamine hydrochloride?

as follows. It is patently ridiculous to regard every toxic chemical and precursor as being so threatening to a country's national security that the Convention must be implemented so as to provide protection against all of them. In fact rather few chemicals possess the particular combination of physical and chemical properties which can make them aggressive enough, not only to provide usable weapons, but also to provide weapons which are comparable in effectiveness with ones already available, whether chemical or non-chemical. (If they are less effective, they would be less likely to be chosen, in which case the chemicals concerned would not seriously threaten the Convention.) It is these rare chemicals, not the myriad others as well,

[204] Editors' note. This table was 'Table 5.2' in the original.

which need to be the focus of the Convention. Not all of them can be specified – either because they are secret, or because they have not yet been discovered, or because the baseline technology of warfare is changing – which is why the Convention had to use the General Purpose Criterion in the first place. So it is on these uncommon chemicals that the national monitoring systems, too, must somehow focus. Areas of scientific inquiry, industry, commerce and other business must be identified in which the chemicals are likely to occur; and these areas must then be monitored. This is not an easy burden to lay upon the National Authority of any industrialized country. But it surely has to be borne.

Where exactly the areas are is a matter for consideration. Some people might think that pharmaceutical research laboratories, as potential fount of novel chemical immobilizing weapons, should be the primary focus. Others might be more concerned about chemical manufacturing industry, noting the possibility of legitimate civil chemicals of sufficient toxicity (either on their own or in conjunction with binary-munition reactants) to stimulate abuse at diminished risk of falling foul of challenge inspection.

So it can be seen that one substantial practical implication of the dependence of the Chemical Weapons Convention on the General Purpose Criterion is that the Article VII National Authorities, as keepers of the domestic compliance-monitoring systems, should be strong rather than weak. The National Authorities are, by default, the principal protectors of the Criterion. Unprotected, the Criterion could easily fall into contempt and neglect, and the treaty-regime itself might not thereafter be able to survive the perils of obsolescence and seeming irrelevance, collapsing, in effect, under its own burden. It would therefore be doing the Convention no service at all to constitute the National Authorities merely as post-boxes*cum*-escort-agencies. This is what very well might happen if, for reasons of administrative convenience, the Schedules came to be treated as the primary arbiters of scope.

A strong National Authority would inevitably be more costly than a weak one, not least in terms of the transparency and accountability mechanisms that would have to be introduced as safeguard against abuse of powers. So the compensating benefits should not be undervalued. Foremost among them are the benefits identified earlier that flow from the General Purpose Criterion: the protection that it provides for legitimate industry involving dual-use chemicals, and its protection of the Convention against technological change.

Those, however, are not the only benefits. The recent sorry events in Japan involving sarin[205] indicate further possible advantages of active national compliance monitoring systems. Above all, such systems could greatly enhance and extend what is the chief means of defence against terrorist acts involving chemical weapons: alert intelligence machinery.

[205] Editors' note. This is a reference to the use of sarin on the Tokyo subway by Aum Shinrikyo in 1995.

7.9 Can University-based Research Promote International Cooperation Against the Spread of Chemical and Biological Weapons?[206]

Editors' note. This paper comprises speaking notes from a 'research-in-progress' seminar hosted by the International Relations [IR] department at the University of Sussex on 12 October 1998. Julian reveals that he was on a list of people who might inadvertently have helped Iraq's illicit weapons programmes, and uses this anecdote as a starting point to explain his views on the responsibilities of universities in preventing the misuse of science. He situates the discussion within wider considerations about the secrecy concerns of national intelligence communities, and the dangers of excessive secrecy.

This paper is much less formal than the other papers in this volume; we have included it to showcase the variety of Julian's outputs and styles as well as to present the content itself, and to give an insight into his presentation style.

Can university-based research promote international cooperation against the spread of chemical and biological weapons?

Question expecting the answer "yes" because that is the research on which I have been engaged for the past quarter-century at Sussex.

Anecdote/reminiscence. What we're up to in the IR end of SPRU.

For my point of departure, I take the visit which the Security Service (MI5) made to this campus earlier in the year. Two officials spoke to a select audience of faculty from the main elements of the University. It was in fact part of a tour the Service was making around many UK universities to spread the message of VIGILANCE.

Vigilance against potential students who might abuse the knowledge we are in the business of purveying. Using it to place weapons of mass destruction into the possession of countries that could become adversaries. Technology transfer of the intangible sort – as contrasted with the tangible transfers evident in international commerce. Teaching as opposed to trade.

High on the list of such countries is Iraq. A country whose innermost weapons-related secrets are being ferreted out by UNSCOM [United Nations Special Commission].

Explain UNSCOM. SCR 687 ceasefire resolution of 1991.[207] Future currently in flux, like the Inter Allied Control Commission in Germany in, say, 1925.

---

[206] Editors' note. The full reference for this is 'Can university-based research promote international cooperation against the spread of chemical and biological weapons?' 12 Oct 1998, Speaking notes for UoS/IR research-in-progress seminar, 12 Oct 98, 8pp. It is listed in Chap. 9 as item 367. Unless otherwise indicated, the remaining footnotes for this paper are Julian's originals. Apart from correcting obvious typos and renumbering headings and tables in keeping with the requirements of this book, we have retained Julian's formatting and spelling throughout the paper and its notes.

[207] Editors' note. UNSCOM – the United Nations Special Commission– was established by the UN Security Council in April 1991 to oversee the elimination of Iraq's weapons of mass destruction and long-range missiles, as stipulated in UNSCR687, the ceasefire ending the 1991 Iraq War. See United Nations Special Commission (UNSCOM) (webpage), https://www.un.org/Depts/unscom/General/basicfacts.html [Accessed 31 March 2025.

UNSCOM has been uncovering the nuts and bolts of how chemical and biological weapons had, in the case of Iraq, spread beyond the elite association of original possessors. An association whose dean was once the United Kingdom.

Hitherto the mechanisms of that spread had largely been matters for speculation, guesswork by the intelligence services, who really weren't much interested in CBW in those Cold-War and therefore nuclear-dominated days. Inference from scattered and often unreliable data, directed by the general tenor of analysis at the time, in other words the institutional prejudices.

UNSCOM enabled movement away from the instinctive explanations of earlier times of how Iraq had managed to acquire the chemical weapons it used against Iran in the 1980s (on which we in the West turned a blind eye, at least in public, in the interests of our favoured regional hegemon and, no doubt also, vested business interests).

Thus, Iraq got them from the USSR (the Cold War explanation); or Egypt helped Iraq in the interests of pan-Arabism (an Israeli explanation).

Thanks to UNSCOM, we were able to see the real enemy here, and it was mostly us: Western business and Western academia lured by gigantic bags of oil-derived gold.

Take for example the Iraqi biological-weapons program which is now so worrisome for us. At that interface of science and government that was crucial for the future of the programme stood, according to UNSCOM, the directing influence of one Dr Rihab Taha – mother of two, wife of a general who is the central figure in the Iraqi military-industrial complex: a biologist who had learnt much of her biology in England, her doctorate coming from East Anglia in 1984.

MI5's message to us was: watch out for that sort of thing here; Sussex has already, in its innocence, transgressed. If you need any advice at all on whether to admit dodgy looking foreigners into your courses, phone us for guidance. Yes, we know the pressures on you to keep admissions up, but you have a duty. If you would prefer not to associate with us, there is always the scheme for voluntary vetting of potential students which the Foreign Office runs in association with the Committee of Vice Chancellors and Principals.

And they went off to talk with other universities, some of which had invited them down for a second hearing.

Now against that background of the Security Service pointing a finger of blame at Sussex, I have a confession to make. UNSCOM keeps several databases to remind itself of what it has learnt on Iraq's now-proscribed WMD programmes and thus to guide itself in its future investigations. (I suspect, though cannot say for sure, that these information hoards are also an element of barter with non-Iraqi sources of information.) The databases embrace individuals who have helped Iraq in one way or another – the best known of these individuals being those several shady businessmen and company directors who have been convicted by courts in Germany in recent years for violating the, it has to be said, hastily introduced export-control legislation. I too am on that list.

I am on it I am told for two reasons. First, because of publications of mine that describe the technicalities of poison-gas and germ warfare – especially those from my

SIPRI days. Such SIPRI publications (whose stereotype reader we were told should be the Egyptian technical adviser to the Geneva Disarmament Conference) were apparently found in profusion within the Iraqi CBW establishments. Second – and this is the one that certain people have taken particular pleasure in telling me about – a video, heavily labelled with Iraqi security-classification markings, found among the so-called "chicken farm" documents soon after the mid-1995 defection from Iraq of General Hassan Kamal, who blew the gaff especially on the biological-weapons programme.

What on earth could this video have been? Answer: a mid-1980s BBC Panorama documentary in which Robinson can be discerned explaining at some length certain basic principles of chemistry. For example. To get to product chemical A, you may have to react chemical B with chemical C, and to get to chemical B you may have to react chemicals D and E, and to get to chemical E you may have to go right back to raw material, such as special rock you may be able to dig out of your own terrain. If you can import supplies of any of the intermediates (or "precursors" as they have become known in the business) then you don't have to do so much industrial chemistry, which could overstretch your chemical industry and chemical economy. But if you can't import it, perhaps because supplier countries have imposed export controls, then you'll have to make it yourself, or else choose a different production pathway that involves more easily accessible precursors.

This is all self-evident and basic stuff, points anyone with the merest smattering of chemistry can make. The only novelty in my rendering was that instead of chemical A, I spoke of isopropyl methylphosphonofluoridate, otherwise known as the nerve gas sarin (the killer of Halabja in 1988 and Tokyo subway travellers in 1995), and instead of B and C, I spoke of methylphosphonyl difluoride and isopropanol; and so on. Now there is nothing remotely secret in the chemical name of sarin or of most of the precursors needed to make it. That knowledge has been in the scientific literature since the late 1940s.

Such, however, was the reason why, so I learnt in 1989, that video was being used as a training aid within certain sectors of UK officialdom concerned with the proliferation of chemical weapons, for my simple fragmentation of a CW-agent production programme into factories for A, B and E showed that a chemical-weapons acquisition effort did not need to rest on some special purpose chemical industry. It could instead exploit the multipurpose character of much of existing chemical industry, to say nothing of its international interdependence.

And that also was why, apparently, the video was discovered labelled with Mukhabarat security stamps among the chicken-farm documents. Or so I'm told. The people telling me then always adding the rider "and that's exactly what the Iraqis did".

Now at SPRU we have a perverse attitude towards publication of our work. The grand refereed academic journal is not what usually comes to mind when we think how best to disseminate our research findings. We are after all among the artisans of the academic world, untenured contract researchers most of us, so we look for more immediate returns on our investment of intellectual capital. Though for the sake of the Research Assessment Exercise (in which SPRU usually does well) we try to control ourselves, we nevertheless have a deep-seated preference, when publishing, for the newspaper article, the private conference paper, even the occasional soundbite on television. It is POLICY research that we do, so this makes sense. BBC Panorama ranks rather high as an influential outlet and, speaking personally, I am always rather chuffed when they call me up.

But in this case? How ought I to be looking back on it now?

In what I said to camera about the fungibility of chemicals, I was trying to make the argument that supply-side export controls can at best be only an interim measure for suppressing chemical-warfare armament pending the establishment of a proper international cooperative regime that can combine both supply-side and demand-side restraint. For a whole host of reasons, export controls can have only a transient effect on CBW proliferation. But I was being told that instead of doing what I thought I was doing, namely promoting what later became the 1993 Chemical Weapons Convention, I was actually promoting the chemical weapons programme of Iraq.

A nasty hint, redolent of blacklisting (given the quarters from which it came), was thus being directed at me. I am not sure that it was meant seriously. But if it was, why? I can only think that, in mocking export controls, I was impugning an orthodoxy which was then just starting to establish itself in Western CBW policy circles. The orthodoxy is one which, for reasons one may wish to speculate upon, prefers supply-side measures – such as the Australia Group (AG), itself modelled on the London Suppliers Club, and which in turn became the model for the MTCR and the Wassenaar Arrangement – over the unfamiliarities, unpredictabilities and uncontrollabilities of new-fangled multilateral disarmament regimes.

It is an orthodoxy which places in formal [sic] alliance at the forefront of international action against proliferation that really deadly combination of our times – the intelligence community and, fronted by ministries of international trade, the business community. For it is these people from whom the national delegations to AG meetings are drawn. The foreign policy and arms control communities have only walk-on parts to play.

I do not, as I say, seriously believe that those hints of blacklisting were a consequence of my Panorama pinprick of the supply-side orthodoxy. But I do draw a lesson. That one should never, in international-relations research, belittle the influence upon events of the intelligence community.

It is rarely easy to figure what that influence is. As far as the UK community goes, my experience is that you rarely get a chance. Its members are institutionally standoffish. I – the Harvard Sussex Program – run a Chatham-House-rules invitation-only seminar in London on CBW issues intended to intermingle government and nongovernment folk. When they choose to accept their invitations, they sit silently at the back taking notes, rarely speaking.

In America things are rather different on the surface but underneath much the same. I will move into another anecdote to illustrate what I mean. I will relate briefly the story of the Yellow Rain, and one of the finer achievements of the Harvard Sussex

Program. The story up to 1987 is told in a paper of ours in the Fall 1987 issue of FOREIGN POLICY,[208] but the story continues.

- Carter admin: international collaborative investigation by scientists.
- Reaganauts: leaf and twig = BWC violation. Nonscience masquerading as science.
- The issue we saw was inept integration of scientific advice into foreign-policy processes. Difficult even under unpoliticized circs: peer-review vs line-of-command: horizontal vs vertical traditions.
- Chicago AAAS [American Association for the Advancement of Science] AGM [Annual General Meeting]1983 – venue for postulation of an alternative hypothesis, to take its place among the several required by proficient investigation of phenomena.
- But it (beeshit) was the hypothesis that survived the data. And that, not communist germ warfare, became the orthodoxy within the scientific community, even the defencescience end of it.

But not the US intel community, which was stacked up around the communist-germ-warfare theory. Had we made no impression on them? We had embarrassed them slightly – to the point at least when they embarked upon silly measures against the Harvard end of the programme: disinformation propagated (agents of influence), attempt to lift clearances. But they came to nothing. So did the official view change? No. Yellow Rain, regarded as T-2 mycotoxin, remains an official threat to US security. Defence against T-2 continues, a nice little earner for a bit of the US defence industry. Nor is it formally possible to change the NIE [National Intelligence Estimate] because the NIO [National Intelligence Officer] has now moved on. Such is the influence of the IC [intelligence community] that that absurd intelligence assessment remains graven in stone. And in the meanwhile the word is being put about – entirely falsely in fact – that the newly opened Soviet archives reveal the Reaganauts to have been right all along. The battle is still being fought (from our side too, I hasten to say). Its lesson is that academics shouldn't mess lightly with the intelligence community: they should be prepared for more than intellectual argument.

Why is the IC so powerful?

A political explanation is the success of the IC in portraying itself as the victor of the cold war – notwithstanding all the idiocies that are now evident in its cold-war behaviour. See especially some of the findings from the Cold War International History Project of the Woodrow Wilson Institute. The so-called "intelligence wars" that peaked in the 1950s and 1960s were the closest that the cold war came to actual warfare. Ergo, the Western IC are the heroes of the cold war, to be treated respectfully and deferred to.

---

[208] Editors' note. See '7.7 The Yellow Rain Affair: Lessons from a Discredited Allegation' in this volume.

Functional explanation: (a) secrecy (b) continuity. Secrecy makes IC argument virtually immune to outside criticism. Continuity means that IC protagonists, when challenged, are much too knowledgeable and experienced to defeat easily.

Which can be a powerful advantage in bureaucratic politics. Witness the influence of ACDA [Arms Control Disarmament Agency], per se a powerless player, on US CBW policy prior to the Reaganauts and even during and after them. Always the same ACDA person on the interagency working groups; after a decade of this it was he, and he alone, who could remember the minutiae and where the bodies were. So people tended to defer to him, unless expressly briefed otherwise.

Likewise UK dominance in 1980s Western CBW intelligence affairs: the same UK person on the UK-US working groups since the year dot, but with new Americans rotating through every two or three years.

Academics can in principle offer a countervailing continuity. Nub.

We may not have access to the secret sources, but we can remember what happened over a period longer than that of the institutional memory of most foreign-policy desks. The papers there usually get retired to storage in no more than three years.

And we can come up with arguments and analyses which, if people within the structures are so minded, they can point to as indications of support. We – if we are so minded – can thus be coopted as cheer-leaders for particular policies, and we can in time generate support for those policies in areas into which government can reach only with difficulty.

This long drawn out process of mutual, and rarely acknowledged, support between GO [government] and NGO [non-governmental organizations] I would point to if asked to explain why, as is now unquestionably the case, the UK has so conspicuously become the product champion for the BWC and the CWC.

Contrast the situation in 1998 with that in 1968 – when arms control was seriously seen in most of Whitchall and Westminster as a communist plot in which participation was of value really only in terms of cold war rhetoric. In 1968, after all, the UK still had a secret chemical rearmament programme running, albeit not very well, forcing our armscontrol people in the Geneva talks into familiar British-type hypocrisy.

OK, so the cold war's over, so of course things are now different from what they were then. But that UK attitude change happened well before Gorbachev. Cognoscenti date its beginnings to the "Mason minute" of about 1973.

It hasn't surfaced yet in the PRO [Public Records Office]. Christopher Mason some of you will know as a Glasgow University academic, a teacher of Politics and IR [international relations]. He is an historic figure – as Whitehall's first ever in-and-outer on the US model.

Besides Glasgow on the list of universities whose research activities unquestionably promoted that UK support for the emergent international CBW control regime I would add LSE [London School of Economics and Political Science], Surrey and Leeds, as well as Sussex.

It is a special sort of research. It succeeds only if it (a) has enough current awareness to know what is or soon will or should be exercising officials, and (b) has enough contact at policy levels to know which the relevant mid-level officials are and which ones are pointing in the same direction as the researchers.

So, what of the post-Cold-War world? The two treaties I have mentioned, the BWC and the CWC, are Cold War productions. Can the international cooperation which they represent be sustained, even developed, in the future? And what is the future role of academic research here?

The UN Security Council is now, finally, back in the business of international security, which should change the focus of our type of research, shouldn't it?

The 1992 Summit statement about the proliferation of WMD being a threat to international peace and security (thereby in principle mandating collective use of force against it under Chap. 8 of the UN Charter).

What are these WMD? 1947/48 – nonconventional: UN Charter Article 26 system for the regulation of armaments.

But today's situation is very different. For example, WMD are no longer hi-tech, but low-tech: accessible via routine civil industry. Once so simple, the concept of WMD is getting to be a real muddle. The UN Security Council Summit statement is likewise an awkward guideline for future policy making. Two illustrations:

(1) A current US court case in which three apparently deranged men in Texas are charged with seven counts of threatening government officials with weapons of mass destruction, viz cactus needles. BWC implementing legislation.
(2) Immobilizing chemicals. These are covered by the CWC so must we regard them as WMD? They are portrayed as nonlethal weapons, which the more simple-minded among us equate with good, instead of the bad that can follow from less politically costly force.

Personally I do not see WMD as a useful concept any longer. Better to disaggregate.

A powerful reason is that new life-science knowledge is opening up possibilities for terrible new forms of coercion and repression. As we start to understand more about processes of development, cognition, inheritance.

Age of biotech lies before us. Like earlier ages characterizable through its dominant technology – nuclear, electronics, metallurgy ... – the technology will no doubt be exploited for hostile purposes. And we may have to live with those terrible new instruments of violence and subjugation. But maybe not – maybe in fact history has armed us against it.

I refer to the taboo against fighting with poison: Manu laws, Roman injunctions, Saracen code, Grotius, the C19 codifications …

Here at Sussex we continue our work on the spread of CB weapons within a collaboration with the BCSIA [Belfer Center for Science and International Affairs] in the Kennedy School of Government at Harvard. HSP has been running for a good many years now, looking to build up a community of people thinking seriously and constructively about CBW, which we do by publishing a journal, organizing national and international workshops, maintaining a unique research archive, encouraging DPhil students into our midst, and offering modest job opportunities – insofar as our precarious funding permits.

We have a number of research projects. I will mention one – our CBW criminalization treaty project in which we have the increasingly enthusiastic participation of a number of senior international lawyers from different countries.

It is stimulated by the UNSCOM experience that indicates rather clearly the parts that traditional arms control, pitched at the level of the state actor, cannot reach:

- the greedy businessman
- the desperate company director
- the evil professor

Multilateral treaty drafting exercise aimed at strengthening individual responsibility for compliance with the existing international CBW regime.

Cp [compare with] Piracy, Highjacking, Protection of Diplomats, Torture – we are looking to define a new international crime – a treaty that will make it an international crime for any person .....

The treaty also creating an obligation upon SPs either to prosecute or to extradite.

That HSP project brings me back to where I started: the MI5 visit to this university.

Two different approaches to much the same problem, transfers of tangible and intangible technology to potential abusers of it.

On the one hand, the intelligence-community-driven national measures of keeping those potential abusers out of the country, away from well-meaning academics.

On the other hand, the option of the international community seeking agreement on a new treaty that would define a new international crime (CBW armament) and create a universal jurisdiction for it.

Are these competing approaches, or are they complementary? (There is actually a middle way: the CWC Art VII extraterritoriality requirement)

In fact HMG [Her Majesty's Government] is now putting the pressure – moving away from the voluntary screening/vetting advocated by MI5 to a compulsory system backed by the force of a new law now being proposed in the wake of the Scott inquiry.[209] This is one of the things being proposed in the DTI [Department of Trade and Industry] *Strategic Export Controls* white paper.

Snuck out in the summer while the universities weren't looking. I hope someone here at Sussex is aware of it.

[209] Editors' note. The Scott Inquiry, led by Lord Justice Scott, was established by the UK government in 1993 to investigate exports of weapons and dual-use equipment to Iraq.

7.10 A Draft Convention to Prohibit Biological and Chemical Weapons Under International Criminal Law[210]

Editors' note. The Biological Weapons Convention and Chemical Weapons Convention operate at the inter-state level – that is, they are formulated *by states, for states,* and they seek to prevent the treaty's states parties from developing, using and proliferating chemical and biological weapons (CBW). Julian and Matthew Meselson recognized that this framework does not fully address development, use, or proliferation by *individuals*. They worked to address this gap by preparing a draft convention that makes efforts to acquire and use CBW an international crime, analogous to global law against torture, which would provide the means to hold individuals to account. Here, they explain this work and present one version of the draft convention.

Any development, production, acquisition or use of biological or chemical weapons is the result of decisions and actions of individual persons, whether they are government officials, commercial suppliers, weapons experts or terrorists. The international conventions that prohibit these weapons, the BWC and the CWC, being directed primarily to the actions of states, address the matter of individual responsibility to only a limited degree.

Article IV of the BWC requires each state party to prohibit the development, production, stockpiling, acquisition and retention of biological weapons anywhere within its territory. Article VII of the CWC requires each state party to enact penal legislation applicable to acts committed in the territory of that state and also to acts committed by its nationals anywhere.

However the BWC and the CWC do not attempt to make the development, production, possession or use of biological and chemical weapons an international crime for which states establish jurisdiction over prohibited acts regardless of the place where they are committed or the nationality of the offender, nor do these treaties contain provisions dealing with the extradition of suspects.

Neither are these deficiencies remedied by the provisions applicable to biological and chemical weapons in the Convention for the Suppression of Terrorist Bombings, opened for signing in January 1998, or in the Statute of the International Criminal Court signed in Rome in July 1998. The Bombing Convention does not apply to the activities of military forces in the exercise of their official duties or to internal state acts – such as the use of CBW weapons by a leader against a population within his own state. Nor does the scope of either of these agreements extend beyond the actual use of CBW weapons to include, as do the BWC and the CWC, their development, production and possession.

---

[210] Editors' note. Julian co-authored this with Matthew Meselson. It was first published as Matthew Meselson and Julian Robinson. A Draft Convention to Prohibit Biological and Chemical Weapons Under International Criminal Law, *The CBW Conventions Bulletin*. 42 (December 1998). Different versions of the paper are listed in Chap. 9 in item 422. Unless otherwise indicated, the remaining footnotes for this paper are Julian's originals. Apart from correcting obvious typos and renumbering headings and tables in keeping with the requirements of this book, we have retained Julian's formatting and spelling throughout the paper and its notes.

What is needed is a new treaty, one that defines specific acts involving biological or chemical weapons as international crimes, like piracy or aircraft hijacking, obliging states either to prosecute or extradite offenders who are present in their territory. Treaties defining international crimes are based on the concept that certain crimes are particularly dangerous or abhorrent to all and that all states therefore have the right and the responsibility to combat them. Certainly in this category, threatening to the community of nations and to present and future generations, are crimes involving the weaponization of disease or poison and the hostile exploitation of biotechnology.

The Harvard Sussex Program, with advice from an international group of legal authorities, has prepared a draft convention that would make certain acts involving biological and chemical weapons crimes under international law. The proposed convention would make it an offence for any person, regardless of official position, to order, direct or knowingly to participate or render substantial assistance in the development, production, acquisition, stockpiling, retention, transfer or use of biological or chemical weapons or to threaten the use of such weapons or to create or retain facilities intended for the production of such weapons. Any person who commits any of the prohibited acts anywhere would face the risk of apprehension, prosecution and punishment or of extradition should that person be found in a state that supports the proposed convention.

The proposed convention would oblige each state party: (i) to establish jurisdiction with respect to the specified crimes extending to all persons in its territory, regardless of the place where the offence is committed or the citizenship of the offender, and (ii) to prosecute or extradite any such offender found in its territory or any other place under its jurisdiction. Decisions regarding sentencing, including consideration of the severity of the offence and of any mitigating circumstances, are left to individual states parties.

The same obligations, to establish criminal jurisdiction and to extradite or adjudicate, *aut dedere aut judicare*, are included in international conventions now in force for the suppression and punishment of international crimes including aircraft hijacking and sabotage (1970; 1971), crimes against internationally protected persons (1973), hostage taking (1979), theft of nuclear materials (1980), torture (1984) and crimes against maritime navigation (1988).

The proposed convention defines biological and chemical weapons as they are defined in the BWC and the CWC, on the basis of the general purpose criterion – and its definitions of prohibited acts are modeled closely on the definitions in these treaties. Commission of a prohibited act is defined as a crime only if committed "knowingly" and it is an admissible defence that the accused person "reasonably believed" that the conduct in question was not prohibited. The proposed convention also includes provisions requiring states parties to cooperate in investigations and to provide legal assistance to one another in the adjudication of offences.

One way forward would be for a group of states to submit the proposed convention or a similar draft in the form of a resolution for consideration by the UN General Assembly, seeking its referral to the UNGA Sixth (legal) Committee for negotiation of an agreed text. If the negotiated text receives the commendation of the General Assembly, the convention may then be opened for signature and ratification, leading to its entry into force. Adoption and widespread adherence to such a convention would create a new dimension of constraint against biological and chemical weapons by applying international criminal law to hold individual offenders responsible and punishable wherever they may be and regardless of whether they act under or outside of state authority. Such individuals would be regarded as *hostes humani generis*, enemies of all humanity. The norm against chemical and biological weapons would be strengthened, deterrence of potential offenders would be enhanced, and international cooperation in suppressing the prohibited activities would be facilitated.

The present draft of the proposed convention, below, was prepared by a working group consisting of James Crawford (Cambridge University), John Dugard (Leiden University), Philip Heymann (Harvard University) and the directors of the Harvard Sussex Program (Matthew Meselson and Julian Robinson). It is based on an earlier HSP draft and on discussions at HSP workshops held in January 1997 at Harvard University and, in association with the Common Security Forum, at the University of Cambridge Lauterpacht Research Centre for International Law in May 1998.

7.10.1 Draft Convention on the Prevention and Punishment of the Crime of Developing, Producing, Acquiring, Stockpiling, Retaining, Transferring or Using Biological or Chemical Weapons

The States Parties to this Convention,

Recalling that States are prohibited by the Geneva Protocol of 1925, the Biological Weapons Convention of 1972 and the Chemical Weapons Convention of 1993, and other international agreements, from developing, producing, stockpiling, acquiring, retaining, transferring or using biological and chemical weapons, and that these prohibitions reflect a worldwide norm against these weapons;

Recognizing that any development, production, acquisition or use of biological or chemical weapons is the result of the decisions and actions of individual persons, including government officials, and that these activities are within the capability not only of States but also of other entities and of individuals;

Affirming that all persons and entities should be prohibited from engaging in these activities, and should be subject to effective penal sanctions, thereby enhancing the effectiveness of the Geneva Protocol, the Biological Weapons Convention and the Chemical Weapons Convention;

Reaffirming that any use of disease or poison for hostile purposes is repugnant to the conscience of humankind;

Considering that biological and chemical weapons pose a threat to the well-being of all humanity and to future generations;

Resolving that knowledge and achievements in biology, chemistry and medicine should be used exclusively for the health and well-being of humanity;

Desiring to encourage the peaceful and beneficial advance and application of these sciences by protecting them from adverse consequences that would result from their hostile exploitation;

Determined, for the sake of human beings everywhere and of future generations, to eliminate the threat of biological and chemical weapons;

Have agreed as follows:

Article I

Any person commits an offence who knowingly:

(a) develops, produces, otherwise acquires, stockpiles or retains any biological or chemical weapon, or transfers, directly or indirectly, to anyone, any biological or chemical weapon;
(b) uses any biological or chemical weapon;
(c) engages in preparations to use any biological or chemical weapon;
(d) constructs, acquires or retains any facility intended for the production of biological or chemical weapons;
(e) assists, encourages or induces, in any way, anyone to engage in any of the above activities;
(f) orders or directs anyone to engage in any of the above activities;
(g) attempts to commit any of the above offences;
(h) threatens to use biological or chemical weapons.

Article II

1. Nothing in this Convention shall be construed as prohibiting activities that are permitted under:

    (a) the Convention on the Prohibition of the Development, Production and Stockpiling of Bacteriological (Biological) and Toxin Weapons and on their Destruction, of 10 April 1972, or
    (b) the Convention on the Prohibition of the Development, Production, Stockpiling and Use of Chemical Weapons and on their Destruction, done at Paris on 13 January 1993, or that are directed toward the fulfillment of a State's obligations under either Convention and are conducted in accordance with its provisions.

2. In a prosecution for an offence set forth in Article I, it shall be a defence that the accused person reasonably believed that the conduct in question was not prohibited under this Convention.
3. It is not a defence that a person charged with an offence set forth in Article I acted in an official capacity, under the orders or instructions of a superior, or otherwise in accordance with internal law.

Article III

For the purposes of the present Convention:

1. Biological weapons means:
   (a) microbial or other biological agents, or toxins whatever their origin or method of production, of types and in quantities that have no justification for prophylactic, protective or other peaceful purposes;
   (b) weapons, equipment or means of delivery designed to use such agents or toxins for hostile purposes or in armed conflict.
2. Chemical weapons means the following, together or separately:
   (a) toxic chemicals and their precursors, except where intended for:
      (i) industrial, agricultural, research, medical, pharmaceutical or other peaceful purposes;
      (ii) protective purposes, namely those purposes directly related to protection against toxic chemicals and to protection against chemical weapons;
      (iii) military purposes not connected with the use of chemical weapons and not dependent on the use of the toxic properties of chemicals as a method of warfare;
      (iv) law enforcement including domestic riot control purposes as long as the types and quantities are consistent with such purposes.
   (b) munitions and devices, specifically designed to cause death or other harm through the toxic properties of those toxic chemicals specified in subparagraph (a), which would be released as a result of the employment of such munitions and devices;
   (c) any equipment specifically designed for use directly in connection with the employment of munitions and devices specified in subparagraph (b).
3. Toxic chemical means any chemical which through its chemical action on life processes can cause death, temporary incapacitation or permanent harm to humans or animals. This includes all such chemicals, regardless of their origin or of their method of production, and regardless of whether they are produced in facilities, in munitions or elsewhere.
4. Precursor means any chemical reactant which takes part at any stage in the production by whatever method of a toxic chemical. This includes any key component of a binary or multicomponent chemical system, that is to say, the precursor which plays the most important role in determining the toxic properties of the final product and reacts rapidly with other chemicals in the binary or multicomponent system.
5. Person means any natural person or, to the extent consistent with internal law as to criminal responsibility, any legal entity.

Article IV

Each State Party shall adopt such measures as may be necessary:

(a) to establish as criminal offences under its internal law the offences set forth in Article I;
(b) to make those offences punishable by appropriate penalties which take into account their grave nature.

Article V

1. Each State Party to this Convention shall take such measures as may be necessary to establish its jurisdiction over the offences set forth in Article I in the following cases:

   (a) when the offence was committed in the territory of that State or in any other place under its jurisdiction as recognized by international law;
   (b) when the alleged offender is a national of that State;
   (c) when, if that State considers it appropriate, the alleged offender is a stateless person whose habitual residence is in its territory;
   (d) when the offence was committed with intent to harm that State or its nationals or to compel that State to do or abstain from doing any act;
   (e) when the offence involved the intentional use of biological or chemical weapons and a victim of the offence was a national of that State;
   (f) when the offence involved the intentional use of biological or chemical weapons against any persons, irrespective of their nationality.

2. Each State Party shall likewise take such measures as may be necessary to establish its jurisdiction over the offences set forth in Article I in cases where the alleged offender is present in its territory and it does not extradite such person pursuant to Articles VII and VIII.
3. This Convention does not exclude any criminal jurisdiction exercised in accordance with internal law, including any internal law giving effect to Article I.
4. Jurisdiction with respect to the offences set forth in Article I may also be exercised by any international criminal court that may have jurisdiction in the matter in accordance with its Statute.

Article VI

1. Upon receiving information that a person who has committed or who is alleged to have committed an offence as set forth in Article I may be present in its territory, a State Party shall take such measures as may be necessary under its internal law to investigate the facts contained in the information.
2. If it is satisfied that the circumstances so warrant, a State Party in the territory of which an alleged offender is present shall take that person into custody or shall take such other measures as are necessary to ensure the presence of that person for the purpose of prosecution or extradition.
3. Any person regarding whom the measures referred to in paragraph 2 are being taken shall be entitled to:

(a) communicate without delay with the nearest appropriate representative of the State of which that person is a national or which is otherwise entitled to protect that person's rights or, if that person is a stateless person, the State in the territory of which that person habitually resides;
(b) be visited by a representative of that State;
(c) be informed of that person's rights under subparagraphs (a) and (b);

4. The rights referred to in paragraph 3 shall be exercised in conformity with the laws and regulations of the State in the territory of which the offender or alleged offender is present, provided that the said laws and regulations must enable full effect to be given to the purposes for which the rights accorded under paragraph 3 are intended.
5. When a State Party, pursuant to the present Article, has taken a person into custody, it shall promptly notify, directly or through the Secretary-General of the United Nations, the States Parties which have established jurisdiction in accordance with Article V, paragraph 1, and, if it considers it advisable, any other interested States Parties, of the fact that such person is in custody and of the circumstances which warrant that person's detention. The State which makes the investigation contemplated in paragraph 1 of the present Article shall promptly inform those States Parties of its findings and shall indicate whether it intends to exercise jurisdiction.

Article VII

1. The offences set forth in Article I shall be deemed to be included as extraditable offences in any extradition treaty existing between States Parties. States Parties undertake to include those offences as extraditable offences in every extradition treaty subsequently concluded between them.
2. If a State Party which makes extradition conditional on the existence of a treaty receives a request for extradition from another State Party with which it has no extradition treaty, it may, if it decides to extradite, consider this Convention as the legal basis for extradition in respect of the offences set forth in Article I. Extradition shall be subject to the other conditions provided by the law of the requested State.
3. States Parties which do not make extradition conditional on the existence of a treaty shall recognize the offences set forth in Article I as extraditable offences as between themselves subject to the conditions provided by the law of the requested State.
4. The offences set forth under Article I shall be treated, for the purpose of extradition between States Parties, as if they had been committed not only in the place in which they occurred but also in the territories of the States required to establish their jurisdiction in accordance with paragraph 1 of Article V.
5. The provisions of all extradition treaties and arrangements between States Parties with regard to offences set forth in Article I shall be deemed to be modified as between States Parties to the extent that they are incompatible with this Convention.

Article VIII

The State Party in the territory of which the alleged offender is found shall, if it does not extradite such person, be obliged, without exception whatsoever and whether or not the offence was committed in its territory, to submit the case without delay to competent authorities for the purpose of prosecution, through proceedings in accordance with the laws of that State. Those authorities shall take their decision in the same manner as in the case of any other offence of a grave nature under the law of that State.

Article IX

1. States Parties shall afford one another the greatest measure of assistance in connection with investigations or criminal or extradition proceedings brought in respect of the offences set forth in Article I, including assistance in obtaining evidence at their disposal which is necessary for the proceedings.
2. States Parties shall carry out their obligations under paragraph 1 in conformity with any treaties or other arrangements on mutual legal assistance that may exist between them. In the absence of such treaties or arrangements, States Parties shall afford one another assistance in accordance with their internal law.
3. States Parties may request assistance from competent international bodies in connection with investigations or criminal or extradiditon proceedings brought in respect of the offences set forth in Article I, as may be authorized by such bodies.

Article X

None of the offences set forth in Article I shall be regarded, for the purposes of extradition or mutual legal assistance, as a political offence or as an offence connected with a political offence or as an offence inspired by political motives.

Accordingly, a request for extradition or for mutual legal assistance based on such an offence may not be refused on the sole ground that it concerns a political offence or an offence connected with a political offence or an offence inspired by political motives.

Article XI

Nothing in this Convention shall be interpreted as imposing an obligation to extradite or to afford mutual legal assistance, if the requested State Party has substantial grounds for believing that the request for extradition for offences set forth in Article I or for mutual legal assistance with respect to such offences has been made for the purpose of prosecuting or punishing a person on account of that persons race, religion, nationality, ethnic origin or political opinion or that compliance with the request would cause prejudice to that person's position for any of these reasons.

Article XII

States Parties shall cooperate in the prevention of the offences set forth in Article I, particularly by:

(a) taking all practicable measures to prevent preparations in their respective territories for the commission of those offences within or outside their territories;
(b) exchanging information and coordinating the taking of administrative and other measures as appropriate to prevent commission of those offences.

Article XIII

1. Each State Party shall inform the Secretary-General of the United Nations of the legislative and administrative measures taken to implement this Convention. In particular, each State Party shall notify the Secretary-General of the United Nations of the jurisdiction it has established under its internal law in accordance with paragraph 3 of Article V. Should any change take place, the State Party concerned shall immediately notify the Secretary-General.
2. Each State Party shall, in accordance with its national law, promptly provide to the Secretary-General of the United Nations any relevant information in its possession concerning:

   (a) the circumstances of any offence over which it has established its jurisdiction pursuant to paragraph 1 or paragraph 3 of Article V;
   (b) the measures taken in relation to the alleged offender, and, in particular, the results of any extradition proceedings or other legal proceedings.

3. The State Party where an alleged offender is prosecuted shall communicate the final outcome of the proceedings to the Secretary-General of the United Nations, who shall transmit the information to the other States Parties.
4. Each State Party shall designate a contact point within its government to which other States Parties may communicate in matters relevant to this Convention. Each State Party shall make such designation known to the Secretary-General.

Article XIV

Any dispute between States Parties concerning the interpretation or application of this Convention which is not settled by negotiation shall, at the request of one of them, be submitted to arbitration. If within six months from the date of the request for arbitration the parties are unable to agree on the organization of the arbitration, any one of those parties may refer the dispute to the International Court of Justice.

Article XV

1. Ten years after the entry into force of this Convention, or earlier if it is requested by a majority of Parties to the Convention by submitting a proposal to this effect to the Secretary-General of the United Nations, a Conference of States Parties

shall be held at [Geneva, Switzerland], to review the operation of the Convention with a view to assuring that the purposes of the preamble and the provisions of the Convention are being realized.

2. At intervals of seven years thereafter, unless otherwise decided upon, further sessions of the Conference may be convened with the same objective.

Article XVI

1. This Convention shall be open for signature by all States from [DATE] until [DATE] at United Nations Headquarters in New York.
2. This Convention is subject to ratification, acceptance or approval. The instruments of ratification, acceptance or approval shall be deposited with the Secretary-General of the United Nations.
3. This Convention shall be open to accession by any State. The instruments of accession shall be deposited with the Secretary-General of the United Nations.

Article XVII

1. This Convention shall enter into force on the thirtieth day following the date of the deposit of the [NUMBER] instrument of ratification, acceptance, approval or accession with the Secretary-General of the United Nations.
2. For each State ratifying, accepting, approving or acceding to the Convention after the deposit of the [NUMBER] instrument of ratification, acceptance, approval or accession, the Convention shall enter into force on the thirtieth day after deposit by such State of its instrument of ratification, acceptance, approval or accession.

Article XVIII

The Articles of this Convention shall not be subject to reservation.

Article XIX

The original of this Convention, of which the Arabic, Chinese, English, French, Russian and Spanish texts are equally authentic, shall be deposited with the Secretary-General of the United Nations, who shall send certified copies thereof to all States.

IN WITNESS WHEREOF, the undersigned, being duly authorized thereto by their respective Governments, have signed this Convention, opened for signature at United Nations Headquarters in New York on [DATE].

7.11 Preventing the Hostile Use of Biotechnology[211]

Editors' note. The Biological Weapons Convention (BWC) faced a critical juncture in 2001-2002, with the treaty's states parties in disarray following a USA decision to withdraw support for an Ad Hoc Group mandated to develop a legally binding instrument to strengthen the Convention. The USA announced its changed position on 25 July 2001, when it reported that it had come to believe that the Ad Hoc Group's approach to treaty verification was not feasible. The USA was not alone in its view, although a very large majority of states parties remained in favour of developing a protocol to strengthen the BWC. Nevertheless, the US position meant that the Ad Hoc Group would never be able to adopt a text by consensus (which was a requirement of its negotiating mandate) and so the Ad Hoc Group negotiations came to an abrupt halt.

The US actions generated consternation amongst the states parties, and the resulting concerns dominated the BWC's Fifth Review Conference scheduled for 19 November-7 December 2001. Then, on what had been expected to be the last day of the BWC's Fifth Review Conference (7 December 2001), the USA proposed ending the mandate for the Ad Hoc Group, thereby putting into question future efforts to strengthen the Convention. In the event, the states parties responded by suspending the Conference, and it resumed the following year (11–22 November 2002).

Accordingly, this paper from September 2002 was written in a context where new approaches were needed to continue work to strengthen the Convention. In it, Julian and his co-authors examine how the international community might act to reduce the risk of biological weapons development and use, with particular emphasis on the responsibilities of states. They argue that while voluntary measures have value, legally binding commitments with effective implementation and verification are needed to address the most serious threats.

Two months before the scheduled resumption of the Fifth BWC Review Conference,[212] the United States let it be known that it favours "a very short Rev Con, if any … with the sole purpose and outcome of agreeing to hold a Rev Con in 2006". Should that view prevail when the states parties assemble in Geneva in November, there could be little serious attempt to strengthen the BWC regime for a further four years. At the same time, the International Committee of the Red Cross (ICRC)

---

[211] Editors' note. Julian coauthored this with Malcolm Dando and Matthew Meselson. It was first published as Malcolm Dando, Matthew Meselson and Julian Perry Robinson. "Preventing the hostile use of biotechnology: the way forward now", *The CBW Conventions Bulletin*, no 57 (September 2002), pp 1–2. It is listed in Chap. 9 as item 416. Unless otherwise indicated, the remaining footnotes for this paper are Julian's originals. Apart from correcting obvious typos and renumbering headings and tables in keeping with the requirements of this book, we have retained Julian's formatting and spelling throughout the paper and its notes.

[212] Editors' note. The BWC's fifth Review Conference took place 19 November–7 December 2001 and resumed in 11 to 15 November 2002. See United Nations Fifth Review Conference of the States Parties to the Biological Weapons Convention [online] available at https://disarmament.unoda.org/fifth-review-conference-of-the-states-parties-to-the-biological-weapons-convention/ [Accessed 9 May 2025]. For more information, see Marie Isabelle Chevrier, "BWC Report: Waiting for Godot or Saving the Show? The BWC Review Conference Reaches Modest Agreement", *Disarmament Diplomacy* 68, December 2002–January 2003, available at https://acronym.org.uk/old/archive/dd/dd68/68bwc.htm [Accessed 9 May 2025] and Nicholas A Sims "Biological Disarmament Diplomacy in The Doldrums: Reflections After The BWC Fifth Review Conference", Disarmament Diplomacy 70, April–May 2003, available at https://www.acronym.org.uk/old/archive/dd/dd70/70op2.htm [Accessed 9 May 2025].

launched an appeal titled *Biotechnology, Weapons and Humanity*, in which it warned of the growing dangers to humanity from the possible hostile use of biotechnology and urged that strenuous efforts be made to prevent such a menace.

It is important to understand the basis of the ICRC's concerns. As has been noted previously in this *Bulletin*, all major technologies have in the past been applied for both peaceful and hostile purposes. If effective action is not taken, the same thing will surely happen to biotechnology. Should this be allowed to occur, the very nature of human conflict could begin to change, as biotechnology comes to be exploited not only to produce weapons that are lethal but also ones designed to affect cognition, behaviour, development and perhaps even inheritance – for hostile purposes. That would indeed be a 'brave new world' to bequeath to future generations.

In deciding where to place our priorities in order to prevent any drift toward such an inimical outcome we have to be realistic about where the principal long-term hazards of biotechnology lie and where they do not. Much concern has been expressed recently over the possible hazards of peaceful research (for example, in regard to the mousepox experiment in Australia), but it is most unlikely that a monster plague could accidentally result from such activities. The further development and application of workable guidelines and peer review of certain sorts of peaceful research are prudent measures but must not distract us from the necessity of dealing with the far greater menace of deliberate use of pathogens, existing or yet to be developed.

More effective surveillance and cooperation between national law enforcement authorities, for example, can help to foil those who may seek to acquire pathogens as weapons of terror. But history shows that it was in the major state-level offensive biological weapons programmes of the Twentieth Century – especially in the UK, the USA and the USSR – that there was the most technologically advanced and most massive preparation for the use of biological weapons. Preventing such state-level offensive programmes in the future should be a primary concern.

A second major cause of concern should be the proliferation of secret, offensively oriented, biological defence programmes such as have recently come into being in the United States. Elements of this programme clearly go well beyond the limits for defensive research set by President Ford and promulgated within the government by the Scowcrott Memorandum of 23 December 1975...,[213] after President Nixon closed down the US offensive programme. There is a grave danger that secret, offensively oriented, defence programmes will acquire a momentum of their own, proliferating and eventually becoming offensive programmes.

A third principal cause of concern must be the ongoing development, within military establishments in the developed countries, most notably the US, of new weapons based on biochemical incapacitants, for example, those acting on the central nervous system. Although such agents are prohibited by the BWC and the CWC for any but peaceful purposes, some persons have advanced specious claims to the contrary, arguing that the use of such agents might be permissible in certain paramilitary operations. This view flouts international treaty commitments and fails to appreciate the

---

[213] The Snowcroft Memorandum was reproduced after the article. See *The CBW Conventions Bulletin*, no 57 (September 2002), pp 2.

hazard that any hostile use of biological or chemical agents, especially by a major state, by breaching the prevailing norm risks eroding that norm altogether.

The international community has numerous means of discouraging the hostile use of biotechnology in the Twenty-first Century. But we must not let pursuit of useful but less effective measures distract us from the need to take more effective ones. Better disease surveillance, for example, if backed up with appropriate medical and public health measures, is greatly needed for its own sake, and it may also assist in distinguishing a natural from an unnatural outbreak of disease. Yet it must be remembered that only at Sverdlovsk was there a detected unnatural outbreak during the whole of the Soviet and Western offensive programmes. Similarly, better articulated codes of professional conduct are to be welcomed as supportive instruments, but even if 99 percent of the world's biologists conform to stringent new codes, one per cent employed in major state-level offensive biological weapons programmes could still produce the havoc we wish to avoid. Moreover, excessive regulation piled onto the scientific community could stifle the very beneficial advances we all wish to see.

The fact of the matter is that multilateral international agreements designed to implement the BWC regime effectively will be needed to deal with the main problem of precluding major state-level offensive biological weapons programmes. In order to be effective, these agreements must include a system of declarations, visits to declared sites and challenge inspections. This will not, in itself, prevent every potential violation, but it will make prohibited activities more difficult and more risky to any who undertake them. And it will give the international community the unity and political will needed to suppress violations, should they occur. In addition, the BWC regime should be strengthened by the addition of agreed sanctions. A promising approach here, one that could be adopted by like-minded states even before a verification regime for BWC can be established, would be the elaboration and implementation of a treaty to criminalize biological and chemical armament in international law. Modelled on treaties applicable to aircraft hijacking, for example, and torture, this one would provide the courts of its member states with criminal jurisdiction over anyone on their territory, regardless of their nationality or official position, who orders or knowingly gives substantial support to the development, production, transfer or use of chemical or biological weapons anywhere. A draft of such a treaty, prepared by the Harvard Sussex Program with advice from international legal authorities is available on the HSP website…[214]

So whilst we pursue other necessary avenues, such as the development of better professional standards to avoid inadvertently dangerous research, we must not become distracted from the main goal. The BWC regime has to be strengthened and effectively implemented as soon as possible, whatever the prevailing winds in Washington. The world was a better place for the 1925 Geneva Protocol despite Washington taking 50 years to ratify it. Indeed, the pursuit of other goals to the detriment of the strengthening of the BWC would be counter-productive rather than just a distraction from what is really required to prevent the hostile use of biology.

---

[214] Editors' note. A version of the draft treaty is also included in this book. See "7.10 A Draft Convention to Prohibit Biological and Chemical Weapons under International Criminal Law".

7.12 "Non-Lethal" Weapons, the CWC and the BWC[215]

Editors' note. In this article, Julian and Matthew Meselson explain how identifying a category of 'non-lethal' weapons represents a risk to the Biological Weapons Convention and Chemical Weapons Convention, and show that the General Purpose Criterion in both Conventions can mitigate this risk. They examine how developments in biotechnology, particularly the growing interest in so-called incapacitating agents, risk undermining the core prohibitions of both treaties. Drawing on historical precedent, legal interpretation, and emerging scientific trends, the authors make a clear case for reaffirming and rigorously implementing the Conventions' foundational principles.

It is hard to think of any issue having as much potential for jeopardizing the long-term future of the Chemical and Biological Weapons Conventions as does the interest in creating special exemptions for so-called 'non-lethal' chemical weapons. The First CWC Review Conference earlier this year was opportunity to address the issue constructively. But, save in the national statements of New Zealand, Norway and Switzerland, the OPCW chose not to do so. In the programme of Review Conference follow-up work that is now getting under way, there is no mention of disabling chemicals, not even tear gas, still less the so-called 'calmatives' and other such incapacitating agents in which interest is now rapidly reawakening.

The one emerging area of technology that today is most in need of strong and lasting arms control – biotechnology – is exactly where new disabling chemicals are coming from, furnishing potential weapons that are tempting some government agencies to depart from or to seek revision of the prohibitions of the CWC. Notions of incapacitating chemical weapons are not new; but, as investment mounts in emergent 'non lethal weapons' (NLW) technologies, it becomes increasingly urgent that the threat they pose to the CWC and BWC regime be recognised.

In the broadest perspective, as different kinds of technology have arisen, they have also been reflected in the means of war and violence. As new inventions or discoveries come along, they find all kinds of applications, including applications for hostile purposes. The rise of biotechnology, however, poses a special problem, as it will inevitably develop means for manipulating all of the life processes, including cognition, development, reproduction and heredity. Therein lie unprecedented and, in time, widely accessible possibilities for violence, coercion, repression or subjugation.

For example, not so long ago we did not know, but know now, that there are several thousand different kinds of what are called receptors in the human brain. A receptor is a protein molecule within or on the surface membrane of a cell which, when a certain kind of smaller molecule called a ligand binds to it, causes something to happen. Receptors and their ligands are part of the body's immensely complex

[215] Editors' note. Julian co-authored this with Matthew Meselson. It was first published as Matthew Meselson and Julian Perry Robinson. "Non-Lethal' Weapons, the CWC and the BWC". *The CBW Conventions Bulletin* no 61 (September 2003), pp 1–2. It is listed in Chap. 9 as item 425. Unless otherwise indicated, the remaining footnotes for this paper are Julian's originals. Apart from correcting obvious typos and renumbering headings and tables in keeping with the requirements of this book, we have retained Julian's formatting and spelling throughout the paper and its notes.

chemical communication system. One group of receptors in our brains are the opioid receptors. The body manufactures ligands – enkephalins and endorphins – that bind with these receptors, variously alleviating pain, inducing sleep, or reducing anxiety. It turns out that these ligands can be imitated by some plant products, one of which is heroin. Also, certain synthetic chemical analogues of heroin bind to particular opioid receptors, causing diverse effects depending on the chemical, including short-term memory impairment, breathing difficulty and flaccid paralysis. One of these chemicals is fentanyl, which was the basis for the 'knockout gas' used by Russian special forces to rescue several hundred hostages in the Moscow theatre siege of October 2002. The US Army chemical corps was studying fentanyl and related chemicals as candidate disabling weapons as early as May 1963.

There are innumerable other kinds of receptors in the brain, most of which we know almost nothing about. Of the few that have been investigated, we do know that some can mediate temporary blindness, for example, or can cause submissiveness, or extreme fear. Others affect memory or motivation. As time goes on, we will learn how to do many kinds of things to people with chemicals that bind to brain receptors or interact specifically with other life processes. Many applications will be beneficial. It should be possible to conduct surgical operations with greater ease and safety with the newer anaesthetics that may be made possible by this kind of research. But, as with all major new technologies in the past, this, as well as other branches of biotechnology, will also have potential hostile applications. The challenge, then, is to promote the peaceful applications of the new technology while preventing its exploitation for hostile purposes. Fortunately, the Chemical and Biological Weapons Conventions provide a timely opportunity for protecting the peaceful uses while maintaining and strengthening the barriers to hostile ones.

How do the CWC and BWC keep abreast of technological innovation? They do so in the same way as they cope with the problem of 'dual use', through the 'general purpose criterion' set out in Article II.1(a) and Article VI.2 of the CWC and Article I of the BWC. Instead of prohibiting a particular chemical, or family of chemicals, the CWC prohibits all toxic chemicals except for the listed purposes that it does not prohibit. These listed purposes include "industrial, agricultural, research, medical, pharmaceutical or other peaceful purposes; ...purposes directly related to protection against toxic chemicals; military purposes not connected with the use of chemical weapons and not dependent upon the toxic properties of chemicals as a method of warfare; law enforcement including domestic riot control purposes". Toxic is defined in the CWC as anything that is harmful, whether employed under conditions in which it is temporarily incapacitating or lethal. Some toxic substances that have been considered for use as disabling chemical weapons are even more toxic than the chemicals developed for lethal purposes, in the sense that extremely small amounts are sufficient to cause an effect. Lofentanil, for example, which is a derivative of fentanyl, is far more toxic than nerve agent. It will cause anaesthesia at a dose of 0.025 micrograms per kilogram body weight, which is hundreds of times smaller than the estimated lethal dose of VX.

The BWC invokes a similar general purpose criterion, prohibiting in Article I "microbial or other biological agents, or toxins whatever their origin or method of production of types and in quantities that have no justification for prophylactic, protective or other peaceful purposes". The inclusion not only of 'microbial agents' but also of 'toxins', captures such natural substances as those ligands and other naturally occurring bio-active molecules and their chemical analogues that can exert harmful effects on life processes. The two Conventions therefore are mutually reinforcing in their prohibition of toxicity as a weapon of war.

Provided the 'general purpose criterion' is implemented properly, it protects the conventions against obsolescence due to technological advance in biochemistry and biotechnology. When new toxic chemicals are discovered, they automatically come under its purview.

Uniquely, the United States has declared a policy that, in effect, excludes certain chemicals from the 'general purpose criterion', and there is now interest in broadening the exclusion still further. During negotiation of the CWC, some members of the US delegation asserted that 'riot control agents' were not toxic chemicals in the sense of the CWC. This is the position espoused by influential US military lawyers today. And, since 1997, the US Weapons Review Program has been generating the opinion that 'calmative agents' such as the fentanyls are also exempt from the 'general purpose criterion' provided those chemicals are classifiable as 'riot control agents'. What is being lost sight of here are the grave risks inherent in *any* blanket exemption for *any* toxic chemical.

The first is the escalation hazard. In March 1965, during the Vietnam War, after newspaper stories had been published about US use of 'riot control agents', Secretary of State Dean Rusk in a prepared statement said: "We do not expect that gas will be used in ordinary military operations." And he did not. Subsequently, however, 25 different types of weapon disseminating the irritant agent CS, including heavy munitions ranging up to 155-mm artillery shell and 750-pound aircraft bombs, were used in Vietnam. Ultimately more than 10 million pounds of CS were dispensed in these munitions. A post-war analysis of the operational use of CS declassified in 1979 could find no report of its use against non-combatants or to save civilians and concluded that "...the reduction in casualties has not been in enemy or noncombatant personnel but, rather, friendly troops, as a result of using CS to make other fires more effective." Thus the United States, whose military is under civilian control and whose vital interests were not at stake in Vietnam, came to use vast quantities of 'non-lethal' chemical weapons under the pressures of a wartime situation in a manner totally at variance with its initially announced policy.

In World War II, the United States, the Soviet Union and the major European belligerents had around 20,000 metric tons of irritant agents (CA, CN, DA, DC and DM) in their stockpiles but used none in combat. The line against all chemical weapons was understood, and the line held. During the Korean War, the United States had huge stockpiles of tear-gas munitions. They were never used except in non-combat situations, most notably on rioting detainees in prisoner-of-war camps. Again the line held. The United States did not progress to using lethal chemical weapons during the Vietnam War, despite its use of tactical CS weapons. However, in World War I, lethal chemical warfare began with the use of tear gas. In Manchuria, in Ethiopia and in the Yemen, lethal chemical warfare began with the use of tear gas.

In the Iraq-Iran war, it began with tear gas, and half of all the chemical munitions later declared to UNSCOM by Saddam Hussein's Iraq were CS munitions, mainly mortar rounds and aircraft bombs. Every confirmed resort to lethal chemical warfare has started with tear gas. One major concern regarding any NLW exemption, then, is battlefield escalation – both escalation to other agents and escalation to unrestricted employment.

There is another kind of escalation, which is the fostering of the growth and influence of institutions that are dependent upon the development and weaponization of chemical agents. Such institutions and their associated bureaucracies and dependent communities inevitably become a source of pressure for doing more in this area, and for promoting the assimilation of chemical weapons into the structures and doctrine of state forces.

A further danger is that hostile resort to disabling chemicals may lead to a loss of confidence in the CWC regime. Imagine the effect on citizens of the world, and their governments, when they observe war on television and see men fighting in gas masks, see weapons discharging clouds of toxic smoke or sprays of gas. What then will they think of the Convention for the Prohibition of Chemical Weapons, or of the Organization mandated to implement it?

Exemption also risks providing cover for cheating. Inspectors may find a large quantity of toxic chemical munitions and, upon asking what they are for, may be told "law enforcement"; and that, pretty much, would be the end of it. Maybe the inspectors were told the truth, but it does not make compliance verification any easier, and it allows for argument about what is and is not a violation.

Finally, and the most serious difficulty of all, exemption blurs the simple line, *no poisons in war*. The simplicity and lack of ambiguity in this ancient dictum make it uniquely a focus for agreement. The importance of averting the hostile exploitation of biotechnology, with its immense potentials for both benefit and harm, is immeasurably more important than the marginal utility of "non-lethal" chemical weapons in military and paramilitary operations. Instead of logic-chopping arguments intended to relax the prohibitions of the CWC and the BWC, we must look to these international agreements as the basis of a clear and generally agreed firebreak against military use in war of all toxic weapons, existing and yet to be devised.

7.13 David Christopher Kelly, 1944–2003[216]

Editors' note. David Kelly was a UK government scientist who was vital to international efforts uncovering illicit biological weapons programmes in the Soviet Union and Iraq. In 2003, he was 'outed' as a source for controversial reporting by BBC journalist Andrew Gilligan, and the UK government failed to protect him from the intense public scrutiny that followed. He committed suicide near his home, shortly after appearing before a parliamentary

[216] Editors' note. Julian co-authored this tribute with Matthew Meselson and Graham Pearson. It was published as "David Christopher Kelly, 1944–2003", *The CBW Conventions Bulletin* no 61 (September 2003) p 36. is listed in Chap. 9 as item 426.

committee. This obituary shows Julian's deep appreciation of Kelly's skill and contribution to global biological warfare disarmament, and warm respect for his generosity and integrity.

We mourn the passing of David Kelly, a modest and good man, an internationally recognised British authority on biological weapons, and a valued friend of the Harvard Sussex Program [HSP]. He died on 17 July 2003, aged 59. He was then Senior Adviser to the Directorate of Counter Proliferation and Arms Control of the UK Ministry of Defence and to the Counter Proliferation Department of the UK Foreign and Commonwealth Office.

He had become a member of the country's Scientific Civil Service in 1973 upon joining, as a molecular virologist, what is now the National Environment Research Council Centre for Ecology and Hydrology in Oxford. There he contributed much to the field of biological control applicable in agriculture, specifically the use of viruses to attack insect pests. In July 1984 he was recruited by the Ministry of Defence to work at what was then the Chemical Defence Establishment at Porton Down as head of microbiology. He directed the research into improved defensive measures against biological warfare that, during the 1991 Kuwait war, enhanced the protection of UK forces against possible Iraqi biological weapons. He also led the successful decontamination of Gruinard Island, which had been a proving ground for anthrax weapons during the second world war. In 1989 he first became involved as a technical expert in the interpretation of the disturbing data emanating from the Soviet Union, chiefly via defectors such as Vladimir Pasechnik, whom he interviewed early on, that pointed to covert Soviet violation of the 1972 Biological Weapons Convention. In 1990, an initially very private 'trilateral process' had been set in motion by the three co-depositaries of the Convention – the governments of the UK, the USA and the USSR – that sought to resolve the situation through on-site inspections in the three countries. These took place during 1991–94, and it was as a key participant that Dr. Kelly began to acquire his experience and high reputation as a field inspector of possible biological-weapons activities. He recalled his work then in one of his very last publications, a chapter in the 2002 VERTIC [Verification Research, Training and Information Centre] *Verification Yearbook*. When, in 1991, UNSCOM began its work in Iraq, Dr Kelly led the first biological weapons inspection mission; and he led the last, seven years later, after 35 intervening inspection visits to the country. This work brought about his merit promotion, in 1992, from Superintendent of the Defence Microbiology Division at Porton to Deputy Chief Scientific Officer; and in 1996 he was transferred from Porton to the Ministry of Defence. He had, in 1995, become Senior Adviser to UNSCOM on biological weapons, and when, much later, UNMOVIC replaced UNSCOM he served in the training of its inspectors. Among the tributes that have been paid to him recently is that of the head of the UK delegation at the opening of the BWC 'new process' in Geneva: few people in recent decades, the ambassador said, can have done more directly to address and deal with the scourge of biological weapons. This extraordinary "service in relation to foreign affairs" had been acknowledged in 1996 when, in the Honours system whereby the British Crown recognises outstanding public service, David Kelly became CMG, a Companion of the Order of St. Michael and St. George.

It was both through Pugwash and through the University of Sussex that David entered into the work of HSP, always with the concern that we be guided by a proper understanding of what biological armament could be, was and was not, and that we did not get things wrong. The activities and findings of UNSCOM provided a vehicle for this educational task that did not necessitate disclosure of classified information, for in Iraq a real-world case study was on display. It was from his UNSCOM experience that, from 1997 onwards, he entered so constructively into the workshops on chemical/biological warfare disarmament that HSP organises for Pugwash so as to bring together scientists from many countries to talk issues through as professionals, regardless of national or political rivalries. And it was from his experiences as an inspector that he spoke at our seminars in Sussex, London and Cambridge, Massachusetts. Perhaps above all else, we benefited from the time and friendship that he was always ready to give to our research students working on aspects of CBW arms control or technology governance.

Generous and public-spirited people are rare enough anyway; but David had rare knowledge as well. We are saddened and diminished by his death.

7.14 Martin Mark Kaplan, 1915–2004[217]

Editors' note. This obituary warmly describes Martin Kaplan's character and contributions to preventing chemical and biological warfare (CBW). Martin worked for the World Health Organization for most of his life; he also actively supported the Pugwash Conferences on Science and World Affairs' efforts against CBW. As well as gratefully detailing Martin's work, the obituary demonstrates the importance of individuals collaborating through global movements such as Pugwash, and the benefit of combining expertise from different disciplines to understand multifaceted problems and devise potential solutions.

Martin Kaplan, motivator of the Harvard Sussex Program, died in Geneva on 16 October 2004. He was 89.

For much of his life, Dr. Kaplan worked at the World Health Organization, where his commitment and scientific insight provided world leadership in veterinary public health. He emphasized the crucial role of animal reservoirs in certain human diseases and organized a landmark cooperative international survey of animal sera that demonstrated and clarified the role of animal reservoirs in influenza. At the Wistar Institute in Philadelphia, where he worked during sabbaticals and home-leaves from WHO in the 1960s and 1970s, he was part of the team that developed animal and human vaccines against rabies. While at WHO and then in Pugwash, he initiated what has today become a global civil society movement against the weaponization of disease and the hostile exploitation of biotechnology. His outlook on the world can be seen in the following passage from an essay he wrote in 1971, *Science and Social Values*:

[217] Editors' note. This was originally published in *The CBW Conventions Bulletin* no.66 (December 2004) p 5. It is listed in Chap. 9 as item 449. See also the photograph of Martin with his wife Lenna and Matthew Meselson and Julian, in Part II, Chap. 4 of this book.

"And yet we are faced with war, poverty, increasing disorder and social alienation, distorted priorities, declining freedom and individual powerlessness. These are products not of man's inherent evil but of the inexorable grinding of the national machines with their imperatives of growth, profit, and glory."

Martin was born in Philadelphia on 23 June 1915, the youngest of eight children whose parents had emigrated from Russia in the 1880s. After college, he trained as a veterinarian at the University of Pennsylvania and then ran an animal practice. He joined the UN Relief and Rehabilitation Administration during World War II and was stationed in Greece. When UNRRA closed in 1947, he transferred to the first of the UN's permanent specialized agencies, the Food and Agriculture Organization, for which he worked in Poland. He was briefly dean of the School of Veterinary Medicine in Waltham, Massachusetts, at what is now Brandeis University, having been chosen for the post at the urging of Albert Einstein, whose vision had been to create an institution consisting only of professional schools, initially in the medical sciences – a means for countervailing *numerus clausus*, the practice imported from Europe in the 1920s by some US universities to limit admissions of Jewish students. The project fell through when interested parties in Boston successfully urged establishment of a normal university with undergraduate classes. Martin rejoined the FAO and then transferred, in 1949, to the nascent WHO in Geneva to form a veterinary section within the Division of Communicable Diseases. He remained in WHO until his retirement in August 1976. Latterly he was Scientific Adviser in the Office of Director-General Marcolino Candau, and finally Director of Research Promotion and Development. After retirement he continued to serve WHO as a consultant, remaining in Geneva with his family.

Martin's most prominent retirement activity, however, was as Director-General from 1976 until 1989 of the Pugwash Conferences on Science and World Affairs. This international movement was rooted in the Russell-Einstein Manifesto of 1955, which reflected the dismay of scientists at the emergence of nuclear weapons and concern about their role, as scientists, within security structures now being rested upon weapons of mass destruction. It called upon scientists to appraise the perils that had thus arisen. Writing in the London *Guardian* of 24 November 2004, Professor Sir Joseph Rotblat, the leader of Pugwash, recalls how Martin had walked into his office a few days after proclamation of the Manifesto and offered his help in any way needed. So began Martin's association with the movement, initially as an observer from WHO, whose leadership was, and remained, supportive. His first major task was to prepare and organize a conference on chemical and biological warfare (CBW). It took place in Pugwash, Nova Scotia, during August 1959, and was the first ever international gathering of scientists knowledgeable in fields from which CBW weapons derive, coming from both sides of the Cold War to characterize problems that CBW posed for humankind and to consider possible remedies.

Martin was not new to this subject. At WHO he had been closely involved in contingency preparations being made by the Division of Communicable Diseases for conducting, if WHO were so requested, an investigation of the allegations that US forces had been resorting to germ warfare in the Korean War. There was no such request, but WHO thereafter remained sensitive to the risk of germ warfare – to "public health in reverse". It was Martin who, during 1969–70, was responsible for *Health Aspects of Chemical and Biological Weapons: Report of a WHO Group of Consultants* and who stimulated, thirty years later, its second edition, *Public Health Response to Biological and Chemical Weapons: WHO Guidance.* For both editions he was able to draw extensively from international expertise available within Pugwash.

This was a resource that had grown richly following the inauguration in 1964 of a permanent Pugwash CBW study group. Martin had conceived the study group as a means for bringing together – privately, in a sustained fashion, and without the publicity that would inhibit frank discussion – academic, industrial and defence scientists from countries of both East and West, of both North and South, as well as from intergovernmental organizations. The enterprise later expanded to embrace other governmental officials, diplomats and, later still, people from nongovernmental organizations active in the fields of arms control and disarmament. It worked because of Martin's extraordinarily wide range of personal and professional contacts. The principal medium of the study group has now become the two-day workshop convened every six months or so, working to a topical agenda with matched participants. The workshops have by now engaged more than 600 people from some 50 countries. The association of the study group in its three successive embodiments with the negotiation and subsequent implementation of the 1972 Biological Weapons Convention and the 1993 Chemical Weapons Convention has been intimate, a consequence of Martin's indefatigable rallying of friends and allies in high places. His perseverance is an example for those who remain to carry on the struggle.

Martin is survived by his beloved wife of 60 years, Lenna Bouchal; by their children Alexa, Peter and Jeffrey; and by four grandchildren. A person as whole, as creative, as decent, and as giving as Martin lives on wherever he gave his labour. We are fortunate in HSP to have known and worked with him.

7.15 Categories of Challenge Now Facing the Chemical Weapons Convention[218]

Editors' note. The Chemical Weapons Convention (CWC) entered into force in 1997, after which the treaty was implemented nationally by states parties establishing domestic processes to fulfil their commitments, and internationally by states parties supporting the

[218] Editors' note. This paper, listed in Chap. 9 as item 459/Rev1, was a post-workshop revision of the paper for wider distribution dated 23 May 2007. Item 459/Rev2, dated 19 December 2007, was a further revised version submitted for publication that was later published as "Difficulties facing the Chemical Weapons Convention" in *International Affairs* vol 84 no 2 (March 2008) pp 223–39. The full reference is J. P. Perry Robinson 'Categories of Challenge now facing the Chemical Weapons Convention', 12 March 2007, 21pp, discussion paper for 52nd Pugwash CBW Workshop, Noordwijk, The Netherlands, 17–18 March 2007. Unless otherwise indicated, the remaining footnotes for this paper are Julian's originals. Apart from correcting obvious typos and renumbering headings and tables in keeping with the requirements of this book, we have retained Julian's formatting and spelling throughout the paper and its notes.

establishment of the Organisation for the Prohibition of Chemical Weapons which in turn set in motion the treaty verification systems. Julian monitored these developments closely.

Written ten years after the Chemical Weapons Convention entered into force, this paper offers a clear assessment of the treaty's strengths, its vulnerabilities, and the risks that could compromise its future effectiveness. Julian identifies two central concerns: the absence of effective governance over dual-use technology and the persistent tension between principled implementation and political expediency. He characterizes the key challenges facing the CWC as including novel utilities for chemical weapons, chemical weapons proliferation, convergence between chemistry and biology, accommodation of national interests, pernicious ignorance, and creeping legitimization. The paper closes with a call for greater vigilance and engagement – particularly from civil society – in upholding the Convention's core principles and purpose.

The Organization for the Prohibition of Chemical Weapons (OPCW) has now existed for a decade, functioning roughly as it should. So a world free of chemical weapons seems attainable. This paper is about the well-being of the CWC – the 1993 Convention on the Prohibition of Chemical Weapons – as the stockpiles and production facilities finally disappear. It has two themes. One is that, in the absence of effective measures of technology governance, resurgent chemical weapons born out of 'dual use' technology will remain a possibility, even a substantial probability. We need to decide now how the CWC should best be implemented so as to provide that governance. The original negotiators of the treaty fortunately did not disregard this key requirement. The second theme is the tension between principle and political expediency that will continue to affect implementation of the CWC. It is a tension that may well increase once the actual disarmament required by the treaty is complete, which in theory will be by 29 April 2012.

People often point to the CWC as showing that the problem of chemical weapons is now solved. It is not. Nor will it be unless the two themes just stated are properly addressed in implementation policy.[219] Now is the time to act, while the impending Second CWC Review Conference is obliging the OPCW to lift its sights further into the future than the press of day-to-day business has hitherto allowed.

Confronting the CWC are serious challenges, in part a consequence of that tension between principle and practice and in part because of wider political and technological change. The present paper characterizes the main challenges and thus points towards possible countermeasures. Its viewpoint is within the academic sector of civil society, where the Convention perhaps has its staunchest friends and where expectations for the Second Review are high. We begin by examining the purpose of the CWC.

[219] The present paper is more on the second theme than the first, which is being developed further in a paper based on the author's presentation at the last Pugwash CBW workshop (Geneva, 18–19 November 2006), "The Governance Regime for Biological and Chemical Weapons, and the Review Conferences of 2006 and 2008".

7.15.1 Why is the CWC Important?

A yardstick for measuring the value of the CWC today is the protection it gives us against particular weapons of mass destruction (WMD), whether they be in the hands of states or of terrorists. That protection is indeed valuable, but to my mind it is not the main reason why the CWC is important. More important is the protection that the CWC, if implemented as its negotiators envisaged, provides against future chemical weapons that will be capable, not of killing us in great numbers, but of *changing* us: weapons that could thereby confer upon aggressors accessible new means for coercion, repression, demoralization or subjugation: weapons that attack, not our lives, but our humanity.

Both chemical and biological weapons (CBW) share a technical characteristic possessed by no other weapon, which is this. CBW can attack life just as other weapons can, being capable of killing their victims no less dead than can bullets or bayonets; but they work by targeting and then disrupting particular processes that contribute to life, which other weapons cannot do save by accident, not design. The nerve gases, for example, target nerve-signal transmission; the blood gases, cellular respiration mediated through the blood. Bacteria and viruses may go for particular organs. Advocates of CBW have long propagated the notion that the effects of CBW might therefore be tailored to suit different tactical or strategic purposes.[220] But, beyond 'tear gas', the disabling skin burns that mustard gas can cause and the uncertain promise of the early psychotropic 'incapacitating agents', this did not begin to become a real option until the pace of change in applicable science and technology (S&T) accelerated during the latter part of the last century, around the time of negotiation of the 1972 Biological Weapons Convention (BWC) and its chemical follow-on. It was becoming increasingly plain then, to those who followed such things, that advances in the life sciences, coupled with technologies that allow the analysis and construction of complex biologically active molecules, could eventually make it possible to design a chemical that would interfere with *any* life process that could be understood in molecular terms. That could extend to the processes of development, inheritance, reproduction, locomotion, sensation, cognition and any other process that keeps us functioning properly as normal human beings.[221] Microbes that are otherwise non-pathogenic might be modified to function as site-specific carriers for harmful molecules. The potential is there, inasmuch as it is not manifest already, for inducing many different forms of malfunction, maybe even ones that discriminate between ethnic groups. It is this potential for changing human beings – for manipulating at will the determinants of our humanity – and doing so in pursuit

---

[220] See, for example, Brig-Gen J. H. Rothschild, *Tomorrow's Weapons: Chemical and Biological*, New York: McGraw-Hill, 1964.

[221] An early statement of this theme is in: J. P. Perry Robinson, "Chemical, biological and radiological warfare: futures from the past", invited submission to the Independent Commission on Disarmament and Security Issues [Chair: Olof Palme], September 1981.

of who-knows-what aggressive strategy that makes CBW uniquely menacing[222] – weapons, in the extreme, of living death. The existence of the BWC is testimony to our sense that we need protection against such a threat. The CWC can provide it.

Science has a way to go yet before the full horrors can be upon us, but the writing is on the wall. For example, the most obvious (though not the only) motivation that can be imputed to Iraq's programme in the 1980s to weaponize aflatoxin[223] is a desire to harm through latent liver cancer a subsequent generation of anticipated adversaries. As Dutch UNSCOM [United Nations Special Commission] commissioner Jack Ooms put it, the objective must have been a weapon of delayed genocide.[224] Artillery shell charged with aflatoxin are said to have been used during the suppression of the Shi'ite uprising in March 1991.[225] This has not been confirmed.

Others may point to other reasons why the CWC is important. The selectivity of action of chemical weapons that gives them their dehumanizing potential is not their only distinctive feature. Another is that, as destroyers of life on a large scale, some chemical weapons, and biological weapons too, can rightly be called WMD. When first used, on 22 and 23 April 1915, the device that brought chemical warfare out from its prehistory, namely massed cylinders of liquefied chlorine gas that could simultaneously be opened into the wind, reportedly asphyxiated five thousand French and Canadian troops at Ypres in Belgium and harmed a further 15,000.[226] Similar numbers of people are said to have fallen victim to Iraqi mustard and nerve gas in the Kurdish town of Halabja after chemical air raids during 16–18 March 1988.[227] For biological weapons no comparable experiences have entered the established

[222] See, further, Matthew Meselson and Julian Perry Robinson, "Chemical warfare and chemical disarmament", *Scientific American* vol 242 no 4 (April 1980) pp 38–47 at p 47; and Matthew Meselson, "Averting the hostile exploitation of biotechnology", *The CBW Conventions Bulletin* no 48 (June 2000) pp 16–19.

[223] See the entry for 25 August 1995 in the News Chronology section of *The CBW Conventions Bulletin* no 30 (December 1995) p 17.

[224] See the entry for 6 May 1997 in the News Chronology section of *The CBW Conventions Bulletin* no 37 (September 1997) p 16.

[225] See the entries for 7 and 7–20 March 1991 and 12 March 2001 in the News Chronology section of *Chemical Weapons Convention Bulletin* (later *The CBW Conventions Bulletin*) no 12 (June 1991) pp 13–14.

[226] But the records are poor and these commonly and widely quoted numbers have been described as too large, albeit initially only by German commentators. For a recent account, see Gordon Corrigan, *Mud, Blood and Poppycock: Britain and the First World War* (London: Cassell, 2004) at pp 164–65, whose view is that, in reality, the chlorine casualties probably numbered around 1500, with 200 deaths.

[227] See the entries for 18 March 1988 and 11 July 2002 in the News Chronology sections of *Chemical Weapons Convention Bulletin* (later *The CBW Conventions Bulletin*), no 1 (Summer 1988) p 11, and no 57 (September 2002) p 55

historical record,[228] but during 1964–68 the United States conducted unprecedentedly large field-trials over open sea of aircraft bioweapons, each of which was found capable of laying down a cross-wind line source of pathogenic microbial aerosol tens of kilometres long that infected experimental animals at sea level over a distance of several tens of kilometres downwind.[229] That could translate into an infective hazard to every person living within an area on the order of thousands, even tens of thousands, of square kilometres. In other words, it appeared from experimental data that some biological weapons would be capable of producing effects comparable in their magnitude to the life-destroying potential of nuclear weapons. Within NATO shortly before then, defence scientists had also been anticipating, correctly, a new generation of *chemical* weapons having similar area-effectiveness.[230]

So, the CWC and the BWC as protection against WMD – that is indeed an alternative frame within which to assess challenges to the treaty. Somehow it has come into common use since the ending of the Cold War, even though it predates it and even though it is a potentially counterproductive and dangerous idea. For advocacy purposes, it has the benefit that there exists widespread concern about WMD in general, and this can be mobilized to propel matters relating to WMD high up the political agenda, even to the point of setting loose war. Not often do CBW attract such priority. Yet the CBW that are most likely to find actual use today are hardly weapons of mass destruction, while those of the future that we fear because they may change us are not mass killers either. It is, to say the least, short-sighted to condone such weapons by implying that WMD are the real priority.

---

[228] But see, for example, Li Xiaofang (editor), *Qixie Kongsu: Blood-Weeping Accusations: Records of Anthrax Victims*, Beijing: Zhongyang Wenxian Chubanshe (2005), ISBN 7-5073-1862-1. It now seems highly probable that the Japanese Imperial Army conducted acts of large-scale biological warfare in China. There is, in addition, strong evidence of widespread use of biological weapons in Brazil during 1957–63 in order to displace indigenous peoples from areas of the Mato Grosso: see, especially, the 20-volume report of the commission headed by Brazilian Attorney-General Jader Figueiredo to investigate charges of corruption among officials of the Indian Protection Service that was published in March 1968. An account of the Figueiredo Report, which is in Portuguese, can be found in Shelton H Davis, *Victims of the Miracle* (Cambridge University Press, 1977) at pages 10–14. See also: Don McCullin with Lewis Chester, *Unreasonable behaviour: An autobiography* (London: Jonathan Cape, 1990) at pages 150–5, and Seth Garfield, “Brazil and the charges of genocide against its indigenous population”, January 1995, www.ciaonet.org/wps/gas01, a Yale University International Security Studies working paper accessed 25 July 2004

[229] Citing the original field-trial report, Ed Regis, in his *The Biology of Doom* (New York, 1999), writes that in 1964, during Operation Shady Grove off Johnston Atoll in the Pacific, Marine Corps *Skyhawks* laid down demonstrably infective clouds of tularaemia and Q-fever bacteria over a sea-level area of nearly 5000 square kilometres per aircraft.

[230] North Atlantic Treaty Organization, Standing Group, von Kármán Committee. *Future developments in chemical warfare*, from the report of Working Group X on Chemical, Biological and Radiological Defence, March 1961, as distributed to the UK Ministry of Defence Advisory Council on Scientific Research and Technical Development, paper no SAC 1928, 11 February 1969, in The National Archive, Kew, file WO195/16864. Regis, *op. cit.*, citing another US field-trial report, writes that in 1968, during sea trials off Eniwetok Atoll, Air Force *Phantom*s discharging dry-agent spraytanks loaded with staphylococcal enterotoxin B over caged monkeys and other animals were estimated to have secured 30 percent casualty rates over 2400 square kilometres per aircraft.

One other common measure of the importance of the CWC is the protection it affords against certain forms of terrorism. This overlaps with the WMD frame. From the USA we have been receiving, very strongly indeed, the idea that it is only a matter of time before terrorists introduce CBW into their campaigns, and that the consequences for society could be catastrophic. So persuaded is the US Congress of this threat that it is currently authorizing and appropriating thousands of millions of dollars annually for countermeasures against CBW terrorism.[231] There is an associated tendency to judge all CBW policy measures, including support for the BWC and the CWC, in terms of their effectiveness against such terrorism. This alarm obscures an important technical reality: that the practical difficulties of achieving catastrophic results from acts of CBW terrorism render such results exceedingly improbable. This is partly why a seven-country study of the vulnerability of EU society to radiological and CBW terrorism has just concluded that any such acts of terrorism today or in the immediate future are unlikely to achieve more than localized nuisance.[232] Yet risk-assessment in this area remains very poorly developed indeed, at least in public,[233] and much more research into the matter seems needed, given the magnitude of conceivable consequences. Nevertheless, for the time being there appears to be no good reason for giving an especially high priority to counter-terrorism as a directing influence on CBW policy-making, still less for treating the CWC as though it were meant as protection against terrorism. The CWC does indeed offer some valuable counter-terrorist tools if it is implemented as it was supposed to be; but that is not its primary value.

The contemporary US biosecurity literature gives us insight into the US concern about CBW terrorism. The concern seems to be conditioned by a belief that terrorists have, or could have, ready access to science and scientists, and that terrorists are somehow more likely to look to new technology than to traditional instruments of terror for their weapons of choice. This would explain the tendency common among US writers to consider the biosecurity implications of S&T change chiefly in terms of its exploitability by terrorists.[234] Such literature may increase our awareness of

---

[231] For detailed expenditure figures, see Stephanie Chang and Alan Pearson, *Federal funding for biological weapons prevention and defense, Fiscal Years 2001 to 2007*, Washington, DC: The Center for Arms Control and Non-Proliferation, June 2006.

[232] See the *Interim Report* dated 17 October 2006 from European Commission Framework 6 project 502476, *Assessment of the vulnerabilities of modern societies to terrorist acts employing radiological, biological or chemical agents with the view to assist in developing preventive and suppressive crisis management strategies* (ASSRBCVUL). Some particulars are posted on the European Union website at http://ec.europa.eu/research/fp6/ssp/assrbcvul_en.htm. The Final Report, dated May 2007, has just been submitted but, in contrast to the unclassified Interim Report, it is marked EU-RESTRICTED. The accompanying Executive Summary, however, is freely accessible.

[233] See Milton Leitenberg, *Assessing the Biological Weapons and Bioterrorism Threat,* US Army War College Strategic Studies Institute monograph, December 2005.

[234] As, for example, in the so-called 'Fink follow-on' report: Institute of Medicine and National Research Council of the US National Academies, Committee on Advances in Technology and the Prevention of Their Application to Next Generation Biowarfare Threats (Co-Chairs: Stanley M Lemon and David A Relman), *Globalization, Biosecurity, and the Future of the Life Sciences*, Washington, DC: The National Academies Press, 2006.

bioterrorism, but its greater value is surely in informing us of the vulnerability of S&T change to malevolent abuse considered more broadly than abuse by terrorists.

A conclusion, then, is this. The main function of the CWC today is to secure elimination of existing stocks of chemical weapons and factories for them. Once that disarmament has been completed, in theory by 2012, the function of protecting against hostile application of dual-use chemicals and chemistry will become paramount. That is why the CWC is and will remain important. Arguably, the hostile application that should head our list of concerns is the sort that aims to exploit the dehumanizing potential of chemical weapons. The mass-killing and terrorization potentials must be up there too, but they are not unique to CBW. In this conclusion we have the makings of an ordering principle that can be applied in ranking the challenges that confront the CWC.

7.15.2 What Would be a Challenge to the CWC?

When, on 3 September 1992, Ambassador Adolf Ritter von Wagner presented a laboriously negotiated draft CWC for the approval of the Conference on Disarmament (CD) in Geneva, he reiterated a point he had made often during his chairmanship of the negotiating body, the CD Ad Hoc Committee on Chemical Weapons:

> There is no precedent for this global, comprehensive and verifiable multilateral disarmament agreement. The draft convention provides for a cooperative, non-discriminatory legal instrument to eliminate the spectre of chemical warfare once and for all. The unique character of its contents is strengthened by the consistent application of two principles: overall balance and adaptability to future needs. Future States parties are offered a balanced legal instrument providing clarity on the fundamental obligations and, at the same time, enough subtlety on matters of implementation so that, with the consent of States parties, the respective provisions may still mature and evolve in the course of future practice."[235]

The draft treaty, he was telling the CD, was a delicate structure in which compromises – on six central matters: the scope of obligations, verification of compliance, safeguards, disarmament, executive procedures, and international cooperation in chemistry – were balanced against one another without, however, precluding from their future implementation any adaptation, if all states parties agreed, to a changed environment. Potential parties were, in effect, being invited to decide, through their domestic political processes, whether they would be better off inside that package of compromises than outside it, with an assurance that the terms of their engagement were not necessarily immutable.

One hundred and eighty eight states have since signed up to this deal (as of mid-April 2007), 182 of them as full parties. The most conspicuous holdouts are Egypt, North Korea and Syria, which have not signed the treaty, and Israel, which has signed but not ratified it. The assumption has to be that all four holdouts wish to maintain the option of chemical-warfare armament. Of the 182 states parties, 12 have, as required

[235] CD/PV.635, 3 September 1992, p 8.

by the treaty, declared that they possessed factories for making chemical weapons at some time after 1945.[236] Of those 12, five have also declared that they possessed stocks of chemical weapons at the time the CWC entered into force, as has one additional state party that must presumably have imported the weapons rather than making them for itself. The combined declared stocks have amounted to rather more than 70,000 tonnes of CW agent.[237]

A working principle is this. Any development or change that causes a state to question its continuing allegiance to the CWC would be a challenge to the treaty. If major or many states were to start such questioning, the challenge would be serious, requiring a collective response if the treaty were to remain in good order. For each state party the constant question would be whether benefits flowing from the CWC regime continued to outweigh the attendant costs and to compensate for any penalties there might be to the national interest: are we still better off inside the regime than outside it? In fact, of course, there are few if any countries where the relevant decision-makers would address the matter in such abstract or holistic terms. Decision would as always be driven by bureaucratic and domestic politics, in other words by the competing interests that such politics serve to accommodate. The cost-benefit frame, however, provides those involved with a convenient and respectable language in which to debate and present the decision, and us to assess it. For this reason it is used here as a framing concept with which to specify the more important challenges to the CWC.

7.15.3 Categories of Challenge

One set of challenges can be seen as primarily technological in character, being a consequence, mainly, of S&T change. A second set is more obviously political. All of them have the potential to harm the CWC if special effort is not made to counter them. They are discussed in turn.

---

[236] The 12 declared possessors of former Chemical Weapons Production Facilities are Bosnia & Hercegovina, China, France, India, Iran, Japan, Libya, Russia, Serbia & Montenegro, South Korea, the UK and the USA.

[237] The 6 declared possessors of chemical-weapons stocks are Albania (declaring 16 agent-tonnes to the OPCW), India (1044 agent-tonnes, maybe somewhat more), Libya (24 agent-tonnes), Russia (40,000 agent-tonnes), South Korea (1056 agent-tonnes, maybe somewhat less) and the USA (27,800 agent-tonnes). Like the (China-supplied) Albanian stockpile, the Libyan one appears far too small to have had much military significance, but Libya also declared that it possessed thousands of tons of precursor chemicals – a supply sufficient to provide a stockpile of chemical weapons exceeding those of India and South Korea together. The amount of mustard and nerve gas used by Iraq during its 1980–88 war with Iran is believed to have totalled 2540 agent-tonnes. The UK had destroyed its cold-war stockpile of some 62,000 agent-tonnes well before the CWC had been agreed, as had other erstwhile possessors.

7.15.3.1 New Utilities for Chemical Weapons

Disarmament, especially WMD disarmament, is an objective widely seen as beneficial, but armament also can bring benefit, by contributing to security. Under some circumstances that could conceivably include chemical-warfare armament.

Military options forgone through renunciation of chemical weapons might then be significant on the cost side of remaining within the CWC regime. Yet the taboo seemingly associated with chemical and other disease weapons appears to mean that most states are content with the disarmament required by the CWC – except that circumstances may now be creating utilities for chemical weapons not previously considered or accessible.

At least three types of new utility can be discerned, and examples of all three seem evident in recent conflicts or in preparedness for them. The first is a consequence of wider changes in the nature of warfare, rather as the shift from 'massive retaliation' to 'limited war' doctrine towards the end of the 1950s elevated the status of CBW in Western military thinking, causing new utilities to be seen for the weapons, especially in Third World settings. Today, a new type of organized violence is taking the place of, for example, those confrontations between highly disciplined and technologically advanced armed forces that characterized the later Cold War. Conflicts these past two decades in the Balkans, the Caucasus, the horn of Africa, Rwanda, Liberia, Sierra Leone, Angola, Sri Lanka, Afghanistan and post-invasion Iraq have eroded formerly clear distinctions between war, organized crime and large-scale violation of human rights. These new wars are fought by seeking political control through the displacement, or worse, of civilian populations and through the sowing of fear and hatred.[238] Because chemical weapons can lend themselves particularly effectively to such objectives, they may conceivably have a greater affinity to the new wars than they did to the old, so, notwithstanding the CWC, they may have an expanding future.[239] This is reason why the recent chemical warfare allegations, emanating from Sudanese, Israeli, Palestinian, Baluchi, Lebanese and US/Iraqi sources, should not remain uninvestigated and thus uncorroborated. Remaining unresolved, a plethora of allegations could imply treaty failure.

The CWC provides for a compliance-verification system run by an intergovernmental organization (the OPCW) having an international inspectorate that ought in principle to countervail this new-utility challenge. But the routines of that system were designed against Cold-War-period conceptions of utility, meaning that the lists of chemicals and types of industrial facility that the OPCW now has under its immediate surveillance are dictated by the types of chemical weapon that fitted old-war, not new-war, requirements. Basically that meant focussing on toxic chemicals that

[238] See especially Mary Kaldor, *New & Old Wars: Organized Violence in a Global Era* (second edition, Cambridge: Polity, 2006).

[239] This future seems already to have begun. Instances of 'new' chemical warfare include episodes in Iraqi Kurdistan, in southern Africa, in Bosnia and perhaps in Chechnya; see J. P. Perry Robinson, "The General Purpose Criterion and the new utility of toxicants as weapons", unpublished paper presented at the 15th workshop of the Pugwash Study Group on Implementation of the CBW Conventions, Oegstgeest, the Netherlands, 23–24 June 2001.

were so intensely aggressive in their effects that weapons disseminating them would be competitive, in quantitative casualty-producing terms or other such measures of tactical efficacy, with modern conventional weapons. Not a great many such toxicants exist,[240] so their coverage in the CWC schedules that govern routine OPCW verification allowed people to suppose that the main threats had thereby been brought under control. In the new wars, however, it is not so much relative aggressivity that determines the utility of CW but rather such other factors as accessibility or availability and terrorizing potential. A whole host of toxic industrial chemicals (TICs) and other chemicals not hitherto regarded as CW agents might thus find application in new-war contexts, as, most recently, chlorine – that long-obsolete, by old-war standards, killer gas that was briefly weaponized during the First World War – has done in Iraq.[241] The fact that most of these chemicals are not listed in the CWC control schedules does not mean that their use for CW purposes is permitted, nor that the CWC is unavailing against them. It means only that the routine international verification procedures currently run by the OPCW in regard to industrial activities are not directed at them. In fact, any abuse of a toxic chemical for hostile purposes is totally outlawed by virtue of the comprehensive nature of the CWC's prohibitions as embodied in its so-called 'general purpose criterion' – those words in, for example, Article VI.2 of the treaty that oblige the national authorities of states parties to ensure that no toxic chemical within their territories, or in any other place under their jurisdiction or control, falls within the illegal realm of purposes prohibited by the Convention.[242] The challenge to the regime therefore lies in the degree to which such national controls may fail to exert a constraining effect. This, regrettably, is an area in which implementation of the CWC is weak. Only a small minority of CWC states parties have implemented the General Purpose Criterion into their domestic law.

A second major source of new utility for chemical weapons is the propensity of knowledge newly gained in the life sciences for suggesting novel modes of attack that could be the basis for militarily or politically attractive new forms of weapon. For

---

[240] For discussion of this point, see J. P. Perry Robinson, "The chemical industry and chemical warfare disarmament: Categorizing chemicals for the purposes of the projected Chemical Weapons Convention", *SIPRI Chemical & Biological Warfare Studies* no 4 (1986) pp 55–104.

[241] CNN from Baghdad, "Scores choke in poison gas attack", posted 0940 hrs EST 20 Feb 07 on www.cnn.com; UN Department of Public Information, release SC/8963 and IK/557 dated 22 Feb 07, "Security Council press statement on Iraq"; Organization for the Prohibition of Chemical Weapons, "OPCW condemns chlorine gas attacks in Iraq", *Chemical Disarmament* vol 5 no 1 (March 2007) p 11.

[242] CWC Article VI.2 opens thus: "Each State Party shall adopt the necessary measures to ensure that toxic chemicals and their precursors are only developed, produced, otherwise acquired, retained, transferred, or used within its territory or in any other place under its jurisdiction or control for purposes not prohibited under this Convention." Known as the 'Molander chapeau', this obligation prefaces the main provisions of the CWC for its industry control regime. The General Purpose Criterion is also set out in CWC Article II.1(a), which states that all toxic chemicals and their precursors are chemical weapons within the meaning of the CWC's prohibitions "except where intended for purposes not prohibited under this Convention, as long as the types and quantities are consistent with such purposes".

example, if a new molecule is discovered that can exert novel disabling effects on the human body, attempts to weaponize it may well ensue. Albert Hofmann's discovery of LSD in 1943 is a case in point, although half a decade elapsed before weaponeers noticed. Or if a hitherto unknown molecular pathway serving a process of life comes to be identified, chemical agents capable of interfering with that pathway might also become identifiable and thus form the basis for a novel weapon. Of course many considerations other than novelty of effect determine the usefulness of a new weapon, so the new science is not itself the challenge to the regime that is here suggested. But it would be a step towards it; and many such can be envisaged.[243] This prospect is not necessarily remote. We should not, for example, disregard the statement reliably attributed to a "former high-level Defense Department official" commenting on the feasibility of US attack on Iranian underground facilities: "We can do things on the ground, too, but it's difficult and very dangerous – put bad stuff in ventilator shafts and put them to sleep".[244] Again, it is the General Purpose Criterion as used in the CWC that is the international safeguard against this challenge. But it is a safeguard only if it can be activated, and this requires continual monitoring of scientific and technological change for any new development that might challenge the regime. This is a task, it should be noted here, that cannot reliably be left to security authorities alone, simply because their surveillance of new science will always be insufficient: the scientific community at large must also be involved.[245]

A third type of novel utility now becoming manifest is the emerging role of chemical weapons, not in the hands of terrorists or other new-war aggressors as in the first novel utility considered above, but for purposes of counter-terrorism. This utility has demonstrably become a stimulus to rich-country questioning of the CWC.[246] It is rooted in past counter-insurgency applications of toxic chemicals, which reach back through the Vietnam War to British, French, Italian and Spanish use of toxic chemicals in colonial situations – a utility that the CWC was intended to suppress. Its re-emergence in counter-terrorist guise is to be seen in the proliferation of weapons based on Agent CR, evident each year in that part of the OPCW Annual Report addressing the declarations of 'riot control agents' required under CWC Article III.1(e), for the extreme aggressiveness and other properties of CR have caused it to be widely rejected as suited to civil police use. Police forces in

---

243 A particularly rich recent source of information on advances in technology that may be applicable to CBW is the Lemon-Relman report referenced in a footnote above.

244 Seymour M. Hersh, "The Iran plans: would President Bush go to war to stop Tehran from getting the bomb?", *New Yorker*, 17 April 2006. It is not clear whether it was a literal or a euphemistic 'sleep' that was meant.

245 This was a matter explored by the UK CWC National Authority Advisory Committee during the October 2001 Sussex workshop on the General Purpose Criterion. See Julian Perry Robinson, "What should be the scope of the CWC? A workshop report", *The CBW Conventions Bulletin* no 55 (March 2002) pp 1–4.

246 See, for example, the Foreword by USAF Lt-Gen Robert J. Elder, Jr, Commandant of the US Air War College, to N. T. Whitbred IV [Commander, USN], "Offensive Use of Chemical Technologies by US Special Operations Forces in the Global War on Terrorism: The Nonlethal Option", *The Maxwell Papers* [Maxwell Air Force Base, AL: Air War College] no 37, July 2006, pp iii–iv.

the UK, for example, are equipped either with Agent CS or with PAVA for law-enforcement use, and, although the UK has also declared Agent CR to the OPCW as a 'riot control agent', it has issued the agent only to its military forces, for counterterrorism. The growing counterterrorist utility of chemical weapons is further evident in the vigorous advocacy to be heard in some quarters for the arming of counterterrorist forces with more advanced types of 'non lethal' CBW. The readiness with which the US Marine Corps has taken to toxin weapons of this type – devices disseminating Agent OC[247] – seems indication of a trend. So, perhaps, is the absence of any serious criticism of the Russian government for having authorized use of toxic chemicals other than riot-control agents by the *spetsnaz* forces that, on 26 October 2002, liberated 634 of the people taken hostage by Chechen separatists in a Moscow theatre. The other 129 hostages were killed by the toxicant used, which is said to have been an agent "based on derivatives of fentanyl" that had been developed by USSR special services.[248] Comparable in some respects, if very different in others, is a counterterrorist utility for CBW that Israel, for example, has demonstrated in its espousal, following the Munich Olympics outrage in August 1972, of assassination as a major tool in counterterrorism.[249]

Perhaps exacerbating the new-utility challenge is the increasing dependence of some countries, not only on state forces for law enforcement including counterterrorism, but also on the private military contractors who have been providing security services at local, national and even global level. The potential value and, therefore, take-up of 'non lethal' chemical weapons by such contractors, who may be regulated less stringently than military or police forces, is already starting to become a matter of expressed concern.[250]

7.15.3.2 Proliferation of Chemical Weapons

Nowadays when people speak of the proliferation or non-proliferation of chemical weapons it is not always clear what they are talking about. In its normal usage, the

---

[247] See the entry for 22 March 1996 in the News Chronology section of *The CBW Conventions Bulletin* no 32 (June 1996) p 27.

[248] See the entry for 26 October 2002 in J. P. Perry Robinson, "Disabling Chemical Weapons: a Documented Chronology of Events", 1 November 2003, pp 139–42. A recent publication in the medical literature identifies the agent used – without, however, citing any authority for the information – as something called "Kolokol-1 [...] containing carfentanil"; see James Geoghegan and Jeffrey L Tong, "Chemical warfare agents", *Continuing Education in Anaesthesia, Critical Care & Pain* vol 6 no 6, (December 2006) pp 230–34.

[249] See Aaron J Klein, *Striking Back: The 1972 Munich Olympics Massacre and Israel's Deadly Response* (New York: Random House, 2005) pp 104–11 and, describing the CBW assassination of Wadi Haddad in 1978, pp 205–8. On the attempted CBW assassination of Khalid Mish'al, see the entries for 25 September 1997 and 19 February 1998 in the News Chronology sections of *The CBW Conventions Bulletin* no 38 (December 1997) p 29 and no 40 (June 1998) p 23. The CBW agent used in this latter episode is said to have been fentanyl, administered through the ear.

[250] See, for example, Alan Pearson, "Incapacitating biochemical weapons: science, technology, and policy for the 21st Century", *Nonproliferation Review* 13(2):151–88, July 2006.

word 'proliferation' conveys the sense that the weapons it is referring to continue to exist somewhere. This works for nuclear weapons, which are not wholly illegal, but since chemical weapons are outlawed by the CWC, 'CW proliferation' or 'CW non-proliferation' implies that the treaty is somehow ineffective or irrelevant. Yet even friends of the CWC use the expression, so it would appear to have a special meaning.

Indeed it does. The special meaning applies 'proliferation' or 'non-proliferation' to chemical weapons as the CWC defines them not to chemical weapons in the ordinary meaning of the term, which is different. The CWC means chemical weapons not only as tangible objects – special artefacts built for arsenals or military stockpiles – but also as holdings of "toxic chemicals and their precursors" that do not satisfy the General Purpose Criterion. This means chemicals held for purposes other than "purposes not prohibited under this Convention", as the CWC puts it, and which are not of "types and quantities that are consistent with such purposes".[251] In other words, it is *intentions* that the CWC also means, not just weapons in the concrete sense. Now that international chemical disarmament is far advanced, thanks to the OPCW, the term 'CW proliferation' primarily refers to the spread of intangibles, not things; and, above all, it refers to the diffusion of technologies that could be applied to CBW *if their possessors so chose.*

It might well be asserted that anyone who speaks of 'CW proliferation' without recognising the significance of the General Purpose Criterion should not expect to be taken seriously; nor, even more so, should the spokespersons of governments that have not implemented the criterion adequately. Regrettably (to repeat), that means the majority of OPCW member states, even including ones in which the idea of 'dual use' technology is well known.

The challenge that CW proliferation poses to the CWC can be disaggregated into two main forms. One is failure to understand what constitutes dual-use technology and the drivers of its diffusion around the world. Without such understanding, governance of the technology, including the formulation and implementation of antiproliferation measures, is bound to be inadequate. The second form of challenge resides in the existence of state or non-state entities that are ready to aid the exploitation of dual-use technology for purposes of weaponizing toxic chemicals. Such exploitation

---

[251] This CWC definition of chemical weapons is to be found in Article II.1 (see above), with Article II.2 stating that by 'toxic chemical' is meant "Any chemical which through its chemical action on life processes can cause death, temporary incapacitation or permanent harm to humans or animals". The "purposes not prohibited under this Convention" that the definition uses in order to exempt some applications of chemicals from the strictures of the CWC are themselves defined later in Article II, in paragraph 9, thus: "(a) Industrial, agricultural, research, medical, pharmaceutical or other peaceful purposes; (b) Protective purposes, namely those purposes directly related to protection against toxic chemicals and to protection against chemical weapons; (c) Military purposes not connected with the use of chemical weapons and not dependent on the use of the toxic properties of chemicals as a method of warfare; (d) Law enforcement including domestic riot control purposes." The terms used in subparagraph (d) are not themselves defined, but Article II.7 states that 'riot control agent' means "Any chemical not listed in a schedule, which can produce rapidly in humans sensory irritation or disabling physical effects which disappear within a short time following termination of exposure".

assistance might possibly be unwitting. The state entities often considered as proliferators in this sense include states that have been characterized as 'rogue states' or as 'failing states'. They may also include states that have deliberately chosen to maintain chemical-weapons capability. This last, it needs to be observed, is a category that ought to exclude all OPCW member states. Yet, for all most of us can tell, it may not in fact do so. Barring challenge inspection or investigation of alleged use, the OPCW Technical Secretariat is able to see only the chemical weapons that have been declared to it by their possessors; and, even then, the CWC confidentiality regime within which it is obliged to operate may well prove too opaque to furnish adequate assurance to people outside the need-to-know reaches of the Secretariat.

As for proliferators among non-state entities, these could include business corporations heedlessly serving a lucrative marketplace, criminal organizations feeding a black market, and terrorist groups seeking new weapons. This is an especially shadowy area about which even less is known (outside, presumably, the world of policing and security intelligence) than state-level proliferation. Any such greedy business-persons or opportunistic criminals there may rarely surface. Again, most of us just do not have knowledge enough to accept or to reject the received wisdom. As for proliferation of CBW among terrorists, this is, for the largely technological reasons alluded to earlier, probably myth. While it is known that certain terrorist groups have indeed looked at CBW options, such intent as they have had to acquire CBW does not seem to have been translated into significant capability. Thus far, other means for terrorist violence have generally proved more attractive or more accessible to them and, since the aberration of Aum Shinrikyo in the 1990s, only the most footling terrorist attempts to acquire CBW have been observed. This is not to say that it could not happen: the chlorine reports from Iraq and, before that, from Chechnya, Bosnia and Sri Lanka, show that it can happen at the non-WMD end of the scale. And the lesson to draw from the still-unresolved 'anthrax letters' affair in the USA during late 2001 is that CBW can put potential for great harm into the hands of technically competent and skilled individuals.[252]

In measures that have been taken outside of the CWC, such as the activities of the Australia Group and also UN Security Council resolutions 1540 (2004) and 1673 (2006), parts of the international community have put in place CW antiproliferation mechanisms that complement those of the CWC. But these mechanisms, which are chiefly dependent on import and export controls, have to function within a trade and technology-transfer environment that does not favour them, an environment in which globalization is impelling diffusion of industrial and other technologies around the world at what seems to be an accelerating rate. The capability of individual states to acquire chemical weapons, if they so choose, is thereby enhanced, and, if they still need specialized assistance, clandestine procurement networks have now gained increasingly dense cover within which to operate. That networks of this type can indeed spring up to meet demand was clearly shown by the UNSCOM/UNMOVIC

---

[252] See, further, Martin Rees, *Our Final Century: Will the Human Race Survive the Twenty-first Century?* (London: Heinemann, 2003); and Milton Leitenberg, *The Problem of Biological Weapons* (Stockholm: The Swedish National Defence College, 2004), pp 137–55.

investigation of Iraqi CBW acquisition[253] and by the Libyan CW programme. Even states that have no immediate wish to acquire CW may nevertheless move to take advantage of these various possibilities as a hedge against circumstances changing – by, for example, building what could serve as 'break-out capacity' into their industrial infrastructure, rather as the USSR created 'mobilization capability' for manufacturing bioweapons within its biotech industry during 1973–91. Iran, victim on a terrible scale to Iraqi chemical weapons during 1983–88, very probably falls into this category[254] while at the same time being amongst the most vocal supporters of the CWC regime and, especially on the medical side, a proactive participant in its international procedures. Nor is there any clear impropriety in such a position, for all the industrial powers, including ones that menace Iran, have manufacturing industries to which they could turn at short notice for CBW agents (whose full weaponization would, however, be more demanding). The fact nevertheless remains that proliferation of this type is a threat to confidence in the regime and therefore a serious challenge to it for as long as the problem of 'dual use' persists.

Nor should it be forgotten that duality exists in a variety of forms. The emphasis here has been on civil-military duality, exemplified by technology that may contribute to production both of pesticides, say, and of nerve gas. This duality has important variants, such as the applicability of law-enforcement technology based on toxic chemicals like Agent CS (an applicability that expressly qualifies as a purpose not prohibited under the CWC) not only to the counterterrorist purposes addressed earlier but also to other military purposes, such as those displayed during the Vietnam War. And beyond that there is the offence-defence duality, exemplified by the ready applicability to CBW of knowledge and other forms of intangible technology that are directly advanced through the study and development of anti-CBW protection. This last form of duality may be especially likely to create dangerous impressions in other countries that it is concealing or dissembling CBW development.

7.15.3.3 Convergence of Chemistry and Biology

CW, or BW, and CBW are terms that can often be used interchangeably (as they sometimes have been in this paper). The case for keeping CW and BW together is usually stronger than the case for treating them separately. Toxins that do not satisfy the General Purpose Criterion are both chemical weapons in the sense of the CWC and biological weapons in the sense of the BWC. Aerosolization or other dispersion

[253] On UNSCOM investigations, see especially: Tim Trevan, *Saddam's Secrets: The Hunt for Iraq's Hidden Weapons* (London: HarperCollins, 1999); Graham Pearson, *The UNSCOM Saga: Chemical and Biological Weapons Non-Proliferation* (London: Macmillan, 1999) and *The Search for Iraq's Weapons of Mass Destruction: Inspection, Verification and Non-Proliferation* (Basingstoke and New York: Palgrave Macmillan, 2005); and Scott Ritter, *Iraq Confidential* (New York: Nation Books, 2005).

[254] See J. P. Perry Robinson, "Dual technology and perceptions of Iranian chemical and biological weapons", draft of 20 July 2005; and International Institute for Strategic Studies, IISS Strategic Dossier. *Iran's Strategic Weapons Programmes: A Net Assessment* (London: Routledge, 2005).

technology used to disseminate the infective agents of biological warfare can also be used for toxic agents of chemical warfare, and diffusion models used to predict the airborne transport of chemical agents can also be used in estimating the likely effects of airborne biological agents. Respirators can protect against both biological and chemical weapons. Historically, bioweapons development has been an offshoot of CW development, often involving the same institutions and similar research skills.

The same instrument of international law, namely the 1925 Geneva Protocol, prohibits both chemical and biological warfare, being rooted in the same taboo against the weaponization of disease as were the Nineteenth Century prohibitions of poison weapons, and on which now rests the present BWC/CWC regime. It is a taboo that these days needs strengthening – not weakening, as by fragmentation. Perhaps it reflects an instinctive fear of disease, whether infectious (transmissible or not) or non-infectious, that may affect us in ways worse even than death.

That said, it is also true that there are major differences between CW and BW. Germ warfare is not chemical warfare. Efforts have persisted in policy circles to move and keep them apart. Such separation has not invariably attracted advocates. An instructive illustration entered the public domain in October 2000 when a letter written nearly 40 years earlier by the US Deputy Under Secretary of State for Political Affairs to his opposite number at the Department of Defense was declassified. The letter had to do with plans for increased forward deployment of US chemical weapons to Europe by creating stockpiles in France and Italy to augment those already in Germany. Its fourth paragraph was as follows:

> This immediate policy issue in turn opens up a broader set of policy problems. The question of chemical warfare doctrine and capabilities applies to geographical areas outside Europe. The fact that CW retaliatory materiel is now available to U. S. forces on Okinawa and in Korea clearly gives it practical relevance to the Far East. Furthermore, the relationship between chemical and biological warfare and their association in the minds of many suggest the desirability of their being looked at in relation to one another.[255]

Within five years, however, the differences between BW and CW had been pressed to the point of the separate arms-control negotiations that eventually brought us the BWC and then the CWC. It was the UK that initiated and promoted the separation. It did so (state papers now suggest) for largely unrelated political reasons, not because its policy-makers were persuaded that separation was inherently better than continued integration of CBW. The political expediency to which the separation was a response – smoothing an awkwardness in UK-US relations during the Vietnam War – eventually passed, and we have, for better or worse, been left with its legacy ever since.

Science has been moving on, and that legacy may quite soon prove an encumbrance as more and more biology becomes understandable in terms of chemistry and chemicals. Already the convergence is making the BW/CW distinction less sharp than many claim it to be, meaning that the distinction is shedding practical significance. The Lemon-Relman Committee report from the US National Academies last

[255] Letter, U. Alexis Johnson to William P Bundy [Deputy Assistant Secretary of Defense for International Security Affairs] dated 14 November 1963, copy in LBJ Library.

year dwelt on this point when making its recommendation of active support for the BWC and the CWC:

> Such international conventions should not be considered the solution to the issues society confronts today with respect to potential harmful uses of advances in the life sciences, nor should they be cast aside and ignored. Despite their limitations, the Committee appreciates their value in articulating international norms of behaviour and conduct and suggests that these conventions serve as a basis for future international discussions and collaborative efforts to address and respond to the proliferation of biological threats. Important opportunities will arise when states parties conduct their next quinquennial reviews of the operation of the BWC (in 2006) and the CWC (in 2008).
>
> The present report has several times noted that technologies are bringing chemistry and biology closer together. That toxins and synthetic biological agents, including bioregulators, immunoregulators and small interfering RNAs, fall within the scope of both treaties is one such linkable feature. These two review conferences will as always be dominated by political considerations, but, in view of the profound developments now under way in the life sciences, the Committee nevertheless draws attention to the possibilities held out by the 2008 conference for building upon the parallel or linkable features of the BWC and the CWC.[256]

The 2006 BWC Review Conference, although a success in other ways, passed over the opportunity the Academies were here suggesting for building upon the linkages between the BWC and the CWC. But given the fact that the CWC regime is far the stronger and better endowed, it is the 2008 CWC Review that will provide the richer opportunity.

How exactly that opportunity should be used is a matter that needs urgent attention. In particular, should not the OPCW start acquiring more technical capabilities in the overlap area, given that that is where major new weapons are most likely to be encountered? Should there not, at the very least, be a proven capacity in the Designated Laboratories for detecting and identifying a proper range of toxins and other biomolecules? And is it not time that the OPCW itself, not simply a few member states, showed through its actions that it recognises the comprehensive nature of the CWC – that it takes the General Purpose Criterion seriously? None of this is necessarily a matter of amending or changing the treaty. If the opportunity is missed, a real challenge to the CWC could develop as people come to realise that the OPCW has failed to keep adequate pace with S&T change. The challenge would be the challenge of ridicule.

7.15.3.4 Accommodating National Interests

Ambassador von Wagner proposed a treaty whose provisions could "mature and evolve in the course of state practice".[257] That is the CWC we now have, room for evolution being established by language that had deliberately contained 'constructive ambiguities' or that had simply left some issues to be resolved later. The alternative

---

[256] This IoM/NRC report is referenced in the Lemon-Relman Report footnote above. The quotation here comes from the passage in Chapter 4 that explains Recommendation 4a.

[257] CD/PV.635, 3 September 1992.

model was a legal instrument free of all ambiguity and with its obligations unnuanced, setting out exactly what its states parties were and were not to do. Even if desirable, such a cast-in-stone treaty could not have been negotiated within the deadline that the negotiators finally set themselves. We have instead been given plenty of space for "subtlety on matters of implementation", and this has been exploited by the policy organs of the OPCW, most notably during their annual negotiations on future programme and budget. Opportunity for augmenting the process is now presented by the Second CWC Review.

At one level this flexibility is much to be welcomed, for the international relationships within which operation of the CWC is embedded are in a constant state of flux, and this is bound to affect the character of the obligations of the treaty. But at another level it may create serious difficulties for the CWC, for it encourages mere political expediency to oppose underlying principles of the treaty. Above all it admits the ineluctable challenge of having to accommodate the interests of the more influential member states even, perhaps, where these may actually degrade the treaty. A pressing new need finally to resolve a 'constructive ambiguity' left over from the original CWC negotiations, or else to deal with some unanticipated puzzle about interpretation, can become opportunity to advance an interest. Its accommodation may perhaps make the regime more stable, but it may also make it increasingly unprincipled.

The proposition that the CWC must evolve if it is to survive changes in international relations clearly lends itself to abuse or, if not abuse, then to shortcomings in the way the CWC is implemented. Disturbing examples of this may be drawn from historical experience, responsibility for each one being attributable to the domestic or bureaucratic politics of this or that major state party. Here are just a few for one state party, the USA:

- Chemical weapons abandoned in Panama not declared.
- Production facilities not declared for certain toxic chemicals not satisfying the General Purpose Criterion, including the Agent CS that was filled into tactical (as opposed to 'riot control') munitions during the Vietnam War, such as 750-pound aircraft bombs.[258]
- Industry declarations submitted to the OPCW so late – some three years after the deadline specified in the CWC – as to distort development of the industry verification regime.
- The ouster of the first Director General, thereby exposing the OPCW to a judgement by the ILO Administrative Tribunal stating that, in "accordance with the

[258] It is of course likely that, in other respects, the US declarations to the OPCW are as accurate as may be expected. Yet who can now say with certainty whether there were or were not other declarable production facilities for undeclared toxic chemicals that were once authorized as weapon-fills but for which production data are not in the public domain? Agent UC (staphylococcal enterotoxin B) and Agent XR (botulinal toxin type A) come to mind. There had been production of both these agents since 1 January 1946, but data on the scale of their production and on the capacity of the production facilities are unavailable. We are therefore obliged to accept that the capacities did not exceed 2.74 kg per day (one tonne per year) and that, accordingly, these two toxins are rightly absent from the CWC schedules.

established case law of all international administrative tribunals, the Tribunal reaffirms that the independence of international civil servants is an essential guarantee, not only for the civil servants themselves, but also for the proper functioning of international organizations".[259]

- The sustained assertion that riot control agents (RCAs), even ones failing to satisfy the General Purpose Criterion, cannot be 'chemical weapons' within the meaning of the CWC.

My point here is not to show that a special finger of blame can be pointed at the USA, for similar lists can be compiled for other influential OPCW member states. My point is simply to exemplify the tension that may arise between principle and practice, the examples being ones of behaviour that clearly runs counter to the spirit of the CWC, if not necessarily to its letter. In aggregate the examples seem to display a powerful member state doing what it wants to do without being wholly mindful of the greater good. Nor should we be surprised by this fact of international life. It is what happens. We may nevertheless point to two challenging consequences for the CWC. One is an OPCW Secretariat that has to be exceedingly circumspect regarding great-power interests before it can take any sort of initiative.[260] The second and related consequence is an accumulating list of CWC-implementation issues that are in the too-difficult-to-deal-with category.

These two challenges could ultimately, if worse came to the worst, prove fatal to the CWC regime. Can governments, collectively through the OPCW, be relied upon to address them? No: that is how governments behave, and that is the core of the problem. Does this mean, then, that it is down to civil society to attempt something? But what can NGOs do other than observe, record, analyse and perhaps (it could do the individuals concerned no good at all career wise) publicize? Insofar as there is any sort of remedy, it resides in the OPCW Technical Secretariat having a Director-General and other senior staff of altogether exceptional ability and probity.

7.15.3.5 Pernicious Ignorance

There is another twist to this dependence of the CWC regime on the whims of the dominant actors. What happens when an influential state party does not care very much about the treaty or when its representatives are inadequately informed about details or about what the treaty is meant to be doing? For example, was President Bush in sufficient possession of the facts when at a press conference on 18 November 2002 he publicly praised President Putin for having authorized the use of an opiate

---

259 International Labour Organization, Administrative Tribunal, 95th Session, Judgement 2232 in Geneva on 16 July 2003.

260 It is therefore to the great credit of the Secretariat that it has been able, over the past decade, to take the lead on several matters of implementation. See Ralf Trapp, "The First Ten Years", in Ian Kenyon and Daniel Feakes (editors), *OPCW: The Creation of the Organisation for the Prohibition of Chemical Weapons: A Case Study in the Birth of an International Organisation,* The Hague: T.M.C.Asser Press, 2007, pp 261–89.

to end the Moscow-theatre siege during the previous month?[261] This is not to say, of course, that great-power representatives are likely to be any more or less well-informed than other people. In fact it is not at all easy for anyone to grasp CBW, or even just CW, issues adequately without prolonged immersion in the subject, and neither may this always work, given the secrecies and sensitivities that still abound. Maybe the problem is that CBW is a subject that historians have largely ignored, meaning that people wanting or having to know about it have no reliable and readily accessible literature available even for basic chronological information and broad-brush analysis.[262] Different concepts of the same matter can co-exist among groups even of specialists without their members appreciating that the concepts are different. In the absence of good reference works of sufficient breadth and independence of vested interest, misapprehension and ignorance have thus become common in policy-shaping circles, even including those people who might have advised President Bush ahead of that press conference.

A situation in which ignorance prevails may well take on the appearance of rampant heedlessness. The converse may also be true. Here are some examples of where this has become pernicious, endangering the regime – examples of where ignorance itself (or perhaps just heedlessness) is a challenge to the CWC:

- It has become common over the years for representatives of a curiously large number of CWC states parties to assert that there is no such thing as a 'general purpose criterion' in the text of the CWC, and therefore that there is no foundation for the comprehensive nature of the CWC or for arguing that its proscriptions extend to chemicals not included in the Schedules. It is not at all clear how the representatives affecting to believe this have managed to overlook Article II.1(a) and Article VI.2 of the CWC.

---

[261] See the fifth entry for 18 November 2002 in the News Chronology section of *The CBW Conventions Bulletin* no 59 (March 2003) p 16.

[262] Among English-writing historians the names of Donald Avery of Canada, Barton Bernstein and John van Courtland Moon of the USA, and Edward Spiers of the UK stand out as especially notable exceptions. A few other writers have ventured into the area, most successfully Brian Balmer of the University College London Department of Science & Technology Studies (see especially his *Britain and Biological Warfare: Expert Advice and Science Policy, 1935–1965, Palgrave Macmillan, 2001,* and his chapter in Mark Wheelis, Lajos Rozsa and Malcolm Dando (editors), *Deadly Cultures: Biological Weapons since 1945*, Harvard University Press, 2006); likewise Jonathan Tucker, biol ogist turned political scientist and author of the admired *War of Nerves: Chemical Warfare from World War I to Al-Qaeda* (New York: Pantheon Books, 2006). It was the absence, forty years ago, of reliable basic texts that stimulated the six-volume study produced by a collective of authors at the Stockholm International Peace Research Institute during 1967–72: *SIPRI, The Problem of Chemical and Biological Warfare: A study of the historical, technical, military, legal and political aspects of CBW, and possible disarmament measures* (Stockholm: Almqvist & Wicksell, New York: Humanities Press and London: Paul Elek, 6 vols, 1971–76). The available open-domain source material has, like the CBW governance regime itself, expanded hugely since then, meaning that the big SIPRI study is no longer an adequate substitute for decent CBW historiography. Two other general texts that should be noted here are: Robert Harris and Jeremy Paxman, *A Higher Form of Killing: The Secret Story of Gas and Germ Warfare* (London: Chatto & Windus, 1982); and Eric Croddy, Clarisa Perez-Armandariz and John Hart, *Chemical and Biological Warfare: A Comprehensive Survey for the Concerned Citizen* (New York: Copernicus Books, 2002).

- The US imputation that RCAs are not 'toxic chemicals' in the sense of CWC Article II.2 (because the USA holds that RCAs can never be chemical weapons) resonates with people who equate 'toxicity' with 'lethality', regardless of the fact that toxicity may take forms other than life-threatening ones. That the CWC does not limit its concept of toxicity to lethal toxicity is clear from the words used in Article II.2. True, if properly used, RCAs rarely cause "death or permanent harm to humans or animals"; but their whole *raison d'être* requires that they produce "temporary incapacitation". Ignorance of this point has led a surprising number of people to think that the CWC applies only to lethal chemical weapons. In fact 'lethal' and 'lethality' are not words that figure in the CWC at all.
- There was even a formal OPCW occasion recently during which a member-state representative expressed disbelief during corridor conversation that the CWC requires states parties to identify, in declarations to the OPCW, "each chemical it holds for riot control purposes".
- Possibly President Bush was heedless of the fact that the use of chemical opiate authorized by President Putin against Chechen separatists might have violated the CWC. The stance taken by Russian Federation officials was that there was no question of any such violation because, as Health Minister Yuriy Shevchenko "officially" declared on television, "chemical substances which might have fallen under the jurisdiction of the international convention on banning chemical weapons were not used during the special operation". Was this yet another instance of high-level ignorance of the General Purpose Criterion – ignorance of the fact that the CWC bans only purposes not chemicals *per se*? Other Russian officials have been more circumspect. The then director-general of the Munitions Agency, Zinoviy Pak, spoke as follows at a conference in Moscow a fortnight later:

> [I]n dealing with this terrorist act Russia did not depart an iota from the Convention. Russia fulfilled that mission in full accordance with the Convention. Namely, the convention allows the use of chemical agents to restore order in a country. These chemicals must be declared by the state to the international organization. Russia has done all that. The chemicals have been declared and Russia uses them. But in this situation we were talking about a law-enforcement action, a unique law-enforcement action that the Convention also allows for [...] So, there are neither legal nor actual grounds of suspecting Russia. Russia has complied with everything scrupulously. And the special services can share with you what this substance was.[263]

In other words it was not because the opiate lay outside the scope of the CWC, as Shevchenko had apparently asserted, but because the action in which it was used constituted 'law enforcement' in the exemptive sense of CWC Article II.9(d). There is an interpretation of this provision, which many people believe to be the best interpretation, which holds that the only toxic chemicals that may be used for law-enforcement purposes are RCAs as defined in CWC Article II.7.[264] Although

---

[263] See the entry for 11–12 November 2002 in the News Chronology of *The CBW Conventions Bulletin* no 59 (March 2003), pp 13–14.

[264] See, especially, Adolf von Wagner, "Keynote Address: The handling of toxic chemicals for law enforcement purposes, including domestic riot control purposes within the Chemical Weapons Convention", *Open Forum on the Chemical Weapons Convention* [The Peace Palace, The Hague, 1 May 2003], www.sussex.ac.uk/Units/spru/hsp/OpenForumCWC.pdf.

this interpretation had been adopted by a number of CWC states parties, Russian behaviour in ending the Moscow-theatre siege did not attract their condemnation, some of them resorting, however, to heavy ambiguity in public statements,[265] this perhaps reflecting the peculiar circumstances prevailing during the preparations for the US-led invasion of Iraq rather than acquiescence in Russian behaviour. The consequences for the CWC of such a political expediency could be long-lasting.

A surprising number of officials, even in countries that are active in their support for the BWC and the CWC appear to believe that only the former treaty covers toxins, not the latter as well. This is apparently a consequence of the Canada/EU policy of referring to the BWC as the 'BTWC'.

The examples just cited all seem to have ready explanations. One other example, however, does not, which makes it all the more important to understand. South Korea insists on its identity as a declared CW-possessor state being withheld from all OPCW publications. The CWC gives it the right to do this, but why does Korea persist in asserting that right even when its officials have made statements to the press about their country's programme for destroying its chemical weapons?[266] It is hard to believe that this is simply another example of a state party abusing the privacy provisions of the CWC through mere heedlessness or ignorance. Nor, surely, can it be a consequence of US stockpiling of chemical weapons in South Korea.[267]

What this paper is calling the challenge of pernicious ignorance thus takes a variety of forms. Once again the remedy seems to lie with civil society, provided a mechanism can be found for promoting sound and comprehensive historiography in the CBW field. But NGOs would need a lot of help, not just financial help but also practical help – in the form, especially, of enhanced access to some types of information. In this regard, it is important that the OPCW should have workable procedures for reviewing the security-classification of the information its Secretariat holds confidential at the request of the member-states that had submitted it originally; and for releasing declassified information. Now is definitely not the time for the OPCW to perpetuate its culture of opacity – of attenuated transparency – for this conveys a message of disregard for the outside world, including the civil society on which, to a degree, the OPCW will depend for its future well-being.

---

[265] A UK Foreign & Commonwealth Office minister, for example, spoke thus to Parliament: "Following inquiries by the United Kingdom and others, Russian authorities have announced that the gas used was based on Fentanyl, an opium based narcotic. Fentanyl is not a chemical scheduled under the Chemical Weapons Convention. Non-scheduled chemicals are not in themselves prohibited under the Convention for use in law enforcement, including domestic riot control purposes. On 30 October, the Russian Health Minister, Mr. Shevchenko said 'I officially declare that chemical substances of the kind banned under international conventions on chemical weapons were not used'." See the entry for 4 November 2002 in the News Chronology of *The CBW Conventions Bulletin* no 59 (March 2003), p 10.

[266] See, for example, the entries for 8 May, 17 July and 4 August 2000 in the News Chronologies of *The CBW Conventions Bulletin* no 49 (September 2000) pp 26, 41, and no 50 (December 2000) pp 26–27.

[267] See the 1963 US State Department letter quoted on above.

7.15.3.6 Creeping Legitimization

Here we return to what, at the outset of this paper, was characterized as the gravest threat of all those that CBW present: potential for changing us. In the ability of CBW agents to target themselves on particular life processes, there is indeed, as the old advocates used to claim, growing scope to 'tailor' the nature or severity of agent effects to a particular combat objective. In that such tailoring could also open the way to weapons suited to hugely malign purpose, an effective governance regime is essential. That same tailoring can, however, provide weapons that, at first glance, appear to be of an altogether more acceptable character, including ones having effects gentler than most other means of violence. Examples include the 'tear gas' of police forces; the psychochemical weapons that, according to past US Army teaching,[268] would cause the enemy to "linger in overpowering reverie"; and the entirely mythical knock-out agents of "war without death" that have figured in science fiction since the Nineteenth Century. Add to these chemicals the various infective agents that can induce highly debilitating diseases of low mortality, and a category of CBW is created whose features seem far distant from those of WMD, whose possession may appear desirable, and whose constraint by treaty may therefore come to seem a liability, notwithstanding the abyss into which the tailoring could also take us.

A rather wide variety of commercial, political and military interests stand to benefit from exclusion of some or all of these non-WMD CBW from the governance regime. *Sub rosa* campaigning to that end has long been under way, most notably during the final months of the CWC negotiation in mid-1992, when the protagonists of what was then starting to be called 'Non Lethal Warfare' (NLW) came up against governmental officials charged with securing consensus on those parts of the CWC text that dealt with RCAs.[269] The issue turned then on whether RCAs fell within the definition of 'toxic chemicals', subject, thereby, to the General Purpose Criterion that would serve to regulate the duality of their application either in warfare (prohibited) or in law-enforcement (permitted). For reasons that remain unclear to this day, the USA favoured exclusion but, finding itself isolated in this position within the Western Group, secured a compromise[270] in which the CWC expressly prohibited use of RCAs "as a method of warfare" but remained silent on the toxic character of RCAs, thus perpetuating a semblance of ambiguity on whether the CWC did or did not capture RCAs. The way became open for determined NLW protagonists to argue that, if tear gas was not proscribed by the CWC, then neither should the more modern varieties, for which they coined the category-label 'Advanced RCA Technology' (ARCAT).

---

268 US Army Chemical Center and School, Fort McClellan, "New chemical agents and incapacitating agents", Lesson Plan LP6075, undated [*ca* 1965].

269 A close account of these and related events is to be found in J. P. Perry Robinson, "Solving the problem of 'law enforcement' ", a paper presented at the 19th Workshop of the Pugwash Study Group on Implementation of the CBW Conventions, Oegstgeest, the Netherlands, 26–27 April 2003, especially in the version currently being revised for publication as "Law enforcement, 'non lethal' warfare and the norm against biological and chemical armament", 10 May 2003.

270 Proposed not in fact by the USA but by a group of eleven Neutral and Non-Aligned States looking also, and in sharp contrast to the USA, for a CWC prohibition of herbicide warfare.

Subsequent ARCAT development projects funded by the US government have included work on the fentanyls and other such intensely toxic chemicals. The process that can be seen here is a surreptitious equation of toxicity with lethal toxicity. In this attempt to loosen the CWC constraint on the weaponization of other forms of toxicity we have started to see a creeping legitimization of non-WMD CBW, which is a most serious challenge to the regime. A situation in which some types of toxic weapon are allowed but not others is certain to be unstable. And it is certain to bring closer a realisation of the dehumanizing potential of CBW.

The instruments of creeping legitimization include, not only 'public diplomacy' and other more hidden pressures for exemption, but also national legislation. In the USA the 'Ensign Amendment' of the 2006 Defense Authorization Act asserts that "riot control agents are not chemical weapons".[271] Fortunately no other state party to the CWC has adopted such a position, nor even commented publicly on what the USA has done. It remains to be seen whether the topic will be kept off the agenda of the Second CWC Review, which is where it clearly ought to be. It should not be supposed that this is a purely state-level problem. Noted earlier in this paper was the possible utilization of ARCAT-type CBW by private military contractors. Any such privatization of CBW is certain to accelerate the creeping legitimization.

7.15.4 Conclusion

The foregoing categorization of challenges to the CWC has been intended simply as an aid to thought about possible countermeasures. A way of taking it forward, short of actually proposing countermeasures (which is not the purpose of this paper), is to think of complementary categories, first of the damage that the challenges might cause, and second of safeguards against such damage, in terms both of actions and of actors. From the matrices that might then be constructed from the categories, and through application of the ordering principle suggested earlier in the paper, types of remedy and their respective priorities would stand out. Such an analysis will be presented in a follow-on to the present paper. For the moment, here are some interim conclusions

[271] In its section 1232, the US National Defense Authorization Act for Fiscal Year 2006 states: "It is the policy of the United States that riot control agents are not chemical weapons and that the President may authorize their use as legitimate, legal, and non-lethal alternatives to the use of force that, as provided in Executive Order 11850 (40 Fed.Reg.16187) and consistent with the resolution of ratification of the Chemical Weapons Convention, may be employed by members of the Armed Forces in war in defensive military modes to save lives, including the illustrative purposes cited in Executive Order 11850." On 27 September 2006, in evidence to the Senate Armed Services Subcommittee on Readiness and Management Support, which is chaired by Senator John Ensign, the Defense Department in the person of Joseph Benkert, Acting Principal Deputy Assistant Secretary of Defense for International Security Policy, testified that the "Administration agrees with [this] statement".

1. There needs to be wider recognition, especially among CWC supporters in civil society, that, throughout the future of the CWC, there will inevitably be tension between political drivers of decision and principled implementation, between political expediency and the wellbeing of the treaty. It is a tension of which we are all aware but, as a challenge, tend to disregard. Sound stewardship of the CWC must be based on minimizing that tension between practice and principle, even, conceivably, to the point of adapting the terms of the treaty to the prevailing political climate. But that is an extreme option, and any such adaptation must never be permitted to endanger the heart of the CWC.
2. The heart of the CWC is the protection it affords against the primary threat that chemical weapons present to us all. This is not the threat of mass killing that chemical weapons share with other WMD. Nor is it the terrorization potential of chemical weapons. It is their potential for changing human beings, perhaps to the point of dehumanization.
3. The General Purpose Criterion is the primary mechanism whereby the CWC provides protection against this danger, and against others too. The criterion is not being implemented adequately. Reasons for this probably lie in the usual mix of ignorance, lack of understanding, incompetence, heedlessness, short-termism, and conflicting political interests, including those interests that are rooted in NLW technology. Action to improve this situation warrants the very highest priority.
4. The OPCW itself ought to be capable of such action, but evidently it is not. The governments of some of its member states seem to have become an intransigent part of the problem, while the Technical Secretariat has been blighted by the tenure policy and the consequent fading of its institutional memory plus, in certain areas, its technical competence, and by the magnitude of the obstacles it must overcome before it can take an initiative. The initiatives that it has been able to take, however, for example on Action Plans, have much enhanced the overall regime.
5. So several actions must now be contemplated. One is concerted, skilled, properly informed and policy-orientated study of those implementation issues that have disappeared into the too-difficult-to-deal-with category, especially issues that bear upon the General Purpose Criterion (e.g., RCAs, the meaning of "chemical action on life processes", changing the Schedules, &c &c). Such study should be put out for scapegoat work within civil society if neither the Secretariat of the OPCW nor member states are inclined to attempt it during the Second CWC Review.
6. We in civil society should not lose sight of the fact that the CWC is an engagement of states parties, of which governments are one element. Governments may represent states parties in the fora of the OPCW, but organs of civil society that are also elements of those same states must not disregard their own responsibilities for the proper implementation of the CWC, especially in areas where our governments appear hamstrung, incapable of action; and especially at the present juncture.

7.16 Bringing the CBW Conventions Closer Together[272]

> Editors' note. Malcolm Dando suggested including this paper. He says, "The reasons I would like to see this included in the selection is obviously first because of its identification of the gap between the CWC and BWC for the agents I spend most of my time worrying about, but more importantly because of Julian's identification of the primary threat from these: CBW not being as WMD but as an existential threat to the human species. As he put it: 'It is the potential for changing human beings and doing so in pursuit of who-knows-what strategy of aggression that makes CBW uniquely menacing: weapons, in the extreme, of dehumanization – a risk, therefore, to the whole human species: an existential risk.' My guess is that few people have yet to grasp this point, but I think the threat must grow as science and technology advances (just think about the growing number of investigations of the mechanisms underlying higher orders of behaviour in Non-Human Primates). So that is where my vote goes".[273]

In 1968, for reasons that are still not entirely clear, the United Kingdom proposed that biological weapons (BW) and chemical weapons (CW) should in future be treated separately at the Geneva disarmament conference, and that the talks should first concentrate on BW. That is what happened, gradually. The international agreement that already existed in the field, the 1925 Geneva Protocol, had taken chemical and biological weapons (CBW) together. Their new separation led to the 1972 Biological & Toxin Weapons Convention (the BWC) and then the 1993 Chemical Weapons Convention (CWC), so in fact both types of weapon became outlawed, and it is possible that neither would have been without that parting of the ways. Yet the divergence led to two very unequal agreements that have what look increasingly like serious gaps in their combined structure. The remedy may require that we appreciate once again the features that BW and CW have in common. It is heartening, therefore, that 'convergence' is appearing as often as it is now doing on international conference agendas. True, the reference is to the convergence of chemistry and biology as, for example, in 'synthetic biology', but that obliges us to wonder for how long it will be prudent to keep the CWC and the BWC as far apart from one another as they are today.

So it is unhelpful that quite so many people should regard BW and CW as too different to be treated in any way other than separately. We need to understand why that view should be as prevalent as it is and listen attentively to the explanations we are given – thus far political ones, chiefly, but also military and technological, though rarely legal. A supposedly clinching argument that has become popular is that BW but not CW are real weapons of mass destruction (WMD), comparable with nuclear weapons. The present paper offers for attention two countervailing considerations.

---

[272] Editors' note. This was originally published as "Bringing the CBW Conventions Closer Together" *The CBW Conventions Bulletin.* 80. September 2008. It is listed in Chap. 9 as item 467. Unless otherwise indicated, the remaining footnotes for this paper are Julian's originals. Apart from correcting obvious typos and renumbering headings and tables in keeping with the requirements of this book, we have retained Julian's formatting and spelling throughout the paper and its notes.

[273] Editors' note. Personal communication from Malcolm Dando to Henrietta Wilson, 12 March 2025.

The first is about the real nature of the danger that CBW present for security, whether that be international, national or human security. The second is about the so-called 'Type C' mode of CW attack that has, in effect, fallen into a gulf that now separates the CWC and the BWC.

7.16.1 CBW as Existential Risk

Some CBW are covered by the established UN definition of WMD, which is as follows: *atomic explosive weapons, radioactive material weapons, lethal chemical and biological weapons, and any weapons developed in the future which have characteristics comparable in destructive effect to those of the atomic bomb or other weapons mentioned above*. There are indeed some CBW that seem capable of causing casualties among living beings – people, other animals and plants – on a giant scale. The *Shady Grove* field trials conducted by the United States off Johnston Atoll in the Pacific Ocean during February and March 1964 are said to have demonstrated that, against caged rhesus monkeys, one single-seat aircraft could establish disease-causing dosages of bacterial aerosol at sea level over nearly five thousand square kilometres. That was a biological weapon, but, as is described below, similar areas of effectiveness were anticipated for future chemical weapons as well. Urban areas of like size – the habitat of maybe millions of people – might be no less vulnerable. So, as casualty agents, there may in principle be some comparability between CBW and nuclear weapons.

Yet, dreadful though that is, it is not the primary risk inherent in CBW. Their main danger is precisely that they need not be weapons of mass destruction, for what is unique about them is that they could in principle serve to subjugate or coerce people, even very large numbers of people, *without necessarily threatening their lives*. A capability for exerting that form of force could become attractive in circles where capability for mass destruction is unattractive.

What is it about CBW that could hold out this allure? Answer: the mechanisms of toxicity and infectivity that characterize CBW are the mechanisms through which we can, if we are so minded, exploit for weapons purposes that most impressive feature of today's science – its accelerating understanding of the processes of life. As we comprehend more and more of how life works at the molecular level, so too may we be able to learn to *manipulate* life at the molecular level. That has been conceivable since the early days of CW. The blood gases, for example, target cellular respiration mediated through the blood; the nerve gases target nerve-signal transmission. So it was, back in the 1950s, that the idea started to take hold that other physiological systems might be disrupted, perhaps including those that define the psyche or the will or govern the ability to function properly. That way could lie potent *disabling* chemical weapons perhaps in great variety.

However, beyond 'tear gas', the disabling skin burns that mustard gas can cause, the uncertain promise of the early psychotropic 'incapacitating agents' such as Agent BZ, and one or two other minor novelties, this way forward did not begin to become a serious option until the pace of progress in applicable knowledge had accelerated during the latter part of the last century, around the time of negotiation of the 1972 BWC and its CW follow-on. It was becoming increasingly plain to those who followed such things then that advances in the life sciences, coupled with technologies that allow the analysis and construction of complex biologically active molecules, would eventually make it possible to design a chemical that would interfere with *any* life process that could be understood in molecular terms. And thoughts could also turn to the idea of modifying the genes of factory or vector organisms so as to obtain such chemicals as gene products. Because of this convergence of chemistry and biology, we may eventually be able to manipulate at will the processes of development, inheritance, reproduction, locomotion, sensation, cognition and any other process that keeps us working as normal human beings. Some of the weapons providing such power could carry the identities of both CW (toxic) and BW (infective). For example, as just suggested otherwise harmless viruses might perhaps be modified so as to produce as gene products during the process of infection particular molecules that could initiate such interference. It is the potential for changing human beings and doing so in pursuit of who-knows-what strategy of aggression that makes CBW uniquely menacing: weapons, in the extreme, of dehumanization – a risk, therefore, to the whole human species; an existential risk.

It seems short-sighted, then, to assess the risks of CBW by reference to their capacities as WMD. Indeed, to think of CBW primarily as weapons of mass destruction may actually be to increase the risk they pose, for it offers legitimacy to those toxic or infective weapons that do not kill on a massive scale: those very biological/chemical weapons that may open the door to the dehumanizing possibilities that now press upon us.

More pernicious even than that: the WMD concept itself is being applied in such a way as to erode one of those central principles of international humanitarian law applicable in armed conflict: the principle of discrimination between combatant and noncombatant, and the consequent illegality of weapons or methods of warfare that are liable to have indiscriminate effects. The focus on the unacceptability of weapons of mass destruction belittles that ancient principle – it offers a sound-bite alternative that distracts attention away from the non-nuclear, non-CBW means and methods that can be just as indiscriminate in their effects but which have become an acceptable component of rich-country arsenals. In the public discourses that are now to be heard on the acceptability or otherwise of weapons, it seems almost as though the principle of discrimination has disappeared. Treating CBW primarily as weapons of mass destruction is to ratify that disappearance.

7.16.2 *Lost in the Gulf*

Other students of CBW history and of technological change may perhaps reach different conclusions about why the CBW Conventions are so necessary. What the foregoing conception does is to point towards molecular dangers from within broad families of toxic chemical that, for the most part, lie outside the CWC Schedules and therefore outside the international CBW control procedures that, thanks to the OPCW [Organisation for the Prohibition of Chemical Weapons], we currently have. Most obviously these families comprise certain bioregulators and other bioactive chemicals of biological origin, as well as their synthetic analogues and also certain conceivable products of biotechnological processes. Yet in fact they are not beyond control. They are all caught by the 'general purpose criterion' of the CWC. And most if not all of them fall within the category corresponding to what the BWC means by 'toxins'.[274] In other words, these most dangerous of CBW are situated within the area where the CWC and the BWC overlap with one another.

Such overlap ought to mean, one might think, that the weapons are well controlled, being subject not just to one but to two international disarmament treaties. In the real world, however, that is not the way it is. That overlap seems simply to have given people involved in implementing one of the two treaties opportunity to relinquish, even deny, responsibility for anything also covered by the other treaty. The area of overlap thus risks becoming a gulf into which things disappear. It looks like this has been happening to toxins.[275]

Closer scrutiny points to that failure of overlap not being just a happenstance of history but a consequence of deliberate policy. Back in 1960, the Military Committee of NATO tasked what became known as the Von Kármán Committee to "develop an estimate of the possible and probable scientific progress to be expected in the next decade."[276] The task was subdivided among 14 working groups of scientific and operational experts that met in March 1961. One of the groups addressed Chemical, Biological and Radiological Defence. Asked to forecast scientific trends to the 1970–75 period and to predict their practical effect on CBR defence, this working group envisaged different categories of possible CBR attack against which defence was required, the forecasting exercise then proceeding through the following eight areas of defence requirement: (1) detection, (2) active neutralization of attack, (3) protective equipments and materials, (4) palliatives, (5) remedial measures, (6) surveillance, (7) large scale decontamination, and (8) means of agent identification. Each area was considered for each of the different categories of attack. What is instructive for the

[274] The BWC meaning of 'toxin' is set out in an international text published by the World Health Organization: Public health response to biological and chemical weapons: WHO guidance (Geneva: WHO, 2004), pp 214–16. It extends to substances that are not called 'toxins' in other specialist communities.

[275] On which see: J. P. Perry Robinson, "The odd neglect of toxin weapons during the 2004 Meeting of Experts", a paper presented at the 21st Workshop of the Pugwash Study Group on Implementation of the CBW Conventions, Geneva, 4–5 December 2004.

[276] This was the start of what NATO later called its Long Term Scientific Studies, the subsequent report from the Von Kármán Committee, in 1961, being described on the website of the NATO Research & Technology Organisation, http://www.rto.nato.int, as "the first multi-national attempt to estimate the impact of scientific and technological advances on military capability".

present paper is not the group's findings but its categorization of attack-modes. For the CW part of the study,[277] there were these three categories:

> *Type A*: "Direct attack on small areas (up to 1 Km^2, say) with respiratory agents. Rapid coverage of the whole area with vapour or quickly-evolved aerosols would be aimed at, in order to achieve maximum surprise against the enemy in the area (surprise is not essential, however, if the enemy is not provided with respirators)."
>
> *Type B*: "Direct attack with agents acting through the skin (areas up to the order of 100 Km^2)."
>
> *Type C*: "Off-target attack (areas up to the order of 100,000 Km^2) with respiratory agents of high toxicity or effectiveness (at least an order higher than the present nerve agents). Surprise against an enemy possessing respirators would be achieved by this method, because the attack is not observed and the agent arrives on target in very small but effective concentrations, which are not detected by the senses nor, quickly enough, by artificial means."

Two questions now arise. Where did the categorization come from? And how sound were the categories as predictive aids, especially the astonishing Type C?

NATO was at that time an alliance possessing chemical weapons: those of France and the United States (but not the UK, which had just got rid of its CW stockpiles and wasn't yet thinking about CW rearmament). It is not unlikely, therefore, that the categorization would have been rooted, if not in actual CW-use doctrine, then at least in an understanding of feasibility. As to the second question, can it be said today that the Type C attack-mode was a real option either in 1961 or during the prospective time period up to 1975? The answer, it seems, is probably yes.

For in 1968 chemical weapons comprising dry-agent spraytanks mounted on American *Phantom* strike aircraft and charged with Agent PG were tested over caged monkeys and other animals at sea off Eniwetok Atoll in the Marshall Islands; this was DTC Test 68-50, and the test data are said to indicate a 30 percent casualty rate over 2,400 km^2 per weapon. If that report is to be believed, not very many sorties would be needed to bring the area of effectiveness up to the order of 100,000 km^2, and it is not inconceivable that, come 1975, there would have been more efficient means of agent-dissemination and even more powerful agents. Agent PG (previously known as UC) is staphylococcal enterotoxin B, an incapacitating agent whose inhalation ED50 in man has been estimated to be at least three orders of magnitude smaller than the corresponding LD50 of nerve-agents such as sarin or Agent VX. Other such superpotent substances falling within the scope of the CWC include botulinal toxins such as the preparation known as Agent XR that was weaponized by the United States as part of its erstwhile chemical/biological-weapons programme, but there is no information in the open domain about their estimated efficacy in Type C attacks. Perhaps Agents PG and XR have by now been superseded as the best exemplars of the category, but, even if they have not, any continuing possibility of Type C weapons would seem to constitute a risk that is not being actively managed by the available tools, including those that have been furnished by the CWC.

---

[277] UK Ministry of Defence Advisory Council on Scientific Research and Technical Development, paper no SAC 1928, 11 February 1969, Future developments in chemical warfare: Extract from the Report of the von Kármán Committee 1961, in The National Archive, Kew, file WO 195/16864.

A year ago, the US State Department published electronically a new volume in its historical series *Foreign Relations of the United States.* It presented a definitive record of US policies towards multilateral arms control during the first administration of President Richard Nixon. It placed in the public domain for the first time hitherto classified documents from the administration's review of CBW policies. One of these documents states that, at the beginning of 1970, the United States maintained production capacity for both PG and XR: 600 pounds per month for the former and 280 lb/mo for the latter.[278]

Those figures correspond to about 3.0 and 1.5 metric tons per year and are therefore greater than the one tonne/year threshold above which CWC states parties are required to declare any CW production facilities they had designed, constructed or used at any time since 1 January 1946. Yet the United States declared neither its PG plant nor its XR plant; nor, so far as is publicly known, was it called to account for failing to have done so.

Perhaps we should not be surprised, then, that neither agent appears on the Schedules of the CWC, meaning that no exemplar of Type C is subject to routine controls. But Type C is a horrendous category, and it is shocking that there should be quite such a gap in OPCW safeguards – the more so given what was supposedly happening during 1974-89 in Project Bonfire, which has been described as an attempt by the USSR to create novel toxin weapons involving human regulatory peptides.[279]

For Agent XR at least, something is known of how this happened. The United Kingdom had proposed during the CWC negotiation that botulinal toxin should be included in the Schedules, but the United States had opposed this. One of the main arguments it used in private to explain its position was that inclusion might compromise the US position that the BWC was unverifiable. The endgame of the CWC negotiation coincided with preparations for the start of VEREX.

7.16.3 Conclusion

Toxins broadly defined present a growing risk to the object and purpose of both the CWC and the BWC even though both treaties prohibit the weaponization of all toxins. This situation seems explicable in terms of reckless pursuit of national interest apparently by at least two states parties combined with heedless *laissez faire* by the other parties. The overlap between the two treaties that toxins occupy has thereby deteriorated into what is effectively a gap. The overlap should surely, as a matter of urgency, be rebuilt through improved implementation of both treaties.

---

[278] See US Policy on Toxins dated 30 January 1970, which is document 177 in US Department of State, Office of the Historian, Foreign Relations of the United States: Nixon-Ford Administrations, volume E-2: Documents on Arms Control and Nonproliferation, 1969–1972 (2007).

[279] Ken Alibek, with Stephen Handelman, Biohazard, London: Hutchinson (1999), pp 154–55 & 163–64.

With commercial interests growing in bioactive chemicals of biological origin, the rebuilding might best take the form of bringing legitimate activities involving such toxins more closely into the transparency-enhancing procedures of the two treaties. There have already been proposals on how such convergence could be achieved. For the BWC, one way forward might lie in developing the existing regime of Confidence Building Measures,[280] though this would presumably have to wait until after the 2011 Review Conference. For the CWC there are the possibilities presented by further development of the verification regime for Other Chemical Production Facilities,[281] though these would work only if the states parties that have done so no longer insist that "produced by synthesis" in the sense of CWC VA IX.1(a) and (b) excludes biotechnological processing.[282] There are surely other options as well.

7.17 Scientists and Chemical Weapons Policies[283]

Editors' note. This paper was nominated by Roy MacLeod as an example of a particular form of Julian's writing that Roy defines as his "essays of public concern" – a term that reflects Julian's distinctive ability to transcend technical expertise, crafting writings that bridge the gap between specialist knowledge and public engagement. The paper echoes many of the themes covered in the other writings chosen for Chap. 7, while emphasizing the shared responsibility of governments, civil society, and individuals in addressing global challenges such as disarmament, non-armament, and the implementation of international treaties.

7.17.1 Abstract

Roles for independent scientists in the formation of public policy on chemical weapons are specified by identifying different types of risk associated with chemical weapons and asking whether independent scientists could contribute usefully to their assessment and therefore to their management. Attention is paid to three sources

---

[280] J. P. Perry Robinson and A. P. Phillips, "Addressing the toxin problem", a submission to the 27th Workshop of the Pugwash Study Group on Implementation of the CBW Conventions, Geneva, 8–9 December 2007.

[281] A. P. Phillips and J. P. P. Robinson, "The CWC and Chemicals of Biological Origin", a submission to the OPCW Academic Forum, The Hague, 18–19 September 2007.

[282] J. P. Perry Robinson, "Toxic biotechnological-process products and chemicals of biological origin", a submission to the 28th Workshop of the Pugwash Study Group on Implementation of the CBW Conventions-, Noordwijk, 5–6 April 2008.

[283] Editors' note. This was originally published as J. P. Perry Robinson. "Scientists and chemical weapons policies", a paper for the NATO Advanced Research Workshop on The Role of Independent Scientists in Assessing the Threat of WMD, Zagreb 13–15 November 2008. It was subsequently revised for publication, and is listed in Chap. 9 in item 466. Unless otherwise indicated, the remaining footnotes for this paper are Julian's originals. Apart from correcting obvious typos and renumbering headings and tables in keeping with the requirements of this book, we have retained Julian's formatting and spelling throughout the paper and its notes.

of novel risk: the changing utility of chemical weapons; changing science and technology; and CW arms control. Having particular regard to the disappointing outcome of the April 2008 conference to review the operation of the Chemical Weapons Convention [CWC] and to the impending shift in the primary focus of the CWC oversight organization from disarmament to non-proliferation, twelve different roles are suggested. These include work on better implementation of the 'general purpose criterion', on expansion of the CWC Schedules, and on convergence of the Biological and Chemical Weapons Conventions.

Key Words. Biological Weapons Convention, chemical warfare, Chemical Weapons Convention, disarmament, Harvard Sussex Draft Convention, non-proliferation, toxin.

7.17.2 Introduction

The Pugwash organization of scientists began in the mid-1950s mainly as reaction to the increasing dependence of security policy on nuclear weapons. As well, its attention to the formation of policy also for chemical and biological weapons has been active and continuous for more than 40 years.[284] In regard specifically to chemical weapons (CW) the present paper takes themes from that past work[285] to address the subject before the present workshop, 'The role of independent scientists in assessing the threat of WMD'.

I start by stating the obvious. 'Assessing the threat' means gauging the capabilities and intentions of people possibly determined to do us harm. It is people, not things, that have to be at the focus of such assessment: inanimate objects can intend nothing and, in themselves, can therefore pose no threat. So, it is not the threat of WMD that this workshop is actually about, but rather the threat of WMD when possessed by enemies. We can thus see at once that feasible roles in threat assessment for independent scientists are limited. Assessing the intentions of potential adversaries is the domain of intelligence and, by definition, people who are 'independent' are kept out of that domain, privy to little of the otherwise unavailable information essential to threat assessment. The reasons for this do not need reciting here.

---

[284] J. P. Perry Robinson, "The impact of Pugwash on the debates over chemical and biological weapons", in Allison L. C. de Cerreiio and Alexander Keynan (editors), *Scientific Cooperation, State Conflict: The Role of Scientists in Mitigating International Discord,* Annals of the New York Academy of Sciences, vol 866 (1999) pp 224–52.

[285] Drawn mainly from an overview by the present author, see J. P. Perry Robinson, "Difficulties facing the Chemicals Weapons Convention", *International Affairs* (London: RUA) vol 84 no 2, March 2008, pp 223–39.

However, threat without agency is 'risk',[286] and risk can most certainly inhere to WMD, chemical weapons among them. Without a contributing assessment of risk – of the damage that CW could cause and the probability of its occurrence – a CW threat could not be defined and would therefore remain unassessable. The role in CW risk assessment of scientists who are independent of salient institutional or other vested interests is, I believe, crucial. For if in fact they had no role in assessing pertinent risks, neither the credibility nor the reliability of the threat assessment could, in the peculiar field of CW, be great. I also think that professionals in threat assessment do their customers no service at all when, as sometimes seems to happen, they disregard the distinction between threat and risk. That way lies exclusion of the input that independent scientists can, should and must make.

Independent scientists have on occasion generated careful WMD risk assessments only to find their work ignored and sometimes even actively suppressed, A common mechanism of suppression is classification – imposing a security marking on the product, perhaps on grounds of 'proliferation sensitivity' or other such impressive-sounding abstraction, thereby depriving the product of the readership for which it had been intended.[287]

7.17.3 Emergent and Changing CW Risks

The utilities of chemical weapons today are not the same as those of the cold war, so, from the standpoint of NATO, the risks that CW now display are also different. And not only has CW utility been changing; so also, to great effect, has applicable technology. If CW policy-making is to remain sound, there is therefore a continuing need, as the context changes, for CW risk assessment. This need is accentuated by developments at the political level. Prominent here are the achievements this past decade of the Organization for the Prohibition of Chemical Weapons (the OPCW), which is the intergovernmental body headquartered in The Hague that is responsible for overseeing implementation of the 1993 Convention on the Prohibition of Chemical Weapons (the CWC). It is longer unthinkable or hopelessly idealistic to envisage a world free of chemical weapons. Of the 71,000-plus agent-tonnes of chemical

[286] This notion I owe to Kai-Bastian Ilchmann, whose doctoral dissertation, *Biothreat and Policy Pathways: Influences upon current bioterrorism policies in the UK,* is nearing completion at the University of Sussex.

[287] Examples include studies of bioterorrism done under EU auspices in 2002 (see EU Joint Research Centre and a consortium of the European Science and Technology Observatory, *Scientific and Technological Advances Relevant to Bioterrorism, and Their Possible Impact on Vulnerabilities in EU Society: A Prospective Study,* IPTS Seville, 29 November 2002) and during 2004–2007 (see M W Leeuw (Project Coordinator), Final Report on European Commission Framework 6 project 502476, *Assessment of the vulnerabilities of modern societies to terrorist acts employing radiological, biological or chemical agents with the view to assist in developing preventive and suppressive crisis management strategies* (ASSRBCVUL), TNO Rijswijk, 4 May 2007.)

weapons declared to the OPCW by CWC states parties,[288] a good 40 percent, plus most of the declared factories,[289] have by now been eliminated under international supervision. So – **utility, technology** and **arms control:** it is important to examine these factors more closely for their likely effects on CW risk. From this should follow a fuller appreciation of roles for independent scientists in CW policy formation.

7.17.3.1 New Uses for Chemical Weapons

We need no reminding that the very nature of warfare has been changing in recent years. Has this meant that chemical weapons are now more suited or less suited to the types of armed conflict for which NATO is obliged to prepare? Either way, CW risks will have changed, necessitating new assessment.

Something of a parallel can be seen in the shift from 'massive retaliation' to 'limited war' doctrine that occurred in the West towards the end of the 1950s. This shift elevated the status of chemical (and biological) weapons in Western military thinking, causing new utilities to be seen for them, especially in Third World settings. The UK, for example, started moving towards a secret CW rearmament programme.[290]

---

[288] Article III.l(a) of the CWC requires each state party to declare to the OPCW detailed information about any chemical weapons it possessed or that otherwise existed on its territory within a month of the treaty having entered into force. Six of the now-184 states parties made such declarations: Albania, India, Libya, Russia, South Korea and the USA. Any state party is entitled to insist that the OPCW keeps some or all of the information that it declares out of the public domain, and some parties have done so, a consequence being that a dependable country-by-country breakdown of the total declared stocks cannot be compiled from open sources. Here, however, are some plausible-seeming estimates made from the best available data: Russia, 56 percent by weight of all the declared chemicals; USA, 39 percent; Libya, 2.0 percent (98. percent of it comprising precursors, however, not actual CW agent); South Korea, 1.5 percent; India, 1.5 percent; and Albania, 0.02 percent. An idea of the military significance of those holdings follows from the total quantity of mustard and nerve gases used by Iraq during its war with Iran in the 1980s. That quantity appears to have been around 2540 agent-tonnes, which is equivalent to about 3.6 percent of all the stocks declared to the OPCW.

[289] Article III.1(a) of the CWC requires each state party to declare any chemical weapons production facility it had possessed or controlled at any time since 1 January 1946 having a production capacity greater than one tonne per year. Twelve states parties have made such declarations: Bosnia & Herzegovina, China, France, India, Iran, Japan, Libya, Russia, Serbia & Montenegro, South Korea, the UK and the USA. The Japanese declaration concerned solely the facility built by Aum Shinrikyo, the cult that was responsible for the terrorist releases of sarin nerve-gas in Matsumoto in June 1994 and in the Tokyo subway in March 1995. Evidently China, France, Iran, the UK and the former Yugoslav republics had destroyed their stocks before the CWC entered into force, as they would otherwise have had to declare stocks under CWC Art III.1(a). Of the states that are not parties to the CWC as of November 2008, those understood to be current CW possessors are Egypt, Israel, North Korea and Syria, with Iraq as a former possessor.

[290] Which, however, it effectively abandoned in 1968 (see Gradon Carter and Brian Balmer, "Chemical and biological warfare and defence, 1945–90", in R. Bud and P. Gummett, editors, *Cold War, Hot Science: Applied Research in Britain's Defence Laboratories, 1945–1990,* Harwood Academic Publishers, 1999, pp 295–338 at 298–99.)

Today, a new type of organized violence is taking the place of those confrontations between highly disciplined and technologically advanced armed forces that characterized the later cold war. Conflicts these past two decades in the Balkans, the Caucasus, the horn of Africa, Rwanda, Liberia, Sierra Leone, Angola, Sri Lanka, Afghanistan and post-invasion Iraq have eroded formerly clear distinctions between war, organized crime and large-scale violation of human rights. These new wars are fought by seeking political control through the displacement, or worse, of civilian populations and through the sowing of fear and hatred.[291] Chemical weapons, with their area effectiveness and terrorizing potential, seem particularly amenable to such objectives, so they may conceivably have a greater affinity to the new wars than they did to the old, and, notwithstanding the CWC, the weapons could therefore have an expanding future. It is a future that seems already to have begun: instances of 'new' chemical warfare include episodes in Iraqi Kurdistan, in southern Africa, in Bosnia and perhaps in Chechnya,[292] and further chemical warfare allegations have been emanating from Sudanese, Israeli, Palestinian, Baluchi, Lebanese, US/Iraqi and, once again, Sri Lankan sources.[293] The emotiveness of CW may mean that, in fact, some or all of the allegations are mistaken or else are false, put out as 'black' propaganda. Investigation by independent scientists has contributed to an understanding of some of these allegations, notably in Kurdistan,[294] Bosnia,[295] and Israel,[296] as it did, too, for the 'Yellow Rain' toxin-warfare allegations in southeast Asia during the late 1970s and early 1980s.[297]

Another area in which independent scientists can contribute to assessment of the risks in emergent new utility is the match between offensive properties of accessible toxic chemicals and the purposes to which such chemicals might be put under realistic scenarios of possible future conflict, having full regard to the practical problems

[291] Mary Kaldor, *New & Old Wars: Organized Violence in a Global Era* (second edition, Cambridge: Polity, 2006).

[292] UK House of Commons (session 1999–2000) Foreign Affairs Committee *Eighth Report: Weapons of Mass Destruction,* London: Stationery Office, 25 July 2000, pp/ 203–206, memorandum dated 17 February 2000 submitted by J. P. Perry Robinson.

[293] 'News Chronology', *The CBW Conventions Bulletin* no 66 (December 2004) through no 78 (February 2008), *passim.*

[294] R. M. Black, R. J. Clarke, R. W. Read and M. T. J. Reid, "Application of gas chromatography – mass spectrometry and gas chromatography – tandem mass spectrometry to the analysis of chemical warfare samples, found to contain residues of the nerve agent sarin, sulphur mustard and their degradation products", *Journal of Chromatography-A,* vol 662 (1994) pp 301–21.

[295] Alastair Hay, "Surviving the impossible: the long march from Srebrenica. An investigation of the possible use of chemical warfare agents", *Medicine, Conflict and Survival* vol 14 no 2 pp 38–73, 1998.

[296] Alastair Hay, Rita Giacaman, Ramzi Sansur and Steven Rose, *Medicine, Conflict and Survival,* vol 22 no 4 (October-December 2006) pp 283–291, 'Skin injuries caused by new riot control agent used against civilians on the West Bank'.

[297] Matthew S. Meselson and Julian Perry Robinson, "The Yellow Rain affair: lessons from a discredited allegation", in Anne L. Clunan, Peter R. Lavoy and Susan B. Martin (editors), *Terrorism, War, or Disease? Unravelling the Use of Biological Weapons,* Stanford University Press, 2008, pp 72–96.

of disseminating chemicals against targets, and to the factors that determine target uptake. During the cold war, the primary criterion of the CW utility of a toxic chemical was whether the chemical was aggressive enough in its effects to make weapons disseminating it competitive, in quantitative casualty-producing terms or other such measures of tactical efficacy, with the modem conventional weapons that might otherwise be preferred. Not a great many such toxicants exist.[298] In 'new wars', however, it is not so much relative aggressivity that would usually determine the utility and therefore the value of chemical weapons but rather such other factors as accessibility or availability of the weapons and their terrorizing potential. A wide variety of toxic industrial chemicals and other chemicals not hitherto regarded as CW agents might thus find application in new-war contexts, as, most recently, chlorine – that long-obsolete, by old-war standards, killer gas that was briefly weaponized during the First World War – has done in Iraq. Relative aggressivity may still be important, but the more central matter for assessment may be psychomotive power,[299] for example, rather than casualty power. A relevant consideration is whether such new-war chemical agents are or are not captured by the Schedules of the Chemical Weapons Convention. Those of them that lie outside the Schedules would of course still fall within the general purpose criterion that sets the comprehensive scope of the Convention, but in that case the practical and administrative difficulties of enforcing the prohibitions could mean that the CWC did not significantly reduce the risk.

Quite another type of novel utility now becoming manifest is the emerging role of chemical weapons, not in the hands of terrorists or other new-war aggressors as just described, but for purposes of counter-terrorism. This utility has demonstrably become a stimulus to rich-country uneasiness with the CWC.[300] It is rooted in past counterinsurgency applications of toxic chemicals, which reach back through the Vietnam War to British, French, Italian and Spanish use of toxic chemicals in colonial situations – a utility that the CWC was intended to suppress. Its re-emergence in counterterrorist guise is to be seen in the proliferation of weapons based on Agent CR, evident each year in that part of the OPCW Annual Report addressing the declarations of 'riot control agents' required under CWC Article III.I(e), for the extreme aggressiveness and other properties of CR have caused it to be widely rejected as suited to civil police use. Police forces in the UK, for example, are equipped either with Agent CS or with PAVA for law-enforcement use, and, although the UK has also declared Agent CR to the OPCW as a 'riot control agent', it has issued the agent only to its military forces, for counterterrorism. The growing counterterrorist

---

298 J. P. Perry Robinson, "The chemical industry and chemical warfare disarmament: Categorizing chemicals for the purposes of the projected Chemical Weapons Convention", *SIPRI Chemical & Biological Warfare Studies* no 4 (1986) pp 55–104.

299 M. W. Leeuw (Project Coordinator), Final Report on European Commission Framework 6 project 502476, *Assessment of the vulnerabilities of modern societies to terrorist acts employing radiological, biological or chemical agents with the view to assist in developing preventive and suppressive crisis management strategies* (ASSRBCVUL), TNO Rijswijk, 4 May 2007.

300 Robert J. Elder, Jr, foreword to N. T. Whitbred IV, "Offensive Use of Chemical Technologies by US Special Operations Forces in the Global War on Terrorism: The Nonlethal Option", *The Maxwell Papers* [Maxwell Air Force Base, AL: Air War College] no 37, July 2006, pp iii–iv.

utility of chemical weapons is further evident in the vigorous advocacy to be heard in some quarters for the arming of counterterrorist forces with more advanced types of 'non lethal' toxic weapon. The readiness with which the US Marine Corps has taken to toxin weapons of this type – devices disseminating Agent OC – seems indication of a trend. So, perhaps, is the absence of any serious criticism of the Russian government for having authorized use of toxic chemicals other than riot-control agents by the *spetsnaz* forces that, on 26 October 2002, liberated 634 of the people taken hostage by Chechen separatists in a Moscow theatre.[301] All but five of the other 129 hostages were killed by the toxicant used, which is said to have been an agent "based on derivatives of fentanyl" that had been developed by USSR special services.[302] Comparable in some respects, if very different in others, is a counterterrorist utility for toxic weapons that Israel has demonstrated in its espousal, following the Munich Olympics outrage in August 1972, of assassination as a major tool in counterterrorism.[303]

Exacerbating the new-utility risk is the increasing dependence of some countries, not only on state forces for law enforcement including counterterrorism, but also on the private military contractors who have been providing security services at local, national and even global level. The potential value and, therefore, take-up of 'non lethal' chemical weapons by such contractors, who may be regulated less stringently than military or police forces, has become a matter of expressed concern.[304]

It is unfortunate that the interest in counterterrorist CW should be growing just as a CW-free world is coming within reach. Risk as well as conceivable benefit attaches to counterterrorist CW. So both now need assessing as a matter of urgency, independently of any interests vested in 'non lethal' weapons technology.

7.17.3.2 Change in Science and Technology

Of the ways in which scientific and technological (S&T) change could affect CW risk, one may think especially of knowledge newly gained in the life sciences suggesting novel modes of attack that could be the basis for militarily or politically attractive new forms of weapon. For example, if a new molecule is discovered that can exert novel disabling effects on the human body at uncommonly low dosage, attempts to

---

301 For a documented account of international reactions to this episode, see J. P. Perry Robinson, "Non Lethal Warfare and the Chemical Weapons Convention", a submission to the OPCW Open-Ended Working Group on Preparations for the Second CWC Review Conference, 24 October 2007

302 A recent publication in the medical literature identifies the agent used – without, however, citing any authority for the information – as something called "Kolokol-1 [...] containing carfentanil", see James Geoghegan and Jeffrey L. Tong, "Chemical warfare agents", *Continuing Education in Anaesthesia, Critical Care & Pain* vol 6 no 6, (December 2006) pp 230–34.

303 Aaron J Klein, *Striking Back: The 1972 Munich Olympics Massacre and Israel's Deadly Response* (New York: Random House, 2005) pp 104–11 and 205–8.

304 Alan Pearson, "Incapacitating biochemical weapons: science, technology, and policy for the 21st Century", *Nonproliferation Review* 13(2):151–88, July 2006.

weaponize it may well ensue. Albert Hofmann's discovery of LSD in 1943 is a case in point, although half a decade elapsed before weaponeers took notice. Of course, many considerations other than novelty or intensity of effect determine the usefulness of a new weapon, so the new science is not itself the source of risk that is here suggested. But it would be a step towards it; and many such can be envisaged.[305] Again, it is the general purpose criterion of the CWC that is the international safeguard. But it is a safeguard only if it can be activated, and this requires continual monitoring of scientific and technological change for any new development that might challenge the CWC regime. This is a task that cannot reliably be left to security authorities or to international civil servants alone, simply because their surveillance of new science is unlikely to be sufficient. The scientific community at large must also be involved.[306] Such a role for independent scientists is explicitly recognised in the CW policies of some countries. In the UK, for instance, the National Authority established in accordance with CWC Article VII.4 has an independent Advisory Committee mandated to "advise on technical developments which may have potential application in chemical weapons."[307]

Technical developments of that kind may originate in causes that have nothing to do with CW. Whether new or old, a particular piece of scientific knowledge may find more than one practical application, meaning that technologies embodying the knowledge may be multipurpose. The chemical dimethyl methylphosphonate is a valuable flame retardant much used to reduce the flammability of garments or of insulating materials in buildings, yet it can also be made into nerve gas. Thiodiglycol is a solvent used in ballpoint-pen inks and in the dyeing of textiles, but it is also just one chemical step short of mustard gas. Chlorobenzalmalononitrile is a sensory irritant widely used by civil police forces in CS Spray and other such devices for law enforcement, but it can also be used for offensive battlefield purposes as payload for artillery shell or aircraft bombs. All three chemicals are thus manifestations of 'dual use' technology of which one application is in chemical weaponry and the other for valued civil purposes. Controlling dual use technology is the primary function of CW anti-proliferation and arms-control measures, but these can work only if the duality is properly recognised. Can this be achieved without the involvement of independent scientists?

---

[305] A particularly rich recent source of information on advances in technology that may be applicable to chemical as well as biological weapons is the Lemon-Relman report of 2006 from the US National Academies. See Institute of Medicine and National Research Council of the US National Academies, Committee on Advances in Technology and the Prevention of Their Application to Next Generation Biowarfare Threats (Co-Chairs: Stanley M. Lemon and David A. Relman), *Globalization, Biosecurity, and the Future of the Life Sciences,* Washington, DC: The National Academies Press, 2006.

[306] Alastair Hay, "Surviving the impossible: the long march from Srebrenica. An investigation of the possible use of chemical warfare agents", *Medicine, Conflict and Survival* vol 14 no 2 pp 38–73, 1998.

[307] http://www.berr.gov.uk/whatwedo/energy/non-proliferation/cbw/national-authority/naac/terms-of-reference/page40779.html, accessed 5 November 2008.

To conclude this brief treatment of S&T change, it is instructive to look at how NATO has approached such risk-assessment problems in the past. One way has been to assemble working parties of scientific and operational experts and ask them to extrapolate into the future from a foundation of present knowledge and experience. An early instance was the precursor of the subsequent Long-Term Scientific Studies, namely the Von Karman Committee that was tasked by the NATO Military Committee in 1960 to "develop an estimate of the possible and probable scientific progress to be expected in the next decade". The Committee's report in 1961 is described on the website of the NATO Research & Technology Organisation[308] as "the first multinational attempt to estimate the impact of scientific and technological advances on military capability". It had 14 working groups that met in March 1961, one of which addressed Chemical, Biological and Radiological Defence. Asked to forecast scientific trends to the 1970–75 period and to predict their practical effect on CBR defence, this working group chose to envisage different categories of possible CBR attack against which defence was required, the forecasting exercise then proceeding through eight areas of defence requirement: (1) detection, (2) active neutralization of attack, (3) protective equipments and materials, (4) palliatives, (5) remedial measures, (6) surveillance, (7) large scale decontamination, and (8) means of agent identification. Each area was considered for each of three categories of attack, which, for the CW part of the study,[309] were these:

> *Type A*: "Direct attack on small areas (up to 1 Km^2, say) with respiratory agents. Rapid coverage of the whole area with vapour or quickly-evolved aerosols would be aimed at, in order to achieve maximum surprise against the enemy in the area (surprise is not essential, however, if the enemy is not provided with respirators)."
>
> *Type B*: "Direct attack with agents acting through the skin (areas up to the order of 100 Km^2)."
>
> *Type C*: "Off-target attack (areas up to the order of 100,000 Km^2) with respiratory agents of high toxicity or effectiveness (at least an order higher than the present nerve agents). Surprise against an enemy possessing respirators would be achieved by this method, because the attack is not observed and the agent arrives on target in very small but effective concentrations, which are not detected by the senses nor, quickly enough, by artificial means."

That categorization of attack modes would have been rooted, if not in the actual CWuse doctrine or existing CW capability of possessor states, then at least in study of feasibility. Type C is the category that surprises some people today, for there is a tendency to portray CW as of only limited effect, hardly even WMD at all. Was it unrealistic, therefore, of the working group to anticipate the possibility of Type C attacks during the coming 10 to 15 years? The answer is probably not, for in 1968 chemical weapons comprising dry-agent spray tanks mounted on *Phantom* strike aircraft and charged with Agent PG were tested over caged monkeys and other animals at sea off Eniwetok Atoll in the Marshall Islands; and the test data indicated

---

308 http://www.rto.nato.int, accessed 31 October 2008.

309 UK Ministry of Defence Advisory Council on Scientific Research and Technical Development, paper no SAC 1928, 11 February 1969, *Future developments in chemical warfare: Extract from the Report of the van Karman Committee 1961,* in The National Archive, Kew, file WO 195/16864.

a 30 percent casualty rate over 2,400 Km^2 per weapon.[310] Agent PG (previously known as UC) is staphylococcal enterotoxin B, an incapacitating agent whose median effective inhalation dose in man has been estimated to be at least three orders of magnitude smaller than the corresponding median lethal dose of nerve-agents such as sarin or Agent VX.[311] Other such superpotent substances falling within the scope of the CWC include botulinal toxins such as the preparation known as Agent XR that was weaponized by the United States as part of its erstwhile chemical/biological-weapons programme, but there is no information in the open domain about their efficacy in Type C attacks. That being the case, should not independent scientists now revisit that 47-year-old categorization in order to form a view on the relevance of Type C chemical weapons to today's risk assessments?

Perhaps Agents PG and XR have by now been superseded as the best exemplars of the category, but, even if they have not, any continuing possibility of Type C weapons would seem to constitute a risk that is not being managed with all the available tools, including those that have been furnished by the CWC.

7.17.3.3 CW Arms Control

The CWC is the cornerstone of today's array of international arms-control measures applicable to CW, alongside the 1925 Geneva Protocol and such plurilateral initiatives as the Australia Group, as well as UN Security Council resolution 1540 (2004) and its successors. Taken together, these measures have made CW increasingly difficult to acquire, develop, produce, retain or use. One may think, therefore, that CW risks have, in general, been reduced by CW arms control. Yet by the same token, shortcomings in the CWC regime may themselves constitute novel forms of CW risk and therefore require careful assessment. Here is another possible role for independent scientists.

What is the most serious risk of this type has already been alluded to: shortcomings in applying the 'general purpose criterion'. The scope of the CWC's prohibitions is impressively comprehensive, for, as well as embracing the familiar CW agents and production intermediates listed in the CWC Schedules, it extends to dual use chemicals without at the same time restricting their civil application, as well as to other toxic chemicals that, for reasons of history, novelty or secrecy, remain unlisted in the Schedules. All of this has been made possible through use of the general purpose criterion, whereby all toxic chemicals, and precursors from which they can be made, are subject to the treaty's provisions "except where intended for purposes not prohibited under this Convention, as long as the types and quantities are consistent with such purposes", as CWC Art.II.1(a) puts it. But, obvious though the benefits of this comprehensiveness are, two great obstacles are being allowed to get in its way.

---

[310] USA, Department of the Army, Fort Douglas, *DTC Test 68-50,* two vols, Deseret Test Center, March 1969, as reported in: Ed Regis, *The Biology of Doom,* New York: Henry Holt, 1999, pp 204–206.

[311] F. R. Sidell, E. T. Takafuji and D. R. Franz (editors), *Medical Aspects of Chemical and Biological Warfare,* a volume in Part I of the *Textbook of Military Medicine* series, [US] Department of the Army, Office of the Surgeon General, 1997, at pages 141 and 622.

One is that, as a matter of practical administration, it is not at all easy to devise and execute effective policies for applying the general purpose criterion adequately, and only a small minority of CWC states parties have yet managed to implement it into their domestic law. The other obstacle is the evident reluctance of some CWC states parties, notably India, to acknowledge that the scope of the treaty is not set just by the Schedules. The final document of the recent Second CWC Review Conference (7–18 April 2008) gave some recognition to the importance of the criterion but failed to envisage ways out of the present predicament. It is not at all clear why those states parties that do so have obstructed proper implementation of the criterion. So central is it to the effective working of the CWC that treaty-wrecking objectives have to be suspected.

The Second CWC Review was a quinquennial opportunity to look further into the future than the OPCW otherwise has time for: opportunity, then, to initiate any necessary repair work and otherwise prepare the ground for actions to countervail emergent risks or other adverse consequences of technical, military or political change. One such change is approaching fast, but, as with the predicament of the general purpose criterion, states parties, not only in the NAM, seemed unconcerned during the Review Conference. On 29 April 2012, according to the CWC and subsequent agreements about implementation, the last of the declared chemical weapons are due to have been destroyed and all the declared CW factories eliminated. That deadline is likely to be missed, but the states parties responsible (Russia and/or the USA) have convincingly demonstrated their continuing commitment, nonetheless, to the disarmament goal. If not in 2012, that goal will certainly be achieved some years later on. When it has been achieved, the OPCW will have discharged what has long been thought of as its primary function, that of overseeing CW disarmament. The Organization will then be set to enter a future that surely needs to be mapped out well ahead of time, since the bulk of its present activities, whether conducted by the OPCW inspectorate or by other headquarters staff, are strongly biased towards verification of stockpile-destruction and capability-elimination. There will inevitably be a tendency to treat the transition as opportunity for cost-cutting and for downsizing the Organization's staff. Yet what will then be the OPCW's primary function will become even more important than it is now: verification of non-production – ascertaining or otherwise building confidence that rearmament is not happening through, especially, abuse of dual-use technology. This will require, above all, effective application of the general purpose criterion, as well as continuation and enhancement of the present industry controls.

If the transition is to be effective in sustaining forward momentum towards a CW free world and not simply regarded as a mere sloughing-off of old tasks, it will be prudent for friends of the CWC to seek an administrative or procedural innovation that would mark the transition, especially one that would have the effect of strengthening the norm against weaponization of toxicity that underpins the CWC. One area for such innovation is machinery for better ensuring compliance through stronger sanctions on non-compliance. Of course sovereign states do not readily commit themselves in advance to punitive or other reactive measures against one another. A possible way forward, then, is a mechanism for holding, not states, but *individuals* personally accountable for contravention of the CWC norm. Such an individual-responsibility approach is already discernible in international affairs. Individuals, or at least sub-state entities, in their identity as persons involved in WMD proliferation, are the main target of UN Security Council resolution 1540 (2004). In Iraq and in the Netherlands, individuals have been indicted, tried and convicted for CW-related crimes.[312] From among the several categories of stakeholder in the development of biotechnology, individual scientists are being made responsible for impeding access to 'dual use' materials thought helpful to bioterrorists. All of this is surely strengthening the norm and could be built upon. Yet the defining feature of a universal norm is not that it should dictate how particular individuals or groups behave, but rather how *everyone* should behave, high or low, head of state or simple citizen. All of us are responsible for the continued wellbeing of the norms against the weaponization of toxicity and other forms of disease, and the onus is therefore upon us all, individually, to uphold them. It is all very well saying that scientists need to think more about conceivable end-applications of their work and be taught about the international treaties that are meant to suppress chemical and biological weapons (CBW). Those treaties – the CWC and the 1972 Biological & Toxin Weapons Convention (the BWC) – codify the norms, but like the 1925 Geneva Protocol that outlaws use of CBW in war, they place their primary constraints on the behaviour of states, not of individuals. If the norm is to be strengthened, the overall regime must surely be developed so that its relevance to the individual as well as to the state becomes clear to all. This needs doing in step with the current moves towards enforceable codes of conduct for scientists, for if it is not those codes may come to be seen as ill-considered acts of discrimination, unpopular and ignored. A more compelling way of inserting the sanction of individual accountability into the regime is needed.

Now that international criminal law is emerging from the constraints of the Cold War, just such a mechanism has become possible. It is a very simple one: a new international convention that would confer on national courts jurisdiction over individuals present in their national territory, regardless of their nationality or official position, who order, direct or knowingly lend substantial assistance to the acquisition, production or use of CBW anywhere. Such a convention might take various forms. One possible text, prepared with advice from an international group of eminent jurists, is the Harvard Sussex Draft Convention.[313] This has already attracted favourable notice from a number of governments and now constitutes the basis for a possible new initiative in the WMD area, perhaps by the European Union. Taken further forward, it could help secure the transition of the OPCW.

---

[312] See Lisa Tabassi and Erwin van der Borght, "Chemical Warfare as Genocide and Crimes against Humanity", *The CBW Conventions Bulletin* no 74 (December 2006) pp 36–44, and, Lisa Tabassi, "The nexus between the OPCW and the International Criminal Court", *The CBW Conventions Bulletin* no 75 (March 2007), pp 1 & 7–12.

[313] Matthew Meselson and Julian Robinson, "A Draft Convention to Prohibit Biological and Chemical Weapons under International Law" in R. Yepes-Enriquez & L. Tabassi (eds), *Treaty Enforcement and International Cooperation in Criminal Matters* (The Hague: TMC Asser, 2002) pp 457–69.

One way or another, the basic norms on which the CWC and BWC both rest need reinforcing, given the climate of potentially adverse change within which they function. Let us remind ourselves of the reasons why CBW continue to carry grave implications for security – international, national and human security. It is not so much because some CBW are 'weapons of mass destruction' within the UN meaning of that expression. It is also because of another distinctive feature of CBW: their ability not necessarily to kill us in large numbers but rather to disable, even *change*, us. These weapons can attack life just as other weapons can, being capable of killing their victims no less dead than can heat, explosive blast, bullets or bayonets; but they work by targeting and then disrupting particular processes that contribute to life, which other weapons cannot do save by accident, not design. The nerve gases, for example, target nerve-signal transmission; the blood gases, cellular respiration mediated through the blood. So the idea started to take hold in the 1950s that other physiological systems might be disrupted, including ones governing the psyche or the will or ability to function properly, thereby furnishing disabling chemical weapons, perhaps in great variety. The old pre-CWC advocates of chemical weapons propagated the notion, which in some quarters persists to this day, that the effects of the weapons might therefore be *tailored* to suit different tactical or strategic purposes.[314] However, beyond 'tear gas', the disabling skin burns that mustard gas can cause, and the uncertain promise of the early psychotropic 'incapacitating agents', this did not begin to become a real option until the pace of change in applicable technology had accelerated during the latter part of the last century, around the time of negotiation of the 1972 BWC and its CW follow-on. It was becoming increasingly plain to those who followed such things that advances in the life sciences, coupled with technologies that allow the analysis and construction of complex biologically active molecules, could eventually make it possible to design a chemical that would interfere with *any* life process that could be understood in molecular terms. It is exactly that form of increased understanding that today's increasing convergence of biology and chemistry is providing. Because of convergence we may eventually be able to manipulate the processes of development, inheritance, reproduction, locomotion, sensation, cognition and any other process that keeps us working as normal human beings. The genes of otherwise harmless microorganisms might be modified to express particular harmful molecules when they infect. The potential is there, inasmuch as it is not manifest already, for inducing many different forms of malfunction, maybe even ones that discriminate between ethnic groups. It is this potential for changing human beings and doing so in pursuit of who-knows-what aggressive strategy of subjugation or coercion that makes CW uniquely menacing:[315] weapons, in the extreme, of living

[314] For example, Brig-Gen J. H. Rothschild, *Tomorrow's Weapons: Chemical and Biological*, New York: McGraw-Hill, 1964.

[315] See Matthew Meselson and Julian Perry Robinson, "Chemical warfare and chemical disarmament", *Scientific American* vol 242 no 4 (April 1980) pp 38–47 at p 47; J. P. Perry Robinson, "Chemical, biological and radiological warfare: futures from the past", invited submission to the Independent Commission on Disarmament and Security Issues [Chair: Olof Palme], September 1981; Matthew Meselson, "Averting the hostile exploitation of biotechnology", *The CBW Conventions Bulletin* no 48 (June 2000) pp 16–19.

death. The existence of the BWC is testimony to our sense that we need protection against that contingency. The CWC, in principle, can provide it and, for this reason, the CWC and its associated regime must be kept in the best shape possible.

Other students of CW history may perhaps reach different conclusions about the object and purpose of the CWC, and therefore different assessments of the risk inherent in convergence. What the foregoing conception does is to point towards molecular dangers from within broad families of toxic chemicals that, for the most part, lie outside the CWC Schedules and therefore outside the routine control procedures run by the OPCW. At first glance (and much more study is needed) these families comprise certain bioregulators and other bioactive chemicals of biological origin, as well as their synthetic analogues and also certain conceivable products of biotechnological processes. This whole category roughly corresponds to what the BWC means by 'toxins'.[316] It includes the CW agents that were mentioned earlier as exemplars of Type C chemical weapons, namely Agents PG and XR that, so conspicuously, are not listed in the CWC Schedules, despite one CWC state party having maintained production capacity for each of them that, in 1970, exceeded one tonne per year.[317]

The convergence of biology and chemistry is an influence that may be moving such substances towards weaponization. May not a good way of reducing the risk of that happening be to promote convergence of the BWC and the CWC? Here is a policy question demanding the input of independent scientists.

7.17.4 Conclusion

Nongovernmental scientists are, by and large, reluctant to engage themselves in public discourse on topics outside their immediate expertise. Most seem disinclined to accept that politicians, decision-makers and opinion formers are often appallingly ignorant of even elementary scientific principles, let alone ones that shape technical aspects of CW risk. Pugwash grew, in part because of this, as a way of stimulating greater participation by scientists in the shaping of WMD policies.

When it comes to arms control, all of us, not only scientists, need reminding that treaties such as the CWC are engagements, not between governments, but between

---

[316] The BWC meaning of 'toxin', which extends to substances that are not called 'toxins' in other specialist communities, is set out in an international text published by the World Health Organization – see, World Health Organization, *Public health response to biological and chemical weapons: WHO guidance* (Geneva: WHO, 2004), pp 214–16.

[317] Interdepartmental Political-Military Group, report to the [US] National Security Council, *US Policy on Toxins* dated. 30 January 1970 (see US Department of State, Office of the Historian, *Foreign Relations of the United States: Nixon-Ford Administrations,* volume E-2: *Documents on Arms Control and Nonproliferation, 1969–1972* (2007), document 177). This document, which records the "existing production capability" for PG as 600 pounds per month and for XR as 280 lbs/mo, did not enter the public domain until well after the USA had made its initial declaration to the OPCW.

states parties. Governments may represent states parties in the fora of the OPCW, but organs of civil society are also elements of those same states, no less responsible for proper implementation of the treaty. If independent scientists can contribute, it would follow that they have an obligation to do so.

The purpose of this paper has been to identify ways in which independent scientists can in fact contribute to implementation of the CWC and, more generally, to the formation of public policies for reducing the risks posed by the actual or possible existence of chemical weapons. A dozen such ways have been identified in earlier parts of this paper. In summary, and in the order identified, they are these:

- Participating in the investigation of allegations of CW use.
- Identifying non-traditional CW agents or precursors for possible inclusion in the CWC Schedules.
- Assessing risks and benefits of counterterrorist CW.
- Monitoring S&T change for technical developments having potential application in chemical weapons.
- Identifying abusable dualities in applications of chemicals or of chemical and other technologies.
- Helping to determine whether the 'Type C' CW attack-mode should still be regarded as plausible, needing recognition in the CWC Schedules.
- Helping to identify, and assess as risks, shortcomings in the CWC regime.
- Contributing to a comprehensive investigation of the problems presented by the CWC's 'general purpose criterion', and of remedies to them.
- Through international discourse with other scientists, advancing understanding of why there is such international variation in perceptions of the risk-reducing capacities of the general purpose criterion.
- Contributing to efforts to build on the opportunities presented by the impending transition of the OPCW's primary function from verifying disarmament to verifying non-armament.
- Helping in efforts to strengthen the norm underpinning the CWC by inserting measures for individual (as well as state) accountability into the overall CWC regime.
- Exploring possible ways for promoting convergence between the CWC and the BWC.

These possibilities can be seen as challenges to independent scientists, both inside and outside Pugwash, who are concerned about chemical weapons and about other WMD.

7.18 Presentation: "Does Arms Control Have a Future?"[318]

Editors' note. From 2012, there were reports that the Assad regime was using chemical weapons in the Syrian civil war in small-scale attacks. Then, in August 2013, an attack in Ghouta resulted in a significantly higher number of casualties, which achieved widespread attention on social media and elsewhere, triggering the UN Secretary-General to investigate. The Ghouta attack was not the last use of chemical weapons in Syria, and international efforts are ongoing to establish what happened in multiple incidents, who was responsible, and to bring justice and accountability to those involved.

In the last years of his life, Julian worked to track, document and understand the use of chemical weapons in Syria, and monitored the international responses. In doing so, he built a set of resources useful to policy-makers, the global chemical and biological warfare (CBW) governance regime, and international work focussed on holding perpetrators to account. As well as this research, Julian actively supported citizens groups in Syria and the UK, working to build links between eyewitnesses, victims and experts that could help their collective efforts to document and analyse evidence needed to understand and demonstrate the nature of attacks, as described in Rim Turkmani's tribute reproduced in Chap. 3 of this book.

These speaking notes set out Julian's views about the successes of the CBW arms control regimes and his conceptualizations of the new utilities of chemical weapons observable in what he calls 'Syrian-mode chemical warfare'. He argues that excessive government secrecy undermines the Chemical Weapons Convention (CWC). The notes were prepared for a closed seminar of government and non-government experts arranged by the UK's Foreign and Commonwealth Office's Arms Control and Disarmament Research Unit (ACDRU).

I was asked to talk for 25 or 30 minutes about "Chemical weapons: past, present and future?". I particularly liked that question mark, and also the licence to roam rather freely. I plan to bring in biological weapons as these can sometimes be seen as a special form of chemical weapon, though I shan't actually say much about them. And I would have liked to speak about the deplorable absence of a comprehensive and dependable history of CBW, but there isn't time enough for that either.

I shall begin simply by asserting that CBW – by which I mean chemical and biological weapons – are an important arms-control success story. I note, above all, the 1972 Biological Weapons Convention (the BWC), the inauguration of the Australia Group in 1985, and the 1993 Chemical Weapons Convention (the CWC). The Foreign Office Arms Control and Disarmament Research Unit – ACDRU – was born at just the time the work that led to all of that was beginning. The work had been set going in 1963 from the Science & Technology side of the United States Arms Control and Disarmament Agency (ACDA), itself only recently born, with which ACDRU swiftly developed what looks from archived state papers in both America and Britain to have been a most constructive relationship. Though ACDA may have led the effort, at least chronologically, I believe it is true to say that, without

[318] Editors' note. These previously unpublished speaking notes are listed in Chap. 9 as item 493, and have the full title "Presentation on CBW by Julian Perry Robinson For ACDRU 50th Anniversary Seminar 'Does Arms Control Have a Future?', (28 October 2015)". Unless otherwise indicated, the remaining footnotes for this paper are Julian's originals. Apart from correcting obvious typos and renumbering headings and tables in keeping with the requirements of this book, we have retained Julian's formatting and spelling throughout the paper and its notes.

ACDRU, there would have been no BWC. An unsung hero of this was Dr Deryck Viney, whom John Walker has identified in his book[319] as the author of the 1966 first draft of ACDRU's pathbreaking concept paper entitled *Arms Control Implications of Chemical and Biological Warfare*. This paper long lay hidden from the reading public behind a TOP SECRET security marking. It can now be read in The National Archives.[320] ACDRU had a somewhat less formative role for the CWC, though my impression as an outsider is that, without ACDRU's very many inputs, the CWC would have been a poor weak thing. As for the Australia Group, that was – is – an altogether different kettle of fish: exclusive not inclusive, and the players who set it going as an informal export-control club coming, not from the arms-control community in defence and foreign ministries, but instead, as I understand it, from the intelligence world and ministries of trade, industry and commerce. There were overlaps of course, still are, between them and the arms-control folk proper.

Am I right to regard the BWC and the CWC as lasting successes, indicative of the power and promising future of arms control? If that is to be believed, I think there are two major counter-indicative questions that we have to confront:

- First, the Biological Weapons Convention is still not much more than a declaration, despite great enterprise by the treaty's friends these past 40 years to give it teeth. Should it not by now have got them? More fundamentally, I would say, we still need to determine whether the BWC actually needs teeth.
- Second, the Chemical Weapons Convention: how well is it really bearing up against the test of Syria?

I had been intending to offer an historical take on those two questions. My argument would rest on a view of CBW history that sees the status of CBW at any one moment as the product of an interacting triad of drivers. The first of the three drivers is the utility seen for the weapons under the circumstances of the time, of course including new science & technology and changing patterns of international relations and of warfare.[321] The second driver is the evolving state of protection against the weapons. This is crucial to their status because the way the weapons work means that, in principle, a far greater degree of protection is available against them than against any other category of warfare weapon, even to the point of negating utilities. The third driver is the prevailing strength of the non-military and non-technical constraints upon use of CBW, this reflecting the near-unique taboo that poison weapons have evidently attracted, a taboo that is codified in the 1925 Geneva Protocol outlawing (first) use of CBW, and now in the 1972 BWC and the 1993 CWC.

---

[319] John R. Walker, *Britain and Disarmament: The UK and Nuclear, Biological and Chemical Weapons Arms Control and Programmes 1956–1975*, Farnham: Ashgate (2012) at page 52. Dr. Viney left ACDRU in June 1967 to head the Czechoslovak section of BBC External Services.

[320] In TNA file FCO 10/184, *Biological and Chemical Weapons – 1967*. TNA files do not, however, indicate how this paper connected to the draft BWC that the UK tabled in July 1969 at the Eighteen Nation Disarmament Conference in Geneva.

[321] There is an overview of these utilities in my Chemical & biological weapons described from an historical standpoint with reference to present-day policy concerns, SHIB O-list item 371, Harvard Sussex Program, 28 January 1999.

I shan't go any further with all that now. It is in any case work-in-progress at the Sussex end of the Harvard Sussex Program. I shall talk instead about the Syria question – chiefly in the hopes of learning more about what the experience these past three and a half years of chemical warfare in Syria really says for the future of the CWC, and therefore also the future of arms control.[322] I shall pay particular attention in so doing to just one of the triad of drivers: the utility of chemical weapons. It is, for the purposes of this seminar, the leading driver, for, if a category of weapon has no great utility or operational value, one may think that applying arms control to it can serve only symbolical purposes, if that.

In that regard one may ask whether – and this is a digression – there has yet been an assessment of the utility of biological weapons that adequately addresses the very many uncertainties and imponderables involved and the absence of real-world experience. I don't believe there has. Since before the Second World War many such assessments of bioweapons were started in Britain and elsewhere, but were any of them actually finished properly? Hence my earlier question about whether teeth should yet be negotiated for the BWC. The answer seems to be obscured by secrecies that prevail even now. The situation might be a lot clearer if at least some of the data gathered during the decade-long programme of field trials and other evaluations that the Kennedy administration in America started in 1961 were finally to see the light of day. That was Project 112, in which I understand Britain participated, an enterprise having just that purpose of CBW utility assessment.

A digression, as I say. Back to Syria, then, where chemical warfare may be continuing even today. Here, to begin with, is a summary of some of its features. Many of you will be very familiar with them, so I should explain now that they are the features I regard as highlights because they pose major questions I think we are now obliged to consider.[323]

During February 2012, against a background in which allegations of chemical warfare had markedly increased elsewhere in the world since the ending of the Cold War,[324] the US State Department began briefing journalists that it was concerned about what would happen, if the conflict in Syria intensified, to the

---

322 What follows is drawn from other work in progress at Sussex, *Binary nerve gas as a stimulus to proliferation of chemical weapons, and its ascendancy during the war in Syria*, the current draft, SHIB O-list Item 484 rev 3, being dated 23 October 2015.

323 My reference sources for the highlights that follow are identified in the three-part documented chronology, *Chemical Warfare in Syria*, Harvard Sussex Program SHIB O-list items 478 rev 2, 481 rev and 490, which is a rolling text.

324 This is suggested (so I wrote in SHIB O-list Item 484 rev 3 at pages pp 5–6) by data held in the Sussex Harvard Information Bank, an archive for which all such reports have long been actively sought by the Harvard Sussex Program. SHIB holdings indicate that use of chemical weapons was alleged in at least 34 different conflicts during the 24 years back to 1990 as compared with CW allegations during 70 previous conflicts back over the preceding 71 years to 1919: 1.42 per year since the Cold War ended as compared with 0.99 during and before it. Of course this apparent growth may simply mean that wars in general have (against appearances) been more frequent lately; or, since very few of the allegations have been properly verified, it could also mean that allegations of CW-use still serve as black-propaganda weapons no less useful than they have long been, or else that they are simply consequences either of genuine misinformation or of inadequate information.

stockpile of chemical weapons Syria was known to have accumulated during the previous decades. Syria had not denied having such stocks, and in November 1996 had publicly explained them as deterrent against Israel. Later in February 2012 there were press reports of the regime using chemical weapons, first incapacitating and then lethal agents, against its opponents in Homs province. In August President Obama announced that such use in the future, or preparations for it, would be "a red line for us". But the allegations continued and, in December, the red-line warning was repeated, by which time chemical warfare had been reported from Daraa and Idlib provinces as well as Homs. This proliferation of reports still continued, and by mid-August 2013 a total of at least 50 episodes of chemical warfare had been alleged in seven of the 14 provinces of Syria. According to records collected and kept by two dedicated human-rights organizations within Syrian civil society, 111 people were killed by chemicals and perhaps ten times that number injured during those scores of episodes. I now think that pretty much all of that reporting was true: that there had been an unprecedented succession of small instances of chemical-weapons employment.

Then, on 21 August 2013, in chemical strikes at Ghouta on the edge of Damascus, the casualty toll greatly increased, according, initially, to the profuse social-media reporting. The team sent by the Secretary-General of the United Nations UN Secretary General, investigating allegations to investigate earlier chemical-warfare reports had just arrived in Damascus to commence its work. The US government announced that 1,429 people had died in the Ghouta chemical strikes. It did not, however, press ahead with the military action implicit in its 'red line' warnings. But over the following month it joined with the Russian government in a successful initiative that obliged the Syrian government to disclose its holdings of chemical weapons and its factories for making them, and to eliminate them all under international supervision – effected by the CWC's implementation authority, the OPCW (Organization for the Prohibition of Chemical Weapons), headquartered in The Hague. The mechanism for achieving this was coercion of Syria's accession to the Chemical Weapons Convention and thus Syria's formal acceptance of the disarmament stipulations and procedures laid down by the treaty. This could be, and was, portrayed as a crowning achievement for that sometimes derided instrument of chemical arms control. By the end of October all the Syrian chemical-weapons production and filling facilities had been disabled pending their final destruction, and by 24 June 2014, albeit later than expected, the 1308 tonnes of chemicals that Syria had declared as chemical weapons had all either been destroyed or taken out of the country for destruction. Those six of the chemicals that were considered the most threatening were later destroyed either by the United States on a special ship in the Mediterranean or else by the United Kingdom in furnaces near Liverpool. Only about 20 of the 1308 tonnes of chemicals were actual chemical warfare agents – in fact mustard gas. The rest were comparatively harmless chemicals from which, given the right equipment, nerve gases could quickly be made.

Yet the growth can also be taken as evidence, if not wholly conclusive, that CW are somehow better suited to new wars than they were to the old.

But then, even as the Syria CW Elimination Programme was still in progress, reports began to be heard, with increasing frequency from April 2014 onwards, of helicopters dropping barrel-bombs charged with toxic industrial chemicals, especially chlorine, on rebel targets in Hama and then in two other provinces. When so used, such toxic industrial chemicals are of course chemical weapons within the meaning of the CWC, captured by its General Purpose Criterion (to say nothing of history, chlorine having been, in April 1915, the first 'weapon of mass destruction'). Before Da'esh forces irrupted onto the scene in June 2014 a dozen more episodes of chlorine barrel-bombing had been reported, with many others subsequently, and a new international investigation was in progress, conducted by a Fact Finding Mission of the OPCW. Although the published evidence clearly pointed to regime responsibility – who else had helicopters? – the regime accused Da'esh. Few, I think, believed this. But Da'esh was now becoming implicated in reports of chemical warfare not involving chlorine barrel-bombs. From July 2014 through at least to last August, these reports included Da'esh use of mustard gas from mortars and artillery, initially against Syrian Kurdish forces in the Kobani canton of Aleppo province, hard by the Turkish border. Meanwhile, regime use of chlorine barrel-bombs in several parts of the country continued to be reported. Then, in a breakthrough on 7 August 2015, the UN Security Council unanimously resolved to establish a full-blown Joint Investigative Mission [JIM] that, in contrast to the earlier UN and OPCW investigations, would seek to attribute responsibility to the perpetrators of any chemical warfare in Syria.[325] How JIM is coming along I do not know (but hope to be told here).

So much, then, for my Syria highlights. I have three broad conclusions or at least arguments to draw from them. *First*, utilities have been found for chemical weapons in Syria that are novel. They are utilities that do not necessitate broadcast dissemination of toxic chemicals over wide areas and in high concentrations, which was a dominant characteristic of Cold-War preparedness for chemical warfare in the European confrontation, and which shaped central provisions of the Chemical Weapons Convention. Instead they are utilities served by small-scale agent releases, closely targeted on key nodes of resistance, and working through impacts on the morale of defenders, not necessarily through killing or injuring them – impacts on what the father of modern chemical warfare, Fritz Haber, when looking back over the World War I experience, called "psychic equilibrium".[326] A label for this, my postulated Syrian mode of chemical warfare, is 'tactical terrorization'. The admittedly sparse open information available from Syria on what exactly the chemical munitions used there were and how they were employed does indeed suggest that a typical operation of this kind required the placing on target of very much less nerve gas than was anticipated during the Cold War. Battlefield planning then envisaged nerve-gas consumption on the order of hundreds of tons per day. Estimates of the amount of nerve gas used in the Ghouta event of 21 August 2013 stand at less than one ton. In both earlier and subsequent Syrian chemical strikes, kilogram quantities seem to have served their purpose.

[325] UN Security Council resolution 2235 (2015).

[326] F. Haber, *Fünf Vorträge* (Berlin, 1924).

Numbers such as these are embedded in concepts that underpin the functioning of the Chemical Weapons Convention. Above all, the idea of a "militarily significant quantity" is what determined the size of the different quantitative thresholds specified in the treaty for the possession, reporting and inspection obligations that it places on states parties. For example, although the CWC is actually a disarmament treaty, it nonetheless allows states parties to possess up to a tonne of, say, sarin nerve gas. Syrian-mode chemical warfare, in which only a few kilograms of agent have been militarily significant in some engagements, therefore raises the question of whether the CWC is still sufficiently fit for purpose.

My *second* conclusion has to do with why that Syrian mode came to the fore in 2012. There is no obvious sign of tactical terrorization with chemical weapons having been deliberately practised in conflicts prior to then. Chemical terrorization for purposes of punishment or retribution or coercing non-combatant populations had most certainly been seen in earlier conflicts, but not I think for purposes of gaining tactical advantage. If anyone believes me to be wrong on this, I do hope they will say why. I now want to suggest that this new utility has a technological explanation.

The OPCW has told us that Syria embarked upon a chemical weapons programme at least as early as 1982. Now, 30 years on, in place of its deteriorated stockpile of nerve gases, Syria had acquired a CW technology that avoided the shortcomings in the way it had manufactured sarin nerve gas. The technology had also, I argue, happened to be one that lowered the barriers to employment of chemical weapons of kinds and in quantities that suited the purposes of tactical terrorization.

The novel Syrian technology I take to be a variant of one for which the United States Army had set a worldwide fashion back in the 1970s: the technology of "binary" nerve gas. Here the chemical munition is loaded, not with nerve gas, but with nerve-gas precursor chemicals that are not allowed to interact until the munition is launched or is being readied for use. This would counter the storage-instability problem that had bedevilled Syrian sarin. The Syrian variant binary was one that Iraq had also been developing, though not in time, it seems, for actual use during its 1980-88 war with Iran. The variant was not the complicated "mix in flight" type of binary favoured by the United States but its rather simpler "mix before use" counterpart. True, this involved severe technological obstacles of its own, but Syria was evidently able to overcome them, or at least live with them.

Most details of the special Syrian technique have not been publicly disclosed. But it became public knowledge at the time of the Ghouta massacre that the technique involved special plant, both fixed and mobile, for mixing the precursor chemicals and loading the still-interacting mixture into munitions a short time before they were needed. I infer from this that such mix-before-use binaries allowed increased flexibility in matching weapons to targets. No longer need that choice be governed by what was on hand in storage – by what had been put into the stockpile years if not decades before. If a need arose for a few munitions better able to exploit particular targets of opportunity, hardware allowing that could quickly be improvised if necessary and uploaded via those mix/fill facilities that the binary munitions had necessitated. Observe, for example, the chemical-warfare reports from Sheikh Maqsoud in Aleppo province during April 2013, which implicated small helicopter-dropped munitions of a kind not seen before as chemical weapons. Other such examples are the varieties of chemical weapon described by the *Le Monde* newspaper people attached to opposition forces in and around Damascus during March–May 2013. Perhaps these small sarin weapons would not even have come into existence if mix/fill plant had not been on hand, particularly mobile plant. In that sense, the Syrian binary technology might be said to have created new utilities for chemical weapons, not simply facilitated old utilities. The tactical terrorization utility seems to have been valued highly enough to be exploited repeatedly. Even after binary nerve-gas had become subject to the international Syria CW Elimination Programme, those morale impacts were apparently still being sought during 2014 and 2015 through the use of chlorine barrel bombs.

So it is that we seem to have been observing a new amenability of chemical weapons to conflicts of the post Cold War era. And we may also be observing a still-running motor of chemical-weapons proliferation driven by technological change. If that is indeed the case, we may have a long way to go yet before we achieve that widely avowed goal of a world free of chemical weapons.

This leads me into my *third and final conclusion* from the Syria highlights. The need for effective arms control has been increased by the developments outlined in my first and second conclusions – by the doubts about the continuing suitability of the various quantitative thresholds in the CWC, and by the new evidence of continuing chemical-weapons proliferation. But the fact that we must once again take seriously the proliferation threatened by binary technology would seem to be making effective arms control still harder to ensure because the technology has opened up a range of potential new chemical weapons that are on the control lists of neither the CWC nor the Australia Group. Bizarrely absent from the Australia Group lists are two of those six most threatening chemicals that were taken out of Syria for destruction elsewhere.[327] I scarcely dare mention also the binary Novichok precursors that post-Soviet Russia was reportedly field testing on a large scale in Uzbekistan during the early 1990s, including precursors not on the CWC schedules either.[328] Some of the fitness for purpose of the CWC, and now the Australia Group too, is thereby again called into question.

Possibly I overstate these dangers, though I don't think so. I conclude that there is now a need to consider whether, in the light of Syria, changes to the rules and regulations of the chemical arms-control regime have become essential. There may also, I am obliged to add, be a need to rebuild confidence in the arms-control regime in the face of those who point to the chlorine barrel-bombs and to the Da'esh mustard gas as evidence of the futility of arms control. So let me say that it is no good our just throwing up our hands at the prospect of chemical arms control failing simply because of our inability thus far to hold to account the perpetrators of the Ghouta massacre and all the other acts of chemical warfare in Syria. The fact that the perpetrators

[327] Namely the sodium salt of O-ethyl methylphosphonothioic acid (EMPTA), and the hydrochloride of diethylaminoethyl chloride.

[328] Namely [deleted] [sic].

are unarraigned war criminals does not mean that the achievements of CBW arms control now count for nothing.

In closing, I would like to point up one further dilemma. It has to do with the obstacle of undue secrecy or, to put it less politely, the consequences of governmental disinformation achieved through the withholding of pertinent information. My Syrian highlights and the conclusions from them originated for the most part in information made available through the efforts of civil society, not government. I have inquired of various officials why so much of the information about how chemical weapons had been used in Syria was so evidently being kept away from public disclosure, and, worse, information about Syrian binaries too. I have been told that it was in order to protect intelligence sources and methods, and because the information was "proliferation sensitive". No doubt both those claims are true – who am I to judge? Yet if there is to be any serious consideration of improving the CWC regime, there will have to be at least some disclosure of that sort of information within the international bodies of the OPCW. How could that ever be achieved while so much continues to be treated as secret – continues to be withheld from civil society? Or must we expect information security always to take precedence over arms control?

So, on that rather combative note, I end my presentation.

Chapter 8
Examples of Julian's Broader Writings

Julian Perry Robinson

8.1 Extract – Possible Impacts of Weapons on Ecosystems[1]

Editors' note. The main focus of Julian's working life was on understanding and preventing chemical and biological warfare (CBW), as represented in the selection of papers in Chap. 7. However, he also wrote about other areas, as demonstrated in the two papers in Chap. 8.

Here, Julian assesses the diverse harms that weapons can inflict on ecosystems, introducing a new theoretical framework for considering ecosystem vulnerabilities. The framework differentiates between direct and indirect stress factors on ecosystems, and allows for assessments of weapon-induced damage to structural components of ecosystems based on how such damage impairs the functional processes that link these components. The extract is from Chap. 4 of a United Nations Environment Programme publication. The original included several references to other chapters from the full report[2]; we have streamlined these mentions, and have included additional information in footnotes where necessary.

[1] Editors' note. This extract is from Chap. 5 of Robinson, JP. 1979. The Effects of Weapons on Ecosystems. United Nations Environment Programme (UNEP) Studies. It is not listed in Chap. 9 as the bibliography there only includes Julian's writings on chemical and biological warfare. Unless otherwise indicated, the remaining footnotes for this paper are Julian's originals. Apart from correcting obvious typos and renumbering headings and tables in keeping with the requirements of this book, we have retained Julian's formatting and spelling throughout the paper and its notes.

[2] Editors' note. For reference: Chap. 1 of the original publication is "Background and Approach", Chap. 2 "Classification of Weapons", Chap. 3 "Ecosystems", and Chap. 5 "Conclusions and Recommendations".

J. P. Robinson (✉)
Brighton, UK

H. Wilson et al. (eds.), *Julian Perry Robinson: Pioneer in the Prohibition of Chemical and Biological Warfare*, Pioneers in Arts, Humanities, Science, Engineering, Practice 32, https://doi.org/10.1007/978-3-032-04064-0_8

8.1.1 Theoretical Framework

The weapons-effects characterizing the weapons treated in Chap. 2 [of the original publication][3] have been described in terms of the damage-agents of the weapons: high kinetic energy fragments, blast, flame, heat flux, toxic and radiotoxic substances, pathogenic microbes and ionizing radiation. Though their actions upon biological systems differ greatly from one another when considered at the level of individual organisms, they have features in common when considered at the higher level of the community. It is possible, therefore, to assess potential impacts of weapons on ecosystems in terms of a relatively small number of stress factors, each one representing the aggregate effect of one or more of the relatively large number of different damage agents.

Several such aggregations are possible. In accordance with the reference points adopted in this study for assessing ecosystem vulnerability,[4] the most useful aggregation is that which bears most directly on the mechanisms of desertification. All terrestrial ecosystems are potentially reducible to deserts or to other such states of low biological productivity, incapable of supporting sufficient plant life to cover more than small fractions of their surfaces. This may be because there is no soil, no water, too much water, or too much salt or other phytotoxicants; because the prevailing temperature is too high or too low; or because the soil structure is inadequate for holding water or nutrients. The conceivable mechanisms of desertification and analogous processes due to warfare follow from this. An indirect example would be oversalination due to inadequate irrigation following damage done to the means of irrigation. A direct example would be destruction of soil structure. The latter might be brought about by repeated or drastic mechanical disturbances, including frequent trampling or passage of heavy vehicles; by extremes of temperature, especially alternating extremes; or by the killing of the microbiota of the topsoil. The capacity of weapons for bringing about desertification would then depend in large measure on their propensity for damaging the soil or killing its biota, effects which would be exacerbated by destruction of the plant cover protecting topsoils from extremes of temperature and from desiccation.

It would seem, then, that the most appropriate aggregation of weapons-effect damage agents is into the three composite stress factors of soil damage, destruction of plant cover, and biocide. All of the weapons under consideration may stress ecosystems in one or more of these three ways. Some will do so more than others. Thus conceived, the stress factors are not independent of one another. Some forms of biocide may, for example, result in a diminution of plant cover which may in turn

[3] Editors' note. Chap. 2 is on "Classification of Weapons". See also the Editors' note at the start of this extract.

[4] Editors' note. The original included a reference to page 32 of the book this extract is from, which lists these criteria for assessing ecosystem vulnerability: (1) the ecosystem's capacity to restore human habitat after weapons-inflicted damage, and (2) the specific form of degradation, in this case desertification. Both are evaluated against the critical factor of recovery time relative to human lifespans.

bring about damage to the soil. Such dependencies may be captured by differentiating direct and indirect stress and considering how the latter may affect resilience towards the former.

As an analytical tool, this notion of stress permits generalizations to be made about impacts of different weapons on each of the different categories of structural component which together … constitute an ecosystem.[5] A similar transformation may be applied to … ecosystem sensitivity characteristics … for these can be conceived as weaknesses in certain of the functional links which together … define an ecosystem.[6] Conclusions about possible impacts of weapons on the different categories of ecosystem under consideration may then be drawn in terms of the degree to which the stresses acting on the structural components are likely to aggravate the functional weaknesses, thereby threatening ecosystemically mediated harm to the human populations of the ecosystems.

8.1.2 Stress Through Soil Damage

The soil is the vital link between the biotic and abiotic components of an ecosystem. It acts as a reservoir for water and the other inorganic substances cycling through the ecosystem. It provides habitat for many of the different populations of decomposer organisms that control the rate and capacity of nutrient cycling. It affords the physical structure in which are rooted most, if not all, of the different populations of primary producers. Soil damage may thus stress an ecosystem in a wide variety of ways. Two broad types of soil damage may be envisaged as a consequence of weapons employment: physical displacement of soil; and alterations in soil structure and composition.

Soil displacement may be a direct consequence of the cratering action of the blast from high-explosive or nuclear weapons. Or it may be an indirect consequence of the destruction of vegetation whose loss subsequently causes increased exposure to the erosive influences of wind and water. The uppermost layers of soil are at once the most vulnerable and the most immediately important for the ecosystem as a whole. They comprise a topmost layer of vegetational litter and other detritus; an intermediate layer of organic matter largely comprising the humus resulting from microbial decomposition of the surface litter; and lower layers of soil having a high

---

[5] Editors' note. The original included a reference to page 29 of the book this extract is from. The structural components of ecosystems mentioned follow Eugene Odum's, and include abiotic elements such as inorganic substances (e.g., carbon and water), organic compounds (e.g., proteins and lipids), and physical climate factors like temperature. The biotic components consist of producers (green plants and algae that synthesize organic matter), consumers (organisms that ingest other organisms or organic matter), and decomposers (bacteria and fungi that break down dead material to recycle nutrients). See: E.P. Odum, *Fundamentals of Ecology* (Philadelphia: W. B. Saunders Company, 1971, 3rd edition), p. 8.

[6] Editors' note. The original included references to Chap. 3, page 30, of the original publication, which consider ecosystem sensitivity characteristics.

porosity as a result of root growth and the burrowing actions of small subterranean creatures. It is this porosity and the crumb-sizing associated with it which enables the soil to retain, according to its chemical and physical nature, supplies of inorganic nutrients in readily assimilable form; in concert with the large water-holding capacity of the litter and humus, it provides the mechanism for infiltration of rainfall into the soil, rather than erosive surface flow. The thickness, content and structure of all these layers will vary greatly from one ecosystem to another. In general, the thinner the layers, the more vulnerable will be the entire ecosystem to weapons that cause soil displacement.

At any given moment, the topsoil present in a particular area represents an equilibrium between natural processes of soil erosion and soil genesis. Therefore, if weapon-effects that cause soil displacement also reduce the natural rate of soil genesis, the result may be an accelerating erosion of topsoil, which is the classic mechanism of desertification. Destruction of vegetation over a wide circumadjacent area will reduce soil-genesis rates. Eroding topsoil becomes wind- or water-borne particulate material which, in large quantities, may menace downwind or downstream areas. The mechanisms of, for example, dust-storms or silting may thus transmit the damaging consequences of weapons from one region to another.

The ecosystemic consequences of soil displacement may not be limited to those of topsoil damage. In some terrains, deep cratering may disturb the water-table and local drainage patterns. If a hardpan is breached, the perched water-table may be lost for many years.

The other major form of soil damage to be considered here comprises a variety of adverse changes in soil structure and composition. Thus, the chemical exchanges which take place in the soil as part of, for example, nutrient cycling occur by way of finely poised chemical and physicochemical equilibria such that small alterations in, say, the sorptive characteristics or the acidity of the soil may tilt the equilibria and block certain of the exchanges. Damage of this type is perhaps more likely to be an indirect than a direct consequence of weapons use: a consequence of, for example, accelerated weathering or other such chemical and physicochemical changes brought about by increased exposure due to cratering or loss of plant cover; or gross damage to constructions such as dykes which then admit increased salinity. Some soil types may be especially sensitive. The latosols, which occur most frequently in the tropical belt within 30 degrees of the equator, are an example. For reasons which remain to be fully elucidated, latosols have a peculiar propensity for assuming a rock-like hardness upon prolonged exposure to the elements. The resultant laterite is a mixture of aluminium and iron oxides and very little else, incapable of supporting either natural growth or agriculture. Such induration is a mechanism of desertification that has confounded ill-conceived attempts at agriculture in a number of forested tropical regions.

Direct changes in soil structure and composition might possibly be brought about by chemical weapons, though it seems improbable that any of the chemical-warfare agents that are prominent today could cause such damage at their likely application densities. It must be noted, however, that the effects of these agents on soil microflora and microfauna do not appear to have been closely investigated. For the present it would seem that thermal stress, rather than chemical stress, is the more serious direct danger. Thus, the heat-flux resulting from the direct action of burning incendiary agents or from thermal irradiation by nuclear explosions, or from spreading fires initiated by either agency, may be of sufficient magnitude to bring about chemical and physical changes in the soil that reduce sorptive capacity and cause collapse of pore structure. Organic particulates may be oxidized and thereby lost from the ecosystem. Habitats for soil fauna may be eliminated. Soil scorching may therefore constitute a major stress on ecosystems.

8.1.3 Stress Through Plant-Cover Destruction

Vegetation is the primary source of utilizable energy for an ecosystem. It is the first link in all the food-chains. It is the means whereby an ecosystem extracts inorganic minerals vital for its biotic components from the geomass. It provides not only food (directly or indirectly) for the consumer population, but also shelter. It stabilizes and moulds the soil, and transforms the prevailing climate regime into a microclimate favouring not only itself but also other biotic components of the ecosystem. The character of an ecosystem is thus dominated by its vegetation, so that damage to the latter will profoundly affect the former.

Destruction of plant cover might be brought about directly by the physical agencies of blast, fragment kinetic-energy, or fire. Or it might be caused indirectly either by soil displacement or by biocidal action upon the component flora. All the weapons under consideration may therefore expose ecosystems to the stress of plant-cover destruction.

The observable consequences of that stress may be diverse, much depending upon the scale of the vegetational destruction. If it is sufficiently large, chains or events will be set in process culminating in a variety of structural changes throughout the ecosystem. Their full range cannot be described here; but the following examples illustrate them.

The impact of plant-cover destruction upon the integrity and structure of the local soil has been suggested [above][7]; increased exposure of the soil to natural erosive forces and to leaching; retardation of soil genesis; and, in latosols, soil induration. Such might be the consequences of complete removal of plant cover. Less drastic reductions in plant cover might also profoundly affect the soil. For example, transpiration of water in an ecosystem would be reduced approximately in proportion to the surface area of damaged foliage. Though some increase in evaporation from the soil surface might occur, the overall evapotranspirative loss of moisture from the soil would in most cases be reduced. One possible consequence would be a generally moister soil and a rise in the local water-table. This could adversely affect the decomposer habitat and the productivity of plant species adapted to the pre-existing lower humidity. Where rainfall is substantial, the increased occupation of the soil's water-storage capacity would also have the effect – well observed by, for

[7] Editors' note. This refers to the above section of this extract, "Stress through soil damage".

example, foresters in temperate regions – of forcing subsequent rainfall out of the affected area via increased surface or subsurface flow. The flood potential of the region would thereby increase and, with it, the possibilities of erosion and massive leaching of nutrients from the soil.

The foregoing effects may be seen as one possible set of consequences of alterations in microclimate due to plant-cover depletion. Such alterations, in addition to decreased humidity, would also include reduced surface carbon dioxide concentrations (leading to diminished local photosynthesis) together with increased surface insolation and wind-velocity. The diurnal fluctuations of temperature gradient across the air layer immediately above the soil would become substantially more intense, and perhaps actively inimical to emergent plant seedlings and small permeant organisms.

Diminished vegetation may alter the reflectivity of the terrain towards solar radiation. Albedo change may provoke microclimate change.

In seeking to gauge the overall impact of stress by plant-cover destruction on ecosystem structure, a major consideration must be the rate and form of likely vegetational recovery. In broad terms, recovery may be seen as a two-phase process. The first phase comprises colonization of the denuded area and the establishment of a pioneer community. The second phase commences with the start of those successional changes[8] … in which the pioneer community gives way to seres of increasing species diversity until the final climax community is reached, one which perpetuates itself through homeostatic mechanisms no longer perturbed by the forces of successional change. The closer the original community to a climax state, the more resilient may it prove, by virtue of its diversity, in the face of any of the stress-factors under consideration. But if the stress is great enough to overload the homeostatic mechanisms – as plant-cover destruction might be if it occurred on a massive scale – the net effect would[9] … be reversion to an earlier and more primitive sere. The more highly evolved the ecosystem, the longer will it take to recover from such a reversion. In any event, the rate of recovery would depend upon the robustness of the ecosystem concerned, and the process could not commence without the initial phase of recolonization. Unless the plant-cover destruction were slight, the new vegetation that came to establish itself in the area would generally be different from the original vegetation, having regard to the alterations that would have taken place in microclimate, nutrient budget, soil moisture-holding capacity, etc. It would establish itself through invasion by *Fusaria* (slime fungi), microscopic algae and bacteria, followed in succession by lichens, mosses and vascular plants. The new plant cover would consist of tenacious and undemanding grasses and similar species. Its overall character would be one of much lower biomass and nutrient-retention capacity. If the stress-factors acting upon the ecosystem had rendered it peculiarly inhospitable, revegetation might be delayed indefinitely, a consequence manifest as desertification. How likely such an eventuality might be would of course depend upon circumstances.

[8] Editors' note. The original included a reference to page 30 of the book this extract comes from. It says "[b]efore an ecosystem attains its climax state of maximum stability, it will have evolved through a succession of intermediate stages or 'seres'."

[9] Editors' note. The original included a reference to page 31 of the book this extract comes from.

The early colonisers would be windborne organisms; the extensiveness of the devastated area would control the appearance of the later ones. As noted,[10] ... recovery is observable in certain arid and tropical-island ecosystems subjected to the massive insult of nuclear explosions. But by way of contrast, there is now the existence of the huge expanses of muddy wasteland that were once, prior to military herbicide operations, lush mangrove forests around the southern coastline of Viet-Nam. Precisely why little if any recolonization of these mangrove sites has yet occurred, at least in the form of vegetation, remains obscure: Testimony at once to the undeveloped state of ecological knowledge and to the gravity of possible ecosystemic consequences of plant-cover destruction.

8.1.4 Stress Through Biocide

All weapons are capable of killing: of exposing living organisms to various kinds of biocidal action. Some are not in fact designed to do so, despite being capable of it. What the military call "antimaterial" weapons, for example, are designed to destroy inanimate structures or equipment, but the stresses which the weapons must exert to do so are invariably of such a kind and magnitude as to be biocidal for some organisms. Different species will differ in their robustness or fragility towards a given weapon-effect; and some types of weapon-effect will kill a wider variety of species than others. Hence, across the broad spectrum of living organisms, weapons will vary in the bandwidths of their biocidal activity. The precise extent of that bandwidth will be a factor of major significance in the degree to which the biocidal action of a weapon will stress ecosystems. Existing knowledge in this area, is, however, very incomplete.... In addition to the dominant primary producers, other organisms among the biotic components of an ecosystem may be especially important to the ecosystem as a whole. It may happen by chance, and could happen by design, that the more vital organisms fall within the biocidal bandwidth of a weapon. The likelihood of this happening by chance would hinge on the extent of the bandwidth – i.e. on the inherent characteristics of the weapon – and on the nature of the ecosystem. For example, an inverse correlation is to be expected with the species-diversity of the ecosystem. The likelihood of it happening by design will no doubt increase as knowledge in systems-ecology advances. Chemical and biological weapons, in particular, offer a disturbingly accessible species-selectivity in their biocidal actions.

The importance of a species of organism within an ecosystem may be assessed by reference to function. For example, the more a species influences the energy throughput of an ecosystem, or the more strongly it affects the environment of all other species, the more important will it be. In these respects, particular species within particular ecosystems may be identified as "ecological dominants". On the criterion

[10] Editors' note. This refers to an earlier discussion on the Mohave Desert test area and Pacific-island test sites.

of energy throughput, the ecological dominants within any one trophic group will be those that have the greatest productivity.

For any ecosystem, the most obviously vital category is that of the primary producer organisms: the green plants which, through the mechanism of photosynthesis, are the ecosystem's principal means of converting solar energy into the chemical energy necessary for sustaining life and growth in its other biotic components. Thus, the stress of biocide upon an ecosystem may be especially strong if a significant range of plant species falls within the biocidal bandwidth of the weapon concerned. … we have indicated the magnitude of some of the possible consequences.[11]

The consequences may assume their gravest proportions in the case of forest destruction. So powerful is the momentum which forests may impart to the cycling of nutrients, and so strong may their influence be on the hydrological and meteorological conditions of a region, that their importance within an ecosystem, and hence their loss, may be felt over an area far exceeding that which is actually forested. There have been innumerable instances in history, which continue to repeat themselves even now, of deforestation being succeeded by widespread land degradation and impoverishment of the region, amounting, in some cases, to desertification and other forms of ecological disaster.

Ecological dominants tend to become fewer in number as geographical latitude increases. In a northern forest, for example, 90 percent or more of the total stand may be made up of one or two species of tree, whereas in a tropical forest it may comprise a dozen or more species. In this particular respect, tropical forests may thus be vulnerable to a stress of narrow biocidal bandwidth than a northern forest.

As to the consequences of biocidal stress upon species at higher trophic levels, it is not yet possible to identify with precision depopulations that would inevitably have an adverse impact upon ecosystems as a whole. Within several ecosystems more narrowly defined than those of the present study, dependencies of one species upon another, as in the foodchains or the pollinating insects, can, of course, be specified; and there are many reported observations of the elimination of one population in an area provoking strong oscillations in other populations.

8.1.5 Relationships Between Scales of Weapons-Use and Magnitudes of Stress and Strain

The strain which an ecosystem may experience when subjected to any of the foregoing stresses will depend upon the characteristics of the ecosystem. It will also depend upon the magnitude of each stress, which in turn will depend upon the nature of the weapons concerned and the scale on which they are used. For each possible weapon-ecosystem impingement, the strain may be conceived in terms of a threshold below

[11] Editors' note. See the above section on "Stress through plant-cover destruction".

which the ecosystem vulnerability ... is insignificant.[12] For some types of weapon and ecosystem, this threshold may be dangerously low. For others, the threshold may be so high as to be unsurpassable during military operations of a plausible kind and scale.

The scale on which weapons may be used will depend upon a range of circumstantial factors of which the limiting one is, ultimately, the availability of resources to the belligerents. A more immediate limiting factor is the number of committed delivery systems available and suitable for the weapons in question. A very rough idea of possible magnitudes of each type of stress that might realistically be anticipated may therefore be gained by estimating for one type of common weapon-delivery-system the area over which the different stresses might be created were a full complement of each of the different weapons to be delivered by it. The resultant area-estimates may then be converted into minimum numbers of weapon-delivery-systems that could stress an area of a given size. Some such estimates, for an area of 500 km^2, are set out in Table 8.1. The delivery-system chosen is a ground-attack aircraft having a payload of up to 6 tonnes of ordnance, its munitions delivery capacity being configured so that it could carry two munitions of 1000-kg rating. Such aircraft are to be found in the air forces of many countries, though in widely differing numbers.

... Above,[13] a theoretical approach has been suggested which could be used to relate the rough measures of stress suggested by Table 8.1 to the thresholds of significant strain envisaged above. The approach requires that damage caused by stress due to weapons acting upon the structural components of an ecosystem be assessed in terms of the extent to which the functional processes that couple the structural components are thereby affected. The degree to which overall function is impaired then appears as the measure of strain. If the strain is manifest as harm to human populations...,[14] the threshold of significant strain will have been crossed. The ecosystem will have proved brittle, not resilient, in the face of the weapon insult.

It has to be said, however, that the full capacity of such an approach for differentiating the likely ecosystemic impacts of the various types of weapon hinges upon knowledge that does not yet appear to exist. Much is known about particular aspects of both structure and function in a variety of ecosystems. But the ignorance remaining is such that any attempt to *correlate* structure and function must deal in hypotheses which will rarely, on the data at present available, be either substantiable or refutable,

---

[12] Editors' note. The original included a reference to page 32 of the book this extract is from, which explains that ecosystem vulnerability is the ability of ecosystems to recover from disturbances based on the rate of adaptation, the extent of change to equilibrium, and the impact on human habitats.

[13] Editors' note. The original includes a reference to page 47 of the book this extract is from,

[14] Editors' note. The original includes a reference to page 32 of the book this extract is from, which explains that the concept of "harm to human populations" mentioned here refers to the assessment of *ecosystem vulnerability*. This vulnerability is defined by the rate at which an ecosystem recovers from perturbation and how different the new equilibrium is from the old. Human survival and development serve as reference points for measuring damage, with particular emphasis on desertification and its long-term impacts on human life-spans.

Table 8.1 Ecosystem-stressing potentials of different weapons[15]
Rough estimates of the numbers of aircraft (a) carrying each type of weapon that might expose 500 km² of terrain to different types of ecosystemic stress

Weapon		Stress factor				
Class	Type	Soil damage		Plant-cover destruction(c)	Biocide	
		Displacement	Scorching(b)		Producer organisms(d)	Consumer organisms(e)
High-explosive	General-purpose HE	340 000(k)	–	7 200(l)	?	7 200(l)
	Fuel-air explosive(f)	–	?	2 300	?	?
Incendiary	Napalm	–	14 000	<42 000(m)	?	?
Chemical	Nerve gas(g)	–	–	–	–	1 100(n)
	Herbicide(h)	–	–	1 700(o)	600(p)	–
Biological	Plague bacteria	–	–	–	–	>30
	Wheat-rust fungi(i)	–	–	–	>3(q)	–
Nuclear	20 kt standard fission(j)	27 000	10(r)	70(s)	9-300(t)	9(u)
	1 Mt fission-fusion-fission(j)	2 000	2(r)	3(s)	1-3(t)	1(u)

(a) Ground-attack aircraft each capable of delivering up to 6 tons of munitions. It is further assumed that, for the high-explosive and incendiary munitions, the payload would comprise bombs of 200–350 kg rating; that the chemical and biological agents would be delivered from spraytanks comparable in size to bombs of 1000 kg rating, two per aircraft; and that no more than two nuclear munitions would be delivered by each aircraft
(b) Refers to soil insufficiently vegetated for wildfire propagation
(c) Refers to forest or other woody vegetation
(d) Refers to death of crop-plant species
(e) Refers to 50 percent mortality among populations of man or similar species
(f) Refers to hypothetical third-generation fuel-air explosive munitions
(g) Refers to o-ethyl S-2-diisopropylaminoethyl methylphosphonothiolate
(h) Refers to a formulation comprising equal parts by volume of the n-butyl esters of 2,4-dichloro- and 2,4,5-trichlorophenoxyacetic acids
(i) Refers to uredospores of *Puccinia graminis* tritici
(j) Fused for surface-burst
(k) Delay-fused bombs
(l) Quick-fused bombs
(m) Lower figure if a spreading wildfire is initiated
(n) A lethal hazard may persist in the area for several weeks or more
(o) Refers to at least 60 percent defoliation lasting for 4–14 weeks of multi-layered deciduous-forest canopy with peak defoliation within 10–24 weeks
(p) Refers to broad-leaved food-crop cultivations
(q) Refers to 85 percent loss of wheat yield if weapons are used under conditions most favouring epiphytetic spread
(r) Refers to thermal irradiation
(s) Refers to 90 percent tree-blowdown due to blast
(t) Refers to nuclear radiation, the range reflecting radiosensitivities of different crop-plant species
(u) Refers to 30-day mortality from nuclear radiation

[15] Editors' note. This table was 'Table 4' in the original.

For the concluding sections that now follow, in which are presented overall impact appraisals guided by the foregoing analytical approach, this is a major caveat.

8.1.6 Impacts on Tropical Ecosystems

... [A] major characteristic of tropical ecosystems, most apparent in the tropical rainforest, is that their cycling nutrients tend in comparison with those of other ecosystems to have longer residence times in the biomass than in the geomass.[16] This is presumably due to a greater rapidity in the heterotrophic utilization of the products of autotrophic organisms. The more the cyclic commodities of an ecosystem are concentrated in the biomass, the more vulnerable may the ecosystem be to biocidal stress. The lesser role of the geomass in nutrient cycling is likely to mean that it will prove incapable, by reason of adaptations among the biotic community, of filling breaches in the cycle caused by biocide. The quality of the soil, for example, is likely to be such that nutrients entering it via the decomposers from a sudden overload of dead plant or animal organisms will not be retained for long, being more or less rapidly leached away by rainfall. There may then occur the phenomenon of "nutrient dumping", already alluded to, in which the entire ecosystem suffers a continuing loss of nutrients. Productivity would fall at an increasing rate unless or until recovery via recolonization started to recouple the broken circuits of the nutrient cycle. In the worst case the ecosystem could suffer desertification.

A tropical ecosystem will be most endangered via nutrient dumping if biocidal stress is concentrated in regions of especially high biomass. Thus weapons may have a particularly strong impact on tropical ecosystems if they are used within heavily vegetated regions and if their biocidal bandwidth encompasses a high proportion of the species comprising that vegetation. The rapidity of the biocide, and its phasing in relation to seasonal changes, including inflexions in rainfall, will be major contributory factors. For example, if the biocide is relatively slow, the increase in the nutrient load on the geomass may be spread over a sufficiently long time for it to be adequately retained. In such cases nutrient dumping may not occur significantly unless either or both of the other two major stress factors operate. Plant-cover destruction, in the sense of soil denudation, would increase exposure of the soil to the leaching and erosive actions of heavy rainfall or winds; so would soil displacement. Nuclear weapons most obviously pose the threat of nutrient dumping to tropical ecosystems. High-explosive weapons might do so, at least in theory, but it seems clear from Table 8.1 that very large numbers of weapon-delivery systems would have to be employed for such effects to be anything other than highly localized. Incendiary weapons would seem to present either a lesser threat or an intermediate one. The latter would be the case if the fires which they initiated took hold and spread of their own accord. ... [T]here would be a substantial probability of this happening in regions where natural

[16] Editors' note. The original includes a reference to page 32 of the book this extract is from.

wildfires are common.[17] Here, however, the ecological dominants would presumably be adapted to survive wildfires, even to prosper in their aftermath. This would not be so in wildfire-rare regions, which would be correspondingly more vulnerable, but here the incendiary attack could have to be of peculiar intensity to initiate a wildfire. Yet given the advances that are now being made in the technology of incendiary weapons, particularly flame weapons, this is a possibility that cannot be discounted.

Herbicidal chemical weapons would also present an intermediate threat. If both biocide and soil-denudation resulted from their use over a large enough area, catastrophic nutrient dumping might well occur. But it would seem from Table 8.1 that such effects would require a large commitment of delivery systems for the requisite chemical weapons. As noted,[18] … the ecological dominants of tropical vegetation – at least of the kind of vegetation against which antiplant chemical weapons might, on military logic, be employed – are relatively large in number. The species diversity would necessitate a wide biocidal bandwidth from the herbicides employed if the soil was to be significantly denuded. If high-explosive or incendiary weapons, especially the latter in the less humid regions, were used in conjunction with herbicides, the danger of nutrient dumping might substantially increase.

The stress factors of soil damage, plant-cover destruction and biocide could operate to menace tropical ecosystems by mechanisms other than nutrient dumping. The sharply regionalized danger of soil laterization and consequent desertification has been noted[19] … Some solace may be derived from the fact that laterization has not been reported from areas of Indochina in which latosols are present and which were subjected to military herbicide operations and other forms of weapons insult. But since it has occurred in certain parts of tropical South America as a consequence of forest clearance, it is a hazard that cannot be ignored.

Other serious consequences of weapons-impact on tropical forest can be envisaged. The risk of soil induration follows from one of a number of fragilities in tropical soil. As noted[20] … forests may shape structure and function in ecosystems, sustaining productivity, over areas distant from those that are actually forested. Such areas may include ones given over to agriculture. In the relatively poor soils common in the tropics, agriculture may be precarious at the best of times, but if adjacent forest vegetation is destroyed, agricultural productivity may seriously decline. Increased erosion, which is always a possibility after deforestation, may act upon lands already virtually denuded by the demands of agriculture and, in especially vulnerable areas, desertification may follow.

Under conditions of high ambient temperature, moisture increases in importance as a limiting factor for most organisms. As the amount of water falls. below the limits of tolerance for forests, trees begin to give way to grassland and the like.

---

[17] Editors' note. This original includes a reference to page 18 of the book from which this chapter was taken, which deals with incendiary weapons as part of Chap. 2's consideration of weapons' characteristics.

[18] Editors' note. See the section on stress through biocide, above.

[19] Editors' note. See the section on stress through soil damage, above.

[20] Editors' note. This chapter section on stress through biocide, above.

[A] labile equilibrium … exists between woody and grassy plant species in tropical savannah.[21] Weapons that expose tropical savannah ecosystems to stresses to which the grassy vegetation is especially sensitive may have the effect of reducing transpiration, thereby tilting the equilibrium so that there occurs an invasion of woody species, initially in the form of scrub or thorn. The value of the land for grazing purposes will then be reduced. If the woody vegetation is harvested for firewood or thorn-fencing, the result will be an aggravated denudation of the land, probably followed by erosion and hence desertification. Weapons possessing the discriminatory biocidal action necessary to induce such a train of events would seem to be limited to chemical and biological weapons having a species selectivity for grasses. All other types of weapon would be likely to affect either both sides of the equilibrium alike, or to act more strongly on the woody species.

8.1.7 Impacts on Arid Ecosystems

As the climate becomes progressively drier in hot regions, the limiting factor of moisture becomes displayed in a diminishing species diversity. Food-chains become shorter, and less variety is available for consumer organisms in their choice of food. This imposes an increasing lability on the equilibrium between the different populations that constitute the biotic community of arid ecosystems. Exposure to the biocidal stress of weapons, especially ones of relatively narrow biocidal bandwidth, may therefore throw the equilibrium into more or less violent oscillation. The indigenous human population of the ecosystem may not have the resources for protection against the consequences and in the worst situation may itself go into rapid decline. The aftermath of the Sahelian drought, in which the grazing lands could no longer support sufficient livestock, is a terrible illustration of such a process.

There are reasonable grounds for doubting whether any use of non-nuclear weapons could induce population oscillations of a kind having human consequences comparable to that of the Sahelian disaster unless that use were on a scale so massive as to serve no other purpose. Only the use of nuclear weapons might result in such an effect unintendedly, and even then the numbers of weapons used might have to be large enough to encompass entire nomadic areas within their overall zone of herbicidal action. In such a case the direct damage to the human population might dwarf the ecosystemically mediated damage.

[21] Editors' note. The original includes a reference to "page 37" of the book this extract is from, which discusses features of tropical ecosystems, where "a labile equilibrium between grassy and woody species, which are antagonistic plant types [, exists]… the equilibrium lies in the different root systems of the two categories of producer organism and their water economy" depending on conditions one or the other type of species will thrive and outcompete the other – at lower annual rainfall the grasses absorb all moisture before it reaches the extensive root systems of woody plants, however, rain above a certain threshold favours woody species that eventually overshadow grassy photosynthesis. These systems are vulnerable to disturbance, whether direct, ecosystemically-mediated, pedological, hydrological or vegetational.

Though it seems improbable, it could be that real dangers of such a magnitude, albeit spread over a longer time period, exist in the more significant role which small consumer organisms appear to play in arid ecosystems as compared with others....[22] Taken together with the lability attendant upon low species diversity, this factor, if indeed it exists, could magnify the overall ecosystemic impact of biocidal weapons specific for such creatures. Such a possibility could presumably come about only with weapons capable of generating fine aerosols of nerve gas or of particular toxins or microbial pathogens. But, unless used at night and on a scale that ensured a sufficiently wide area coverage by the aerosol before the daytime lapse conditions in atmospheric stability set in, such an aerosol would most likely fail to permeate the small-consumer habitat to an adequate extent. If there is any empirical evidence on this matter from arid region weapon-test sites, it has yet to be published.

In arid regions moisture is both an upper and a lower limiting condition upon most plant species. But although an increase in moisture may, because of the abundance of incident solar energy, lead to a rapid colonization by species adapted to greater moisture, a decrease in moisture does not lead to an equally rapid colonization by species less demanding of moisture. Thus,[23] ... soil damage which reduces moisture retention may precipitate a sudden loss of plant cover, which in turn will expose the soil to further degradation by wind erosion. This is a mechanism of desertification which, in some soil types, might be set in motion by heavy use of high-explosive or flame weapons. Since the eroding topsoil could be carried in dust storms, vegetation outside, the affected area might also be destroyed. It is conceivable that such a desertification mechanism might be activated the other way about, with partial plant-cover destruction causing a decline in the moisture-retention of the soil sufficiently great to ensure complete soil denudation, and sufficiently long lasting to deny regrowth of the same moisture-adapted plant species; erosive wind action might then permanently impoverish the soil. Any of the weapons capable of stressing ecosystems by plant-cover destruction or herbicidal biospecificity might set such events in motion. Among non-nuclear weapons, the relatively great area effectiveness of antiplant chemical weapons would be the most threatening; though it would be hard to envisage a military rationale other than ecocide for their use in arid ecosystems.

8.1.8 Impacts on Arctic Ecosystems

Arctic ecosystems have vulnerabilities towards weapons-stress in common with arid ecosystems because of their poor species diversity. Much of what is said in the preceding section may be no less applicable to them. They have additional vulnerabilities that stem from low ambient temperature. Primary productivity is low and the biogeochemical cycling of nutrients is very slow. Biocidal stresses of weapons acting

[22] Editors' note. The original includes a reference to page 36 of the book this extract is from, which considers arid ecosystems.

[23] Editors' note. The original includes a reference to page 37 of the book this extract is from.

upon the primary producer communities of arctic ecosystems as a whole and a very long-lasting one. Recovery periods may stretch over several human generations, so that major losses of vegetation may, for practical purposes, amount to desertification. Tundral forest, once lost, may never recover.

The cold may act to increase greatly the environmental persistence of toxic agents disseminated by chemical weapons in arctic ecosystems. Their biocidal action may thus become considerably extended by virtue of the prolonged time periods over which they place sensitive organisms at risk. The sluggishness of the biogeochemical cycling will reduce the mobility in the ecosystem of any toxic or radiotoxic substances that enter the food-chains. However, since the latter are abnormally short, with scant variety of foodstuffs for consumer organisms, there may nonetheless be vicious manifestations of bioaccumulation in the case of some poisons and radionuclides. Thus, although the lability of the population equilibria defining arctic communities might well be thrown into oscillation by any large-scale use of weapons in arctic ecosystems, it can be expected that the fluctuations of population will be markedly more violent and long-lasting in the cases of nuclear weapons and some types of chemical weapon.

8.2 Qualitative Trends in Conventional Munitions – The Vietnam War and After

Editors' note. This chapter is the second of two examples of Julian's non-CBW writings that we have included in Chap. 8. Julian considers lethality, spread and broader trends of the weapons in use at the time he was writing (1979), especially those employed in the Vietnam War.[24,25] Applying his expertise in identifying, accessing and understanding diverse open sources, he assesses different categories of weapons and their kill mechanisms. He also explores weapons under development and what these suggest about the future of munitions and conflict.

There is now a ferment in the design of routine weapon systems which many commentators believe is revolutionising the nature of conventional warfare. It is providing field commanders with capacities for destruction that have hitherto been available only with nuclear weapons. The most spectacular aspect, and the one that has attracted

[24] Editors' note. This was originally published as Julian Perry Robinson, "Qualitative Trends in Conventional Munitions: the Vietnam War and After". Chap. 3 in Mary Kaldor and Asbjørn Eide (eds), *The World Military Order: The Impact of Military Technology on the Third World*, The Macmillan Press, 1979. It is not listed in Chap. 9, as that bibliography only includes Julian's writings on chemical and biological warfare. Unless otherwise indicated, the remaining footnotes for this paper are Julian's originals. Apart from correcting obvious typos and renumbering headings and tables in keeping with the requirements of this book, we have retained Julian's formatting and spelling throughout the paper and its notes.

[25] This chapter is taken from a longer study, 'Trends in conventional munitions and their implications for qualitative arms limitation', Science Policy Research Unit, University of Sussex, May 1977, currently [sic] being revised for publication in the *SPRU Occasional Papers* series. Editors' note. A manuscript for the latter paper is available via the Sussex end of the Harvard Sussex Program..

Table 8.2 A comparison of the lethalities of successive major weapons[26]

Weapon	Lethality index
Sword	20
Javelin	18
Bow and arrow	20
Longbow	34
Crossbow	32
Arquebus, 16th century	10
Musket, 17th century	19
Flintlock, 18th century	47
Rifle, Minie bullet, mid-19th century	150
Rifle, breechloading, late 19th century	230
Rifle, magazine, World War I	780
Machine gun, World War I	13 000
Machine gun, World War II	18 000
Tank, World War I (armament: two machine guns)	68 000
Tank, World War 11 (one 3-inch gun, one machine gun)	2 200 000
Field gun, 16th century, ca. 12-lb round shot	43
Field gun, 17th century, ca. 12-lb round shot	230
Field gun, 18th century, Gribeauval, 12-lb shell	4 000
Field gun, late 19th century, 75 mm high-explosive shell	34 000
Field gun, World War I, 155 mm high-explosive shell	470 000
Howitzer, World War II, 155 mm high-explosive shell with proximity fuse	660 000
Fighter-bomber, World War I (one machine gun, two 50-lb high explosive bombs)	230 000
Fighter-bomber, World War II (eight machine guns, two 100-lb high explosive bombs)	3 000 000
Ballistic missile, World War II, high-explosive warhead (V-2)	860 000
Fission explosive, 20 Kt airburst	49 000 000
Fusion explosive, 1 Mt airburst	660 000 000

Source Col T. N. Dupuy, 'Quantification of factors related to weapon lethality', annex III-H in R. Sunderland *et al.*, *Historical trends related to weapon lethality*, contract report from the Historical Evaluation and Research Organisation to the US Army Combat Developments Command (Oct 1964)

most of the commentary, is the extraordinary increase in capability for locating distant targets and then accurately directing weapons against them. Less noted have been the developments in what weapon-designers call 'kill mechanisms' and in the ways in which these can be packaged as munitions for weapons. The developments have been major; and it is important, not least for the future of arms limitation, to understand what they are. The present chapter is therefore an attempt to piece together the scattered information which the open literature contains on munition design so as to identify significant trends.

[26] Editors' note. This table was 'Table 3.1' in the original.

The focus is on destructiveness. There is a curious statistic that is to be found in some military writings which can be used as a yardstick: the 'lethality index'.[27] It reflects the area of ground per unit of time over which a weapon can create a mortal hazard to people or destroy whatever other types of target it may be designed against. The parameters of the index allow very different types of weapon to be compared, provided such matters as logistical constraint are ignored. Table 8.2, taken from a study commissioned by the US Army Combat Developments Command,[28] shows anti-personnel lethality indices for successive major weapons from the sword to the hydrogen bomb. We may observe from this that it took pre-twentieth-century weapons designers some three millennia to achieve a thousand-fold increase in lethality, but the next thousand-fold increase came within a bare half-century. Nuclear fission technology provided the final order-of-magnitude jump; but in Table 8.3, which shows anti-personnel lethality indices calculated on exactly the same basis for present-day weapons, we see that the lethality of non-nuclear weapons now overlaps with that of nuclear weapons.

The origins of the present upsurge in the power of conventional weapons may be located in the Korean War.[29] This stimulated in the rich countries a search for technological ways of coping with military-manpower inferiorities on the battlefield. Several of the anti-personnel weapon concepts which, much later, were developed into operational weapons, emerged at that time, primarily – though by no means exclusively – in the United States. But they were overtaken in the R&D laboratories by battlefield nuclear weapons, and pressure to develop them did not recover until the early 1960s when the emphasis in military requirements shifted to 'limited war'. Thereafter the pressure became intense, at least in the West, and the great resources of technological skill and R&D capacity that were invested in support of the US war effort in Vietnam yielded several more weapon concepts, many of which still remain in development. It now seems inevitable that, over the coming years and in different parts of the world, these concepts will become fully reduced to practice for the arsenals of the rich countries, their dependents, and their clients.

8.2.1 Categories of Munition

In the final resolve, battle (of any type) is and always has been a matter of taking and holding ground. The use of foot-soldiers continues to be the preferred way of doing this. By the same token, infantry force remains the basic element in an overall

[27] T. N. Dupuy, 'Quantification of factors related to weapon lethality', annex 111-H in R. Sunderland et al., *Historical Trends Related to Weapon Lethality*, contract report for the Advanced Tactics Project of the US Army Combat Developments Command (Washington, D.C., 1964). See also 'Tactical nuclear combat', *Ordnance* (Nov-Dec, 1968) p. 292.

[28] Ibid.

[29] E. Prokosch, *The Simple Art of Murder: Antipersonnel Weapons and their Developers*, rev. ed. (Philadelphia, .1972); J. S. Tompkins, The Weapons of World War Ill: The long road back from the bomb (London, 1967).

Table 8.3 Lethality indices of some modern weapons[30]

Type of weapon	Specific weapon* for which index is calculated	Lethality index†
NON-NUCLEAR WEAPONS		
Assault rifle	5.56 mm M16	4 200
Light machine gun	7.62 mm M60	21 000
Medium howitzer, HE shell	155 mm M109 with M107 HE projectiles	89 000
Shoulder-fired TPA flame-rocket launcher	66 mm M202 with 4-rocket M74 clip	1 200 000
Medium howitzer, nerve-gas shell	155 mm M109 with M121 GB projectiles	1 400 000
Automatic grenade launcher, HE/frag grenades	40mm XM174 with M406 grenades	1 500 000
Fighter-bomber with napalm firebombs	Phantom with 19 × BLU-1 750-lb firebombs	1 900 000
Main battle tank	M60 with a 105 mm gun, a light MG and a heavy MG	3 200 000
Medium howitzer, HE/frag submunition shell	155 mm M109 with M449 'improved conventional munitions'	3 500 000
Multiple rocket launcher, nerve-gas rockets	115 mm M91 with 45 × M55 GB rockets	6 800 000
Tactical guided missile, HE/frag bomblet warhead	Lance with M251 warhead	7 200 000
Fighter-bomber with HE general-purpose bombs	Phantom with 19 × M117 750-lb bombs	9 600 000
Multiple rocket launcher, HE/frag rockets	140 mm RAP-14 (French) with 21 rockets	12 000 000
Fighter-bomber (1950s) with FAE cluster bombs	Skyraider with 14 × CBU-55 cluster bombs	20 000 000
Heavy bomber with HE general-purpose bombs	B52 with 108 × Mk 82 500-lb bombs	23 000 000
Fighter-bomber with nerve-gas bombs	Phantom with 19 × MC-1 750-lb GB bombs	28 000 000
Bomber with blockbuster light-case HE bombs	Hercules with 2 × BLU-82 15,000-lb bombs	52 000 000

(continued)

[30] Editors' note. This table was 'Table 3.2' in the original.

Table 8.3 (continued)

Type of weapon	Specific weapon* for which index is calculated	Lethality index†
Tactical guided missile, nerve-gas warhead	Lance with E27 GB warhead	91 000 000
Fighter-bomber with HE/frag cluster bombs	Phantom with 19 × CBU-58 cluster bombs	150 000 000
Heavy bomber with HE/frag grenade clusters	B52 with 2 × SUU-24 dispensers for ADU-256 clusters	207 000 000
NUCLEAR WEAPONS		
Tactical guided missile, 'mininuke' warhead	Lance with developmental 0.05 Kt whd, airburst	60 000 000
Tactical guided missile, 1 Kt warhead	Lance with M234 whd, airburst at middle yield option	170 000 000
Medium howitzer, 'mininuke' shell	155 mm M109 with 0.1 Kt shell, airburst	680 000 000
Tactical guided missile, 20 Kt warhead	Pluton (French) with AN-52 warhead, airburst	830 000 000
Fighter-bomber with 350 Kt bomb	Phantom with one B-61 bomb at highest yield option	6 200 000 000
Strategic guided missile, 1 Mt warhead	Submarine-launched M-20 missile (French)	18 000 000 000
Strategic guided missile, 25 Mt warhead	SS-18 (Soviet) intercontinental ballistic missile	210 000 000 000

Conventions: (FAE) fuel-air explosive; (frag) fragmentation; (HE) high explosive; (LC) light case; (MG) machine gun; (TPA) thickened pyrophoric agent, a napalm follow-on

Notes *US weapons unless otherwise indicated

†Calculated from available area-of-effectiveness data according to the method of Col. Dupuy as in Table 8.2

defence against territorial invasion or armed insurrection. This means that weapons for killing troops, or at least putting them out of action, are the primary armament of armed forces. Historically, anti-personnel weapons have indeed formed the bulk of national arsenals. But the counters to firepower are manoeuvre and protection, which therefore enter into the design of the weapons. If a moving force is to be attacked, the choices are either to out-manoeuvre it, to use weapons that can be brought to bear accurately and rapidly on moving targets, or to use area weapons that saturate terrain across which the force is moving. There are two variants on the last alternative: the use of massed delivery systems, each weapon delivering munitions with relatively small areas of effectiveness; or the use of a small number of weapons to deliver munitions with a large area of effect ('area munitions').

If a protected force is to be attacked, the type of protection dictates the choice of weapon. Protection through concealment within natural cover may be countered with one or other form of area weapon, or with special weapons (such as fire or chemical defoliants) that strip away cover. Protection by terrain may be countered with 'indirect-fire' weapons – ones capable of reaching over obstacles to direct fire – used for area effects unless the target is observable. The protection afforded to static forces by field fortifications or to mobile forces by armour, demands special weapons capable of destroying materiel. Just as the sword and the spear are the archetypal anti-personnel weapons, so too are the battering-ram and the sap and archetypal anti-materiel weapons. As weapons technology developed, it came to provide anti-personnel weapons that could also be used to breach light fortifications or armour, simultaneously killing people inside them. This was the primary feature of the first gunpowder weapons with firearms, especially cannon firing exploding-shell projectiles, being among the first of what are now known as 'general purpose' or 'anti-personnel/anti-materiel' ('anti-PAM') weapons. In Europe by the seventeenth century, anti-PAM rather than anti-personnel weapons were becoming the dominant armament. Further shifts came with the introduction of 'high' explosives and machine-guns towards the end of the nineteenth century, for these had the effect of making weapon systems *per se* primary targets of attack. Indirect-fire weapons in particular became increasingly important targets, for with their long reach, their relative invulnerability except to other indirect-fire weapons, their ease of supply and their capacity for saturation attacks on area targets, they had become formidable obstacles to attacking forces, and formidable threats to defending ones.[31] Herein lay part of the stimulus behind the development of tactical ground-attack aircraft and, to a lesser extent, of tanks. The emergence of these and other such weapons created a demand for further counter-measures in the form of special anti-tank and anti-aircraft weapons, these in turn demanding special counter-counter-measures.

As a consequence of all this, the major preoccupation of weapons designers over the last forty years or so has been the development of a large variety of anti-weapon or anti-counter-measure weapons. Three categories can be distinguished. The first comprises updated forms of traditional anti-PAM weapons: multi-purpose weapons considered cost-effective against a wide range of personnel and materiel targets, including lightly protected weapons. They include napalm and the routine types of high-explosive (HE) artillery shell and bomb. The second category comprises anti-PAM weapons in which most of the destructive energy is harnessed for penetrating armour or degrading materiel, the anti-personnel effects being subsidiary. Examples include weapons dispensing the many varieties of anti-tank warhead: heavy projectiles made of tungsten or depleted uranium, HE in squash-head or hollow-charge configurations, and so on. Although they are intended as anti-materiel weapons, and would not be considered cost-effective against unprotected personnel targets, their objective of disabling enemy weapons systems is in fact met by killing the people operating these systems; and for reasons associated with the requirement of prior penetration of protection, their anti-personnel kill mechanisms are often extraordinarily horrible. The third category comprises true anti-materiel weapons, ones that are designed to disable enemy weapons systems or counter-measures, not necessarily by killing their occupants, but by destroying components or machinery. Weapons which are powerful enough to do this will certainly also kill people; and some of them have

[31] F. Scott,. 'To strike unseen', *Journal of the Royal United Services Institute* (June 1971) pp. 30-5.

particular anti-personnel applications which are considered cost-effective. Fuel-air explosive mine-clearing weapons are one example; some types of anti-vehicle or anti-flak incendiary and fragmentation weapons are others.

A good taxonomy should impose order on a subject, but it may also have value in exposing disorder. The foregoing attempt demonstrates that a simple categorisation of modern weapons as either anti-personnel or anti-materiel is not possible. Ways of using weapons may perhaps be so categorised, but not the weapons themselves. All that can be said is that most weapons have anti-personnel effects; a few are designed for this alone (e.g. small-calibre assault rifles or some multiple-flechette warheads); but the majority are intended to have varying degrees and kinds of anti-materiel effectiveness as well. And if weapons are ranked according to the gravity of their effects on the human body, those designed purely for anti-personnel uses may in fact be nearer the less hideous than the more hideous end of the scale.

Within each of the foregoing categories, maximum lethality is provided by indirect-fire weapons. This follows from characteristics already noted, namely range performance, concertability and ease of replenishment, which together enable large weights of munitions to be delivered over a wide front. It is therefore on trends in the design of munitions for indirect-fire weapons – tube artillery, rocket artillery and ground-attack aircraft – that this chapter concentrates, first on the form of the munitions and then on their kill mechanisms.

The cheapest indirect-fire weapons to maintain and use are mortars; the most costly, supersonic ground-attack aircraft. If the latter are to be worthwhile, the effectiveness of their ammunition has a great deal of expense to justify. It follows that of all indirect-fire munitions those dispensed by aircraft are likely to be the most advanced. A pattern may be observed of novel concepts first appearing in air munitions and then, as experience is gained and production costs diminish, filtering down to less elaborate weapons systems. This is true of, for example, cluster munitions, which appeared first as cluster bombs, then as warheads for large battlefield rockets, and now as artillery shells of smaller and smaller calibres.

The latest trends in indirect-fire munitions are obscured from the public eye by military secrecy. We may, however, learn much – as the weapons designers did themselves – from the experience of the Vietnam war. Thus, if we can identify time-trends in the expenditures of different varieties of air munition over Indochina, we may be able to judge in broad outline which way indirect-fire munitions in general have since been developing.

8.2.1.1 Munition Trends During the Vietnam War

We know from US Defense Department releases the annual tonnages of air munitions dropped on targets in Indochina either by US aircraft or by other aircraft carrying US-supplied munitions. Lumsden provides a collection of such data.[32] This tells us

[32] M. Lumsden, *Antipersonnel Weapons* (Stockholm, SIPRI, forthcoming). Editors' note. SIPRI published this in 1978 as *Anti-personnel Weapons*.

nothing about the different varieties of air munition used, but on this we have another source of information, namely what appears to be a complete year-by-year Defense Department record of the numbers (and costs) of all the different munitions procured or programmed for procurement by the US armed services during 1964–73.[33] If the figures on air munitions are extracted from these records (omitting the nuclear, biological and casualty-chemical munitions) and the total tonnage of the different procurements then calculated from what is known, or can be estimated, about the individual weights of the 80-odd different munitions, annual air-munition procurement can be compared with annual air-munition expenditure over Indochina. This is the procedure which underlies Table 8.4. We note from the table that procurement and use ran closely in step, the former exceeding the latter in 1966, 1968–9, and 1973, though not by much. This bears out what may be inferred from other sources and what Defense Department witnesses had testified to Congressional committees at the time: that the initial air operations over Indochina were supplied from war-reserve stocks of munitions of Second World War and Korean War vintage; that these stocks were 'largely expended' by 1967[34]; that standby production-lines for old munitions were then reopened and production-lines for new munitions built; and that not until 1973 did consumption rates slacken so as to allow replenishment of the denuded war-reserve stockpile.

There are many different varieties of air-to-ground munition, and some are more heavily relied upon than others. The US procurement record provides evidence on relative degrees of importance. Five broad categories of munition may be distinguished. The first two are the traditional armament of ground-support aircraft: machine guns, and free-fall bombs dropped individually. Then there are unguided rockets, an innovation (for aircraft) of the Second World War which, in the air-to-ground rather than the air-to-air role, languished after the war and did not fully recover until rockets became popular as a means for arming helicopters, which happened early in the Vietnam war. Munitions dispensing clusters of small free-fall sub-munitions constitute the fourth category; the simplest forms go back to the First World War but, during the 1950s, development programmes commenced which subsequently yielded a wide variety of bomblet and minelet cluster munitions, some of extraordinary complexity. Finally, there is the category of guided weapons: air-to-ground guided missiles and glide bombs fitted with target-homing devices that permit exceptionally accurate delivery. The latter are the 'smart' bombs which came into Increasing use in Indochina from 1967 until the USAF withdrew from direct operations.

Table 8.5 shows annual US Air Force procurements of each of these categories. The data are presented as percentages of total annual procurements, both by tonnage and by dollar cost. There are two major omissions: guided air-to-ground missiles other

[33] US Dept of Defense, Reports of programs for fiscal years 1964-73: 'Procurement of equipment and missiles, Army: Ammunition, P-4800'; 'Other procurement, Navy: Weapons and support equipment'; 'Procurement, Marine Corps: Budget activity 1: Ammunition'; and 'Other procurement, Air Force: Munitions and associated equipment' (30 June 1966-31 Dec 1972).

[34] US Dept of Defense, Congressional testimony. Submission for the record of the House Defense Appropriations Subcommittee, Feb 1974.

Table 8.4 Tonnages of US conventional air-to-ground munitions used over Indochina, and tonnages procured by the US Air Force and US Navy, each year during 1963–73[35]

Category	1964	1965	1966	1967	1968	1969	1970	1971	1972	1973	Total
US Air Force procurements (megadollars, by fiscal year)	58.6	175.9	839.2	1016.0	1063.4	1109.6	730.4	540.4	586.5	664.8*	6784.8
US Air Force procurements (kilotonnes, by fiscal year)	15.4	81.5	565.0	542.5	997.5	1089.5	649.5	540.3	638.7	678.0*	5797.4
US Navy procurements (kilotonnes, by fiscal year)	7.0	51.4	138.8	224.6	322.6	353.6	164.5	63.5	108.8	161.7*	1596.5
Total US procurements (kilotonnes, by fiscal year)	22.4	132.9	703.8	767.1	1320.1	1443.1	814.0	603.8	747.5	839.7*	7393.9
Quantities used in Indochina (kilotonnes, by calendar year)	?	285.8	458.4	845.6	1304.0	1258.5	886.7	692.3	983.7	378.1†	7039.1

Notes *Programmed procurement as of 31 Dec 1972
†January through June only
Sources On procurement: Figures calculated from data contained in US Defense Department annual returns on types, numbers and costs of munitions procured or programmed for procurement during fiscal years 1964–73
On use: M. Lumsden, Antipersonnel weapons (Stockholm: SIPRI, forthcoming)

[35] Editors' note. This table was 'Table 3.3' in the original.

Table 8.5 Relative importance of different categories of air-to-ground munition as indicated by annual US Air Force procurements during fiscal years 1964–73[36]

Category of weapon	1964	1965	1966	1967	1968	1969	1970	1971	1972	1973*	1964-73
PERCENTAGES BY WEIGHT OF MUNITIONS											
Massive bombs and mines	57.5	81.1	91.4	86.9	93.8	92.1	86.4	86.2	90.4	91.6	90.2
Precision guided bombs	np	np	.	0.3	0.1	0.4	1.0	2.5	2.6	0.8	0.8
Machine-gun ammunition	2.6	3.9	2.8	2.8	1.2	1.0	0.7	1.1	1.1	1.0	1.4
2.75 inch rockets	np	5.2	1.7	3.0	1.6	1.6	1.5	0.8	1.3	1.4	1.6
Cluster munitions	39.9	9.8	4.0	7.0	3.4	4.7	10.5	9.4	4.8	5.1	5.9
PERCENTAGES BY COST OF MUNITIONS											
Massive bombs and mines	6.7	32.7	56.2	44.7	63.1	61.3	46.1	49.8	56.8	57.8	53.9
Precision guided bombs	np	np	0.1	3.9	3.0	4.0	5.9	12.3	11.3	6.6	5.1
Machine-gun ammunition	5.3	12.3	14.6	11.9	8.2	4.6	2.7	5.1	5.1	4.5	7.6
2.75 inch rockets	np	12.2	6.5	7.7	6.5	6.8	5.1	3.6	5.6	7.5	6.5
Cluster munitions	88.1	42.9	21.7	31.8	19.3	23.3	40.3	29.2	22.1	23.6	26.9

Conventions: (np) no procurement; (.) less than 0.05%
Note *Programmed procurement as of 31 Dec 1972
Source Figures calculated from data contained in US Defense Department annual returns on types, numbers and costs of munitions procured or programmed for procurement during fiscal years 1964–73

than smart bombs; and the fuzes for those munitions for which fuzes are supplied separately. These omissions probably do not greatly affect the overall tonnages, but they could make a substantial difference to the relative dollar costs per unit weight of each category of munition. Be that as it may, the table clearly indicates that the two categories in which innovation both burgeoned and was applied were those of smart bombs (which are beyond the scope of this essay) and of cluster munitions.

[36] Editors' note. This table was 'Table 3.4' in the original.

8.2.1.2 Cluster Munitions

Table 8.5 indicates that cluster munitions – the 'CBUs' of Vietnam war reporting – were in production as War Readiness Materiel before the US Air Force had begun operations over Indochina. By 1966 requirements for cluster munitions were apparently exceeding supplies, a user-demand which subsequently, it would seem, both increased and was met. Somewhere between 50 and 100 different cluster munitions were studied in US R&D laboratories during the war,[37] and of those more than 30 went into production and were presumably used in Indochina.

The US Air Force procured about 350,000 tonnes of cluster munitions during 1964–73, and the US Navy another 40,000 tonnes. Though this was less than six per cent of the total weight of air munitions procured during that period, it accounted for more than a quarter of the procurement expenditure. This gives some idea of the importance attached to cluster munitions.

The logic behind cluster munitions is that the smaller a munition the greater is the area-effectiveness of a given total weight. This is because the destructive effects of a single munition, whatever its kill mechanism, diminish rapidly with distance from its point of function. Hence, if a munition is to exert destructive effects over an area, kills at the periphery necessitate overkill at the centre. Cluster munitions allow killing agents to be distributed more evenly over a target, thereby reducing net overkill and providing greater economy. They may, in some cases, allow attacks on area targets which, for logistical reasons, would otherwise be impossible to engage.

From the standpoint of design, the chief problem is to ensure that the cluster munition dispenses its sub-munitions evenly over the target. The design of the dispenser and the design of the sub-munitions are the two basic variables. Thus, some of the requisite dispersion may be achieved by having the dispenser forcibly eject the sub-munitions over an area, rather as artillery may be used for pattern fire. Alternatively, the sub-munitions may be fitted with some sort of self-dispersing mechanism so that they distribute themselves over the target area after release from the dispenser. The more sophisticated types of cluster munition combine both methods.

Table 8.6 shows the total US air cluster munition procurement during 1964–73 grouped into different categories according to the type of dispenser. It will be seen that procurement was initially directed towards rearward-ejection dispenser munitions (REDMs). These comprise an externally carried sheaf of tubes, looking rather like a rocket-pod pointing backwards. There are large streamlined versions for use from tactical ground-support aircraft at high speed; and there are smaller, simpler versions for helicopters and light aircraft. Each tube of an REDM – about 70 mm calibre in the early US ones – is loaded with several bomblets which are ejected in a rapid stream at a rearward velocity roughly the same as the forward velocity of the aircraft; they thus fall in a tight pattern along the flight path. Over the 1964–73 time period, procurement of REDMs increasingly concentrated on systems for low-performance aircraft. For high-performance aircraft, downward ejecting systems evidently came to be considered more attractive. Sub-munitions for these DEDMs could be larger than

[37] For a partial listing see Table 6.12 in Lumsden, op. cit.

those for rearward-ejecting systems, for attacks on hard targets, as in the CBU-33 and CBU-38. Or they could be much smaller, thus enabling very wide area coverage. The latter property was exploited in Indochina in support of new mine-warfare techniques, DEDMs being used to scatter enormous numbers of small anti-personnel mines over areas across which passage was to be denied. Thus, the CBU-28/37 Dragontooth, the CBU-34/42 WAAPM, and the SUU-41 Gravel anti-personnel mining systems were extensively used during Air Force Igloo White operations in support of the 'McNamara Wall'. This is now an international growth area in indirect-fire weaponry.

In that DEDMs provided a means for dropping free-fall clusters of sub-munitions in quick succession, they have their antecedents in devices used during the Second World War for dropping small incendiary bombs from bomber aircraft. Among these the Soviet AK-2 Aerial Release Case may be noted. When B-52 and other bombers came to be used over Indochina, modernised versions of these devices, chiefly the SUU-24 Hayes Dispenser, were procured so as to allow the latest types of bomblet and grenade to be used in high-altitude area bombing attacks. The dispensers dropped large canisters – boxes commonly weighing about 150 lbs and with no special aerodynamic shaping – of bomblets or grenades, the boxes opening some way above the target to scatter their contents. The preferred payloads were evidently self-dispersing sub-munitions: spherical bomblets with vanes built into their outer surface shaped so as to impart rotation as the bomblet fell through the air and then Magnus Lift to give the desired lateral motion. Two Hayes Dispensers can be carried in a B-52; each dispenser can hold 72 canisters; and each canister can hold some hundreds of submunitions depending on their size (more than 500 in the case of the small M40 grenades). Table 8.6 shows that procurement of Hayes Dispensers took place during 1965–6, and of canister-clusters for them intermittently during 1965–72. Press reports of B-52 cluster-bombing did not begin to appear in the outside world until the end of this period.

Canister clusters for Hayes Dispensers and the like accounted for about half of the tonnage of cluster munitions procured during 1966, but thereafter true cluster bombs became far the most heavily procured of all cluster munitions. In these, such as the Sadeye and Rockeye, the dispenser does not remain attached to the aircraft, but is dropped as an aimable free-fall unit resembling a general-purpose bomb; the casing opens in mid-air to release the bomblets. In the commonest variety, based on the navy's Sadeye dispenser which opens like a cockle-shell, the bomblets are of the spherical self-dispensing type, no special ejection mechanism being used to scatter them other than the spin imparted to the falling dispenser by its tailfins. The smaller but more elaborate Rockeye dispenser, also a navy innovation, does not use self-dispersing bomblets but appears instead to spin out its payload by means of a rocket motor. Loaded with hollow-charge bomblets, the Rockeye dispenser is the principal US air-delivered anti-tank munition. Cluster bombs of various types are included in US arms exports; certain NATO countries, Iran, and Israel have been among the recipients. Cluster bombs were used in the Israeli air-raid on Damascus

Table 8.6 Cluster munitions procured by the US Air Force and US Navy, 1964–73[38]

Cluster munition	Type of sub-munition used		Annual procurements (weight of munitions purchased, kilotonnes)									
			1964	1965	1966	1967	1968	1969	1970	1971	1972	1973
HIGH-SPEED REARWARD-EJECTION DISPENSER MUNITIONS		TOTAL, percent wt:	*82.6*	*16.9*	*5.8*	*15.7*	*19.7*	*2.9*	*0*	*0*	*0*	*0*
Dispenser w/ anti-personnel bomblets, CBU-1	Fragmentation, BLU-4		1.29	0	0	0	0	0	0	0	0	0
Dispenser w/ anti-vehicle bomblets, CBU-2	Heavy pellet, BLU-3		4.06	0	0.07*	5.81	3.84	0	0	0	0	0
Dispenser w/anti-tank bomblets, CBU-3	Hollow charge, BLU-7		0.77	1.03	0.33	…	0	0	0	0	0	0
Dispenser w/ apers-smoke bomblets, CBU-12 & CBU-13	WP, BLU-17 & BLU-17 + BLU-16 (HC)		0	0.41	0.96	0.56	..	0	0	0	0	0
Dispenser w/ apers-jungle bomblets, CBU-46	Fragmentation, BLU-66		0	0	0	0	3.71	1.92	0	0	0	0
		TOTAL:	6.12	1.44	1.36	6.37	7.55	1.92	0	0	0	0
LOW-SPEED REARWARD-EJECTION DISPENSER MUNITIONS		TOTAL, percent wt:	*0*	*32.7*	*19.1*	*10.5*	*0.6*	*5.6*	*7.0*	*4.0*	*11.2*	*9.7*
Dispenser w/ anti-vehicle bomblets, CBU-14	Heavy pellet, BLU-3		0	2.79	3.44	2.27	..	1.41	2.45	0.72	0	0

(continued)

[38] Editors' note. This table was 'Table 3.5' in the original.

Table 8.6 (continued)

Cluster munition	Type of sub-munition used		Annual procurements (weight of munitions purchased, kilotonnes)									
			1964	1965	1966	1967	1968	1969	1970	1971	1972	1973
Dispenser w/ apers-smoke bomblets, CBU-22	White phosphorus, BLU-17		0	0	0.12	0.29	0.16	0	0	0.14	0	0
Dispenser w/apers jungle bomblets, CBU-25	Fragmentation, BLU-24		0	0	0.92	1.71	0.07	2.30	2.89	1.28	4.09	3.56
		TOTAL:	0	2.79	4.48	4.27	0.23	3.71	5.34	2.14	4.09	3.56
DOWNWARD-EJECTION DISPENSER MUNITIONS		TOTAL, percent wt:	*0*	*12.4*	*8.6*	*4.9*	*5.5*	*3.5*	*4.1*	*1.0*	*10.6*	*0*
Re-usable dispenser, Gladeye	(unloaded)		(..)	(0)	(0)	(0)	(0)	(0)	(0)	(0)	(0)	(0)
Re-usable Tactical Fighter Dispenser (TFD), SUU-41	(unloaded)		(0)	(0)	(0)	(2792)	(2088)	(..)	(0)	(0)	(0)	(0)
Disp w/clustered anti-personnel bomblets, CBU-7	Fragmentation, BLU-18		0	1.06	1.68	…	0	0	0	0	0	0
Strongback w/ clustered apers-irritant blets, CBU-19	CS irritant, BLU-39		0	0	0.13	0.09	0	0	0	0	0	0

(continued)

Table 8.6 (continued)

Cluster munition	Type of sub-munition used		Annual procurements (weight of munitions purchased, kilotonnes)									
			1964	1965	1966	1967	1968	1969	1970	1971	1972	1973
Disp w/clustered apers minelets, CBU-28 & CBU-37	Blast, BLU-43 & BLU-44 Dragontooth		0	0	0.16	1.29	0.85	..	0	0	0	0
Disp w/clustered apers-irritant bomblets, CBU-30	CS irritant, BLU-39		0	0	0	0.13	0.55	0.89	..	0	0	0
TFD w/anti-tank mines, CBU-33	Hollow charge, BLU-45		0	0	..	0.23	..	..	0	0	0	0
TFD w/clustered apers minelets (WAAPM), CBU-34 & -42	Fragmentation, BLU-42 & BLU-54		0	0	0.06	0.27	0.70	1.38	2.57	0	0	0
Dispenser w/anti-PAM bombs, CBU-38	Fragmentation, BLU-49		0	0	0	0	0	0.04	0.55	0.52	3.89	0
TFD w/clustered ?apers jungle bomblets, CBU-43	Fragmentation, BLU-48		0	0	0	0	0	..	0	0	0	0
TFD w/apers bomblets, CBU-44	?		0	0	0	0	..	0	0	0	0	0
TFD w/multi-purpose minelets, CBU-45	?		0	0	0	0	..	0	0	0	0	0
		TOTAL:	0	1.06	2.02	2.01	2.09	2.31	3.13	0.52	3.89	0

(continued)

Table 8.6 (continued)

Cluster munition	Type of sub-munition used		Annual procurements (weight of munitions purchased, kilotonnes)									
			1964	1965	1966	1967	1968	1969	1970	1971	1972	1973
CANISTER-CLUSTERS FOR INTERNALLY-MOUNTED DISPENSERS (FOR B-52 BOMBERS, &c.)		TOTAL, percent wt:	*0*	*13.3*	*50.1*	*9.9*	*0.2*	*0.1*	*4.4*	*0*	*2.8*	*0*
Re-usable dispenser, SUU-24	(unloaded)		(0)	(80)	(84)	(0)	(0)	(0)	(0)	(0)	(0)	(0)
Canister w/ anti-personnel jungle bomblets	Fragmentation, BLU-24		0	0	1.14	..	0	0	0	0	0	0
Canister w/apers grenades, ?ADU-256 (*et al.*)	Fragmentation, M40 (and other grens)		0	0.16	0.05	0.13	0.09	0.07	0	0	0	0
Canister w/anti-vehicle bomblets, ADU-253	Heavy pellet, BLU-3		0	0.98	7.04	2.47	..	0	0	0	0	0
Canister w/anti-PAM bomblets, ADU-272	Pellet, BLU-26		0	0	2.94	1.44	..	0	3.36	0	1.02	0
Canister, anti-PAM, ?BL-29	Napalm-B		0	0	0.58	0	0	0	0	0	0	0
Canister w/anti-PAM bomblets, CDU-22	Fragmentation, BLU-63		0	0	0	0	0	0	..	..	0	0
		TOTAL:	0	1.14	11.75	4.03	0.09	0.07	3.36	0	1.02	0
CLUSTER BOMBS		TOTAL, percent wt:	*17.4*	*24.7*	*16.4*	*59.0*	*74.0*	*87.8*	*84.4*	*95.1*	*75.5*	*90.3*

(continued)

Table 8.6 (continued)

Cluster munition	Type of sub-munition used		Annual procurements (weight of munitions purchased, kilotonnes)									
			1964	1965	1966	1967	1968	1969	1970	1971	1972	1973
125-lb, clustered anti-PAM bombs, M1A2 & M1A4	Fragmentation, M41		0	1.23	2.02	0.56	1.73	3.38	0.75	0.55	0	0
750-lb, clustered anti-PAM bomblets, M36	Incendiary, M126		0	0	0	0	0	2.81	4.26	6.21	..	0
500-lb, Rockeye w/ anti-tank apers bomblets, MK20	Hollow-charge/frag, MK118		0	0	0.12*	0.31*	0.63*	1.90*	3.65*	5.80*	7.65*	4.39*
750-lb, Sadeye w/ anti-PAM blets, CBU-24, -29 & -49	Pellet, BLU-26, BLU-36 & BLU-59		1.29*	0.87*	1.71*	23.16*	25.79*	49.53*	50.73*	22.10	..	0
750-lb, Sadeye w/ anti-PAM bomblets, CBU-52	Frag/incendiary, BLU-61		0	0	0	0	0	0	1.07	1.72	6.64	8.03
750-lb, Sadeye w/ anti-personnel bomblets, CBU-53	Incendiary, BLU-70		0	0	0	0	0.24	..	0	0	0	0
750-lb, Sadeye w/ anti-PAM bomblets, CBU-54	Incendiary, BLU-68		0	0	0	0	0.02	..	0	0	0	0

(continued)

Table 8.6 (continued)

Cluster munition	Type of sub-munition used		Annual procurements (weight of munitions purchased, kilotonnes)									
			1964	1965	1966	1967	1968	1969	1970	1971	1972	1973
500-lb, clustered anti-PAM bombs, CBU-55	Fuel-air explosive, BLU-73		0	0	0	0	0	0.07*	1.42*	0.50*	0.86*	4.12*
750-lb, Sadeye w/ anti-PAM bomblets, CBU-58	Fragmentation, BLU-63		0	0	0	0	0	0	2.28	14.24	12.54	16.42
500-lb, Rockeye w/ discr AT-apers blets, CBU-59	Hollow-charge/frag, BLU-77		0	0	0	0	0	0	0	0.08	0	0
750-lb, Sadeye w/ sub-munitions, CBU-70	?		0	0	0	0	0	0	0	0.13	0	0
		TOTAL:	1.29	2.10	3.85	24.03	28.40	57.69	64.15	51.32	27.68	32.96
		GRAND TOTAL:	7.41	8.52	23.46	40.70	38.37	65.69	75.97	53.98	36.68	36.52

on 9 October 1973 (a reprisal for the earlier Syrian FROG-missile attacks on Israeli kibbutzim).[39]

Several of the air-delivered cluster munitions used over Indochina were still in development at the time of their use, their procurement being for operational evaluation. There were probably others also, purchased out of R&D appropriations (and therefore not showing in the procurement tables): examples from among non-cluster munitions include the big Air Force fuel-air explosive bombs (BLU-72 and BLU-76) described later. Some of these developmental cluster munitions may later have been abandoned, but there is little published information either on this or on new types that have emerged since.

The 1972 and 1973 procurement data give some idea of which of the many cluster munitions were considered the most satisfactory and most appropriate for the post-Vietnam war-reserve stockpile. The US Defense Secretary's 1972 Defense Report states that these were the first two years of a

> 5-year procurement program to build an inventory of more sophisticated, more effective air munitions. When completed, this modernised air munitions inventory will, through a higher degree of effectiveness per sortie, improve the ability of our fighter/attack aircraft to achieve desirable effectiveness with fewer sorties and less exposure to anti-aircraft defenses. This air munitions modernisation program calls for significant additions of new munitions with better target destruction capability than those currently in the inventory.[40]

Of the $187 million sought in the FY 1973 budget for procurement of cluster munitions, 75 percent was to be spent on the stockpile.[41]

Recent writings continue to emphasise the importance of three roles which air cluster munitions played during the Vietnam war: the suppression of anti-aircraft defences; wide-area scatter mining; and tank destruction. For all of these, a new and much larger free-fall dispenser has been developed: the SUU-54, which is in the 2000-lb class. It is used in a variety of munitions that are still in development: examples include the GBU-2 Pave Storm flak-suppression munition; the Gator wide-area anti-personnel and anti-tank mining system; the Grasshopper anti-vehicle mine system; and the Piranha anti-tank mine system for use in shallow water.[42] For direct attack of tanks there is a 1974 report of a conceptual weapon in which the SUU-54 would be used to dispense independently-homing anti-tank sub-munitions.

The United States does not, of course, have a monopoly in cluster munitions. Very little seems to be known about Soviet cluster munitions, except that they exist. Among air-delivered anti-tank munitions, the Rockeye cluster bomb has at least two foreign competitors of similar size. There is the British Mk 1 No 1 Cluster Bomb (the BL755)

---

[39] N. Thimmesch, 'War's Double Standard', *Philadelphia Bulletin* (20 Dec 1973).

[40] US Dept of Defense, Report to the Congress by M. Laird (Secretary of Defense). Annual Defense Department Report FY 1973 (Washington, D.C., 1972).

[41] US Dept of the Air Force, Congressional testimony, W. W. Snavely·(Assistant Deputy Chief of Staff, Systems and Logistics), Statement before the Senate Defense Appropriations Subcommittee (18 Feb 1972).

[42] C. A. Robinson, 'ADTC presses weapon studies', *Aviation Week and Space Technology* (15 July 1974) pp. 265–7, 269.

which was conceived in 1964, shortly after the Rockeye, but which did not enter production until 1971; it dispenses a larger hollow-charge/fragmentation bomblet with, so it is said, greater precision and reliability.[43] Foreign sales of the BL 755 are being actively promoted: customers already include five other NATO air forces and at least one African country (Kenya), and there are hopes of US purchases also. Then there is the French Belouga cluster bomb, which seems similar to the BL755 but was still in development in 1976.[44] Three different types of submunition are available for it: anti-tank and anti-personnel bomblets, and a sub-munition for area-interdiction. There is an Australian cluster bomb, reported to be in development in the 1972 Defence Report. Finally there are West German developmental cluster-bombs within the Strebo programme on fixed and free-fall dispenser munitions.

The West German work on cluster munitions was kept secret until late 1970, when information on three types of sub-munition was released: the Pandora anti-tank minelet, designed to destroy the tracks of a tank but incapable of being triggered by wheeled vehicles or by unwary footsteps; the Medusa, a larger and more destructive magnetic anti-tank (hollow-charge) minelet; and two varieties of Dragon Seed bomblet, one anti-personnel and the other hollow-charge armour piercing.[45] One of the design considerations in the Strebo programme is the fact that Soviet tank-warfare doctrine envisages a massing of tanks into 500 m^2 prior to attack; certain of the developmental Strebo munitions – presumably including cluster bombs loaded with Dragon Seed – are said to be capable of enabling a single aircraft (e.g. a Starfighter or a Tornado) to attack such a square; other Strebo cluster munitions are being developed to give still larger area-coverage.[46] These systems are not confined to cluster bombs. There is what appears to be a DEDM dispensing Medusa minelets; and there is the Streuwaffen, a 19-tube REDM dispensing 409 half-kilogram bomblets (presumably Dragon Seed). The French Giboulée system, in development since 1966, is also an REDM, dispensing 50 mm anti-personnel (fragmentation) or anti-tank (hollow-charge) bomblets.[47]

The progression of cluster munitions from aircraft to missile to artillery to mortar systems is clearly displayed in the US arsenal. The non-nuclear warheads developed by the US Army for its battlefield-support guided missiles and heavy artillery rockets have all been of the clustered-bomblet type: first, since the 1950s, warheads containing spherical self-dispersing bomblets charged with chemical-warfare or biological-warfare agents, and then warheads containing fragmentation bomblets, some of them of types originally developed for air force cluster bombs. The latest version, the 1000 lb M251 warhead for Lance, standardised in 1974, contains 833 BLU-63 bomblets. The unit cost of a Lance fitted with this warhead was stated, in March 1976, to be $207,000. US production has so far been entirely confined to fulfilling foreign military sales orders: 200 for Israel (deliveries began in February

[43] Weapon notes no. 12: BL-7 55 cluster bomb, *Flight International* (14 Feb 1974) pp. 221–3.

[44] R. T. Pretty and D. H. Archer, *Jane's Weapon Systems 1976-77* (London, 1976).

[45] S. Geisenheyner, 'Letter from Europe: Area weapons', *Air Force Magazine* (Feb 1971) pp. 19–21.

[46] 'New weapons to counter land attack', *Aviation Week and Space Technology* (24 Apr 1972) p. 60.

[47] Pretty and Archer, op. cit.

1976), and others for Belgium and the Netherlands. The M251 warhead is optimised for use against personnel and light materiel targets; still in development in 1977 was a warhead effective against personnel carriers and tanks.[48] Bomblet warheads have also been developed for air-to-ground missiles, but little information has been released about them. The Bullpup missile, in its AGM-12E variant developed for the Vietnam war, is said to have a warhead containing 830 BLU-26/36 fragmentation bomblets.

Bomblet and minelet payloads for cannon-artillery projectiles of 8-inch, 155 mm, and 105 mm calibre have been in development since the early 1960s. Bomblet shells have been one of the two main foci in the US Army's 'Improved Conventional Munition' (ICM) programme (the other focus evidently being the development of special high-fragmenting iron alloys for shell casings).[49] Minelet shells are the basis of the closely related Army programme on Artillery-Delivered Target-Activated Munitions (ADTAM), another manifestation of the current interest in new offensive mine-warfare techniques.

Field evaluation of ICM artillery clusters in comparison with current general-purpose HE shell took place in South Vietnam during 1968. It is reported that against like targets – presumably soldiers in the open – up to 55 percent fewer ICM rounds were needed.[50] US Army Congressional testimony in 1974 spoke of the 105 mm ICM round as being ten times more lethal than the normal HE round; and it was said that the follow-on round then in engineering development would be more lethal still.[51] The US Army and US Marine Corps procured about 50,000 tonnes of cluster-shell projectiles during 1964-73.

As to mortar systems, the US Army has standardised an anti-personnel cluster round, the M453, for its 107 mm mortar.[52] There does not appear to have been any major procurement of this prior to 1974.

Outside the United States, it seems that the main interest in ground-launched cluster munitions has to do with offensive mine-warfare systems. Most of the available information is on West German developments: a 1971 report refers to 155 mm artillery rounds loaded with Pandora anti-tank minelets; a 1973 report, to development of warheads containing Pandora, Medusa, or Dragon Seed sub-munitions for the LAR 110 mm multiple rocket launcher (MRL) system.[53] The latter project now seems to have culminated in the Minenstreumittel-Werfer (MSM/W), which was

[48] US Dept of the Army, Congressional testimony. Col. Whalen, statement before the Senate Armed Services Committee (9 Mar 1976); and 'Latest U.S. Army equipment on show', *International Defense Review* (Feb 1977) pp. 101–8.

[49] See E. Prokosch,. 'Antipersonnel weapons', *International Social Science Journal* (1976), Vol. 28, no. 2, pp. 241–358; and E. Prokosch, 'How large is your wound?', unpublished MS.

[50] US Army Materiel Command, 'Program achievements 1962-1972' (23 Mar 1972).

[51] US Dept of the Army, Congressional testimony, J. R. Deane, Statement before the House Armed Services Committee (21 Mar 1974).

[52] US Army Materiel Command (n. 20, above). Editors' note. This corresponds to US Army Materiel Command, 'Program achievements 1962–1972' (23 Mar 1972).

[53] S. Geisenheyner, op; cit.; and 'Tri-lateral RS 80 artillery rocket project', *International Defense Review* (Apr 1973) p. 147.

undergoing final evaluation in 1976[54]; derived from LAR, it comprises a 15-tube truck-mounted launcher, each tube firing a canister containing five AT-II anti-tank minelets to within a range bracket of 800–3000 m. There are signs that the French Army is developing cluster warheads for its own MRL system, the 140 mm RAP-14. Around 1970, the British Foil MRL project was subsumed into the nascent trilateral British~German-Italian RS-80 project on the joint development of a long-range (40–60 km) heavy-calibre (250–75 mm) MRL system; cluster warheads of anti-personnel and anti-armour minelets are envisaged for this. Britain withdrew from RS-80 in 1975.

Cluster munitions are, as described above, primarily intended for area weapons. Their quantitative area-effectiveness is therefore an important factor in decisions about their development and acquisition. Such information is also directly relevant to those questions of discriminateness and indiscriminateness which imbue much of the present international consideration of qualitative limitations on conventional weapons. But military authorities tend to keep this information secret, and it is difficult for outsiders to form a clear picture; there is also much unreliable information in circulation.

Table 8.7 provides estimates in respect of particular cluster munitions and certain other area munitions.

It should also be said in relation to Table 8.7 that the area of effect sought from an area weapon will depend on the type of target the munition is intended to attack and the type of operation in which it is to be used. In some cases only a limited area coverage will be sought.

For an anti-tank cluster bomb, for example, the desired coverage might be defined by the area of ground over which a tank might move during the time-of-fall of the bomb, plus an additional area to compensate for delivery errors. In other cases, maximum area effectiveness will be a primary objective, either as a means for reducing the weight of munitions needed to attack certain types of target or to provide special wide-area attack capabilities. For the latter there seems at present to be a marked growth in military requirements. The proliferation of ground-force precision guided munitions, particularly the new highly mobile anti-tank and anti-aircraft guided weapons, is undoubtedly stimulating greater interest in area-saturation weapons for the defence of armour and ground-support aircraft. Again, the increasing interest, noted earlier, in scatter-mining as an adjunct to offensive manoeuvre has created requirements for new wide-area target-activated weapons.

8.2.2 Categories of Kill Mechanism

The area of effectiveness of a munition is a measure of the area of ground over which the ‘hit-kill probability’ exceeds a specified level. This probability defines the combined probabilities of a target-element at which the munition is directed

[54] ‘German mine-laying techniques’, *International Defense Review* (Aug 1976) p. 648.

Table 8.7 Rough estimates of maximum effective areas of representative area munitions[55]

Type of munition	Effective area coverage (ha)	Actual munition on which the estimate is based
750-lb napalm firebomb	0.2	BLU-1 (US)
750-lb general-purpose HE bomb	0.4	M117 (US)
155 mm cluster shell with apers fragmentation bomblets	0.5	M449 (US)
Light RE dispenser munition with apers frag bomblets	0.6	CBU-25 (US)
600-lb anti-tank cluster bomb	0.7	Belouga (French)
Heavy RE dispenser munition with apers frag bomblets	1.0	Giboulée (French)
2000-lb general-purpose HE bomb	1.1	Mk84 (US)
500-lb fuel-air explosive cluster bomb	2.6	CBU-72 (US)
155 mm battalion fire, HE shell	5.8	M114/M1 (US)
External-carriage DE dispenser munition, apers frag bomblets	6.6	CBU-7 (US)
750-lb anti-PAM fragmentation cluster bomb	7.0	CBU-24 (US)
Anti-PAM frag bomblet warhead for battlefield-support missile	8.7	Lance/M251 (US)
Ground-launched anti-tank minelet cluster	9.0	MSM/W (West German)
Ground-launched apers frag warhead cluster	10	RAP-14 (French)
2000-lb anti-PAM fragmentation cluster bomb	19	Pave Storm (US)
15,000-lb light-case HE bomb	49	BLU-82 (US)
0.1 kiloton airburst nuclear-fission warhead	92	(hypothetical)
Nerve-gas bomblet warhead for heavy artillery rocket	110	Honest John/M190 (US)
Bomb-bay DE dispenser munition, clustered apers frag bomblets	150	SUU-24/ADU-256 (US)

Source: The bases of the estimates, and the published sources on which they rely, are set out in J. P. Perry Robinson, 'Trends in conventional munitions...', op. cit.[56] Actual effective area coverage of course depends on such factors as target-protectedness, delivery mode, etc. The estimates given here are roughly comparable points within ranges varying widely in breadth

being struck (the 'hit probability') and of being put out of action by a strike (the 'kill probability'): a person killed or disabled, or a piece of equipment destroyed.

[55] Editors' note. This table was 'Table 3.6' in the original.

[56] Editors' note. The full bibliographic details are: "Trends in conventional munitions and their implications for qualitative arms limitation", Science Policy Research Unit, University of Sussex, May 1977. A manuscript for this paper is available via the Sussex end of the Harvard Sussex Program.

Older types of munition tended to have either high hit probabilities or high kill probabilities, but not both. The basic anti-personnel weapon of the Second World War, the infantryman's rifle, is an example: during the war something like 50,000 bullets were fired for each hit, but every third or fourth hit killed. Cluster munitions represent an attempt at trading-off hit and kill probabilities against each other so as to achieve a greater overall hit-kill probability.

To this same end, the actual kill mechanism that is built into a munition or sub-munition has offered much scope to the ingenuity of weapons designers. For killing, the military have always preferred to hack, pierce, or shatter. Other possibilities have been less well regarded, but trends of fashion may be discerned, loosely connected with technical change. Flame, for example, has become increasingly popular during the present century; and, in the package of different kill mechanisms proffered by nuclear fission and nuclear fusion, the military are now seeing particular attractions in ionising radiation. In contrast, toxicity evidently has little allure, and infectivity even less. Physical methods of destruction, centred on bullets and explosives, remain in the ascendancy.

Chemical explosives provide two basic kill mechanisms: energy transfer by blast wave or by moving fragments. The latter, when exploited by detonating an explosive within a metal casing, is especially conducive to area effectiveness; the former, to high kill probabilities over limited areas. Fragmentation munitions come in three broad categories according to the type of casing used to surround the explosive. In the original type, going back to the early days of gunpowder, the casings did not incorporate any special aids to fragmentation: they shattered into a broad spectrum of fragment sizes, some large and excessively destructive, others so small as to be harmless. The trade-off between hit and kill probabilities was therefore poor. To improve this, techniques of 'controlled fragmentation' emerged which sought to narrow the fragment spectrum by scoring the casing so as to create a multiplicity of fracture points (a technique which did not in fact work very well until it was appreciated, some time after the Second World War, that the scoring had to be done on the inside surface of the casing, not the outside). Another technique for controlling fragmentation has been to construct the casing of helically-wound notched steel wire or rod. A technique for still greater control, developed during the 1950s, is to pre-fragment the casing into steel pellets, embedding them in a soft matrix. This is a modern twist in a line of development which stretches back through shrapnel shell to the earliest forms of artillery case shot (bullet-filled shell). In fact neither pre-fragmented nor scored casings are suitable for artillery munitions since they are too weak to withstand projection from a gun barrel: their use is limited to aircraft and mortar bombs, grenades, rocket warheads, and sub-munitions. Maximised fragmentation effects from artillery shell are currently obtained either by using shell casings made of special iron alloys whose crystal structure serves to narrow the fragment-size spectrum, or by using modern versions of case shot. These include the sub-munition-filled projectiles noted earlier and canister rounds having a payload of small metal missiles. The Beehive artillery ammunition provided to US forces in Indochina, largely on an experimental basis and mostly for direct-fire guns, is the latest manifestation of the latter: each round ejects some thousands of small fin-stabilised needles, known as 'flechettes'.[57] Quite large numbers of multiple-flechette projectiles were used during the Vietnam war, particularly as air-to-ground rocket warheads. Originally anti-personnel, multiple-flechette munitions are now being developed for armour~piercing uses as well.[58] Depleted uranium is a favoured material for making the flechettes.[59] Its density provides the necessary increase in weight, and it also has the chemical property of reacting exothermally [sic] with iron: the additional heat thus generated during impact with armour creates intense local overpressure which aids penetration, and pyrophoric fragments are strewn about which are capable of igniting fuel and the like.

Just as design techniques are available for maximising the fragmentation effects of explosive munitions, and for optimising the fragment-size spectrum for use against different hardnesses of target, so also may the blast effects be maximised (though at the expense of fragmentation) and optimised. The thinner the casing, the greater the blast. Light-case explosive munitions have been developed which range from the very small to the very large. At one extreme are the 'tea bag' anti-personnel minelets (such as the US Gravel minelet), consisting of a small fabric bag of pressure-sensitive explosive. About 25 grams (more than is contained in a Gravel minelet) of an appropriate explosive detonated at foot level is sufficient to amputate a man's leg. Mines of this type are designed to be scattered by the thousand over a wide area.[60] At the other extreme are such massive munitions as the 7½-ton 'daisy cutter' bombs (BLU-82) employed in Vietnam for clearing helicopter landing-zones and as wide-area anti-personnel weapons.[61]

Optimisation of blast effects for particular targets is achieved by shaping the explosive charge so as to focus the blast or otherwise make it conform to the target. When hollow-charge munitions (also called HEAT or shaped-charge munitions) are used for armour penetration, a conical hollow is shaped into the impact end of the charge and lined with a thin layer of metal, usually copper. The detonating explosive vaporises the copper, and its directional effect squirts the metal forward in a thin jet capable of cutting through armour and injecting an intense pulse of thermal energy; the effects on people inside the tank are terrible. In the medical literature they have recently been characterised in terms of an 'anti-tank missile burn syndrome'.[62] Ten per cent of Israeli casualties during the 1973 Yom Kippur War displayed it: half of them had more than 20 percent of their body surface burnt, and almost a third had

---

[57] US Army Materiel Command, Historical Office, Arsenal for the Brave: A History of the United States Army Materiel Command 1962–1968 (Sep 1969) pp. 155–7.

[58] 'Flechette warhead for 2.75 in rockets', *International Defense Review* (1973) no. 3, p. 376.

[59] 'Ballistics unit tests research concepts', *Aviation Week and Space Technology* (15 July 1974) p. 275; D. Dardick, S. Green, S. Rush and F. Fedowitz, 'Gun/projectile systems', *Space/Aeronautics* (Mar 1967) pp. 92–9.

[60] US Dept of the Air Force, Congressional testimony, R .. D. Anderson, statement before a subcommittee of the Senate Armed Services Committee (19 Nov 1970).

[61] A. H. Westing, 'The super bomb', *American Report* (New York, 8 Sep 1972) p. 3.

[62] N. Ben-Hur and H. Soroff, 'Combat burns in the 1973 October war and the anti-tank missile burn syndrome', *Burns* (1975) no. 1, pp. 217–21.

third-degree burns, down to and through the subcutaneous tissue. In 'squash-head' (HESH) munitions, also used for the attack of tanks and other armoured vehicles, the explosive charge is adapted to flatten itself against the armour prior to detonation (only direct-fire weapons provide enough impact velocity for this flattening to take place). The blast wave transmitted through the metal breaks up the inside surface into a hail of high-velocity fragments, a phenomenon known as 'spalling'.[63] Fuel-air explosive munitions provide another means for optimising blast: the explosive charge – a detonatable cloud of fuel – is spread out like a gigantic pancake immediately above its intended area target.

Blast effects diminish more rapidly with distance (in air) than do fragmentation effects, but within the area over which they remain effective, their hit probability (so to speak) is essentially 100 percent. Other types of casualty agent capable of entirely enveloping an area include fire initiated by incendiaries, and the toxic or infective aerosols of chemical/biological warfare. Weapons exploiting these kill mechanisms have a number of peculiar operational disadvantages, mostly stemming from uncertainties associated with the sensitivity of the kill probability to environmental factors; the uncertainties become translated into poor predictability of target effects and hazards to neighbouring friendly forces, meaning difficulty in calculating munitions requirements for specific operations. These and related technical problems perhaps serve to reinforce military inhibitions about using fire, poison, or disease as weapons, thus going some way towards explaining why chemical, biological, and incendiary warfare have remained on the periphery of military favour. This situation in the case of incendiary warfare has been changing since the Second World War; and there is a case for locating the reasons for the change in the rapid growth in understanding of how to create and direct fire. A similar process of technical advance may in due course come to increase the incidence of chemical warfare. As area weapons, chemical warfare munitions have strong attractions in the abstract, not least in economy of killing: decigrams or more of fragmentable metal in conjunction with a like weight of high explosive are needed to disable a person, but it may take less than a milligram of nerve gas to kill.

8.2.2.1 Kill Mechanism Trends During the Vietnam War

The US air-to-ground munitions procurement record provides evidence of the relative importance attached to the different kill mechanisms for indirect-fire munitions. Table 8.8, to which the same caveats attach as Table 8.5, categorises US Air Force annual procurements of air-to-ground munitions according to kill mechanism. Notable in it, once the procurement/use steady-state had been reached in 1967, is the remarkably consistent distribution of procurement between the three dominant classes of kill mechanism: general-purpose explosive, fragmentation maximised explosive, and flame. The fluctuations, like the erratic procurements of other kill

[63] US Army Materiel Command, Research and development of materiel, engineering design handbook. Elements of armament engineering. Pt. 2: Ballistics (Sep 1963).

mechanisms, presumably stem from special military requirements in different phases of the war.

The procurement record also provides evidence of the extent to which the different categories of kill mechanism were exploited in the different categories of munition. It also displays time trends in their exploitation. Thus, as the Vietnam war progressed, three particular growth areas became prominent. The first was in incendiary cluster bombs (the M36/CBU-52 procurements). The second was in fuel-air explosive cluster bombs (the CBU-55 procurements). The third was the growth in the general area of fragmentation munitions and its increasing concentration in the anti-PAM rather than the anti-personnel or anti-materiel categories. These are trends which are worth exploring in more detail. Since fragmentation-munition design is already discussed rather fully in the open literature,[64] the remainder of this chapter will concentrate on new incendiary weapons and on fuel-air explosives.

8.2.2.2 Heat and Flame: Controlled Fireballs

The trends of the 1960s and early 1970s in the exploitation of fire as a weapon show only faintly in the US procurement data. There are two main categories of fire weapon, though the distinction between them is not clear-cut: ones using flame munitions, with which the damage is mostly caused by direct contact with a burning flame agent; and ones using incendiary munitions, which damage not so much by direct contact as by initiating fires that then propagate through a target. Flame agents are broadcast; incendiary agents are used as intense point sources of fire.

Napalm, which is petrol gelled or otherwise thickened by means of additives, is an example of a flame agent; and many different varieties of it have been developed. The 1964–73 US Air Force and the US Navy procurement record indicates that, although napalm formed a high proportion of total US air-munitions procurement during the earlier part of the Vietnam war, the proportion declined as the war continued. Whether this reflects a similar decline in the use of air-dropped napalm during the war is not clear. The available data on the quantities of napalm dropped over Vietnam show that levels of use and of procurement paralleled each other during 1963-7, but with the latter so much above the former that large reserves would seem to have accumulated above the former that large reserves would seem to have accumulated for later use. In contrast to napalm, incendiary procurement shows a rapid growth during the later period of the war. The growth was in large measure due to the effectiveness of incendiary cluster bombs in destroying vehicles during interdiction operations. When stocks of this type of munition – primarily the 750-lb M36 cluster of 182 magnesium/thermate 4-lb stick bomblets, dating back to the Korean

[64] J. Pearson, R. G. S. Sewell and R. T. Carlisle, assignors to the Secretary of the US Navy, US patent no. 3 566 794, *Controlled fragmentation of multiwalled warheads* (app. 26 Nov 1958, iss. 2 Mar 1971); P. Crevecoeur, 'Anti-personnel hand grenades – some recent European developments', *International Defense Review* (1973) no. 3, pp. 372-4; M. Held, 'Air target warheads', *International Defense Review* (1975) no. 5, pp. 719–24; E. Prokosch, 'How large is your wound?', unpublished MS.

Table 8.8 Relative importance of different categories of kill mechanism as indicated by annual US air force procurements of conventional air-to-ground munitions during fiscal years 1964–73[65]

Category of kill mechanism	Percentage of total annual munitions procurement, by weight of munitions										
	1964	1965	1966	1967	1968	1969	1970	1971	1972	1973*	1964–73
Solid bullets	np	0.6	2.8	2.8	1.2	1.0	0.7	1.1	1.1	1.0	1.4
Explosive: general-purpose	1.3	31.4	62.1	79.0	90.5	85.8	81.6	83.8	89.9	89.3	82.8
Explosive: blast-maximised	np	np	..	0.2	0.1	np	0.2	0.3	0.1	0.5	0.2
Explosive: hollow-charge	5.2	1.2	..	..	np	..	np	0.6	0.2	0.9	0.2
Explosive: fragmentation-maximised	35.0	13.2	5.1	8.9	4.5	5.3	10.1	7.9	5.1	5.0	6.3
Napalm and other incendiaries	58.4	53.6	29.8	8.9	3.7	7.7	7.5	6.3	3.6	3.3	8.9
Chemical irritant (CS)	np	np	..	..	0.1	0.1	np	np	np	np	..

Conventions: (np) no procurement; (..) less than 0.05%
Note *Programmed procurement as of 31 Dec 1972
Source As Table 8.5. [Figures calculated from data contained in US Defense Department annual returns on types, numbers and costs of munitions procured or programmed for procurement during fiscal years 1964–73.]

[65] Editors' note. This table was 'Table 3.7' in the original.

war and suitable for carriage only in bombers or on the slower ground-attack tactical aircraft – had been depleted, in the spring of 1968, production lines were re-opened and development of newer versions hastened.[66] The chosen successor to the M36 was the CBU-52, which dispenses 254 BLU-61 2.2-lb spherical self-dispersing fragmentation bomblets incorporating metallic zirconium as an incendiary agent. This material was known, particularly from work on high explosive-incendiary (HEI) projectiles for aircraft cannon, to be an exceptionally potent incendiary, giving rise to long-burning pyrophoric fragments easily capable of setting fire to fuel tanks and the like.[67] Powders of zirconium, and of other reactive metals such as lithium, have also been under consideration, and perhaps also used, as a means for increasing the heat flux from burning napalm or other gelled-fuel flame agents.[68]

Incendiary cluster bombs were originally developed as a means for burning cities. In the tactical role, their new-found applications were not limited to anti-vehicle uses. In testimony before a Congressional committee, a US Air Force witness sought to justify the plans for increasing procurement of the CBU-52 during 1973 by citing increased military requirements both for anti-PAM weapons due to 'the introduction by the enemy of large ground formations into the Republic of Vietnam' and for anti-aircraft defence suppression weapons due to 'the resumption of bombing in North Vietnam'.[69] Thus, heat as a kill mechanism, at least in conjunction with fragmentation, was evidently on the upsurge for air munitions. The same was apparently true for ground munitions. In its 1973 ordnance-research overview, the US Army Munitions Command wrote: 'Research studies have confirmed that rapid application of heat energy (heat flux) is the most effective method for destroying or damaging military targets'[70] – a judgement, it is to be noted, which refers to 'military targets' in general, regardless of their nature.

This is a remarkable appraisal. If justified, it can scarcely fail to be reflected in the munitions of the future. The performance of incendiary and flame munitions in Vietnam was no doubt part of the reason for it; but more important would seem to have been a particular laboratory development: a successful demonstration of what was called the 'controlled fireball concept'.

It is generally prudent to remain sceptical when, as in this instance, technologists claim a technological breakthrough. Yet on paper the development appears to

[66] US Dept of the Air Force, Congressional testimony (supra, n. 28). Editors' note. This corresponds to US Dept of the Air Force, Congressional testimony, R. D. Anderson, statement before a subcommittee of the Senate Armed Services Committee (19 Nov 1970).

[67] C. C. Balke and W. S. Graff, assignors to the Secretary of the US Army, US patent no. 2 801 590, *Pyrophoric element* (app. 14 June 1951, iss. 6 Aug 1957); G. H. Custard, assignor to the Secretary of the US Army, US patent no. 3 421 439, *Incendiary projectiles* (app. 29 Sep 1961, iss. 14 Jan 1969); Stevenson, assignor to the Secretary of the US Army, US patent no. 2 951 752, *Incendiary composition* (app. 21 May 1958, iss. 6 Sep 1960).

[68] US Army Munitions Command, *Laboratory posture report FY-73* RCS AMCDI-101 (1973).

[69] US Dept of the Air Force, Congressional testimony. W. W. Snavely, statement before the Senate Defense Appropriations Subcommittee (13 Sep 1972).

[70] US Army Munitions Command (supra, n. 35). Editors' note. This corresponds to US Army Munitions Command, *Laboratory posture report FY-73* RCS AMCDI-101 (1973).

be a major one. In particular, it looks as though those peculiarly awful weapons, the flamethrower and the napalm firebomb, may soon be outmoded; and that fire munitions will become increasingly popular for weapons to which at present they are poorly suited, such as artillery systems. The prospect thus seems to be a substantially upgraded status for incendiary warfare so that, in the years to come, fire, despite the repugnance which still attaches to its use as a kill mechanism, may be employed much more commonly in war than it has been so far.

The military allure of incendiaries and flame, if indeed it is growing, may explain the treatment they received at the Lucerne and Lugano conferences of government experts convened in 1974 and 1976 to explore the possibility of prohibiting particular conventional weapons. Burns were the one type of injury caused by conventional weapons which the medical experts present were prepared to regard as exceptional, which is the sort of validation that jurists need if the rule of international law is to be advanced in the field of weaponry. Yet despite the medical evidence, military experts produced a succession of reasons why it would be wrong to ban the use of even some types of fire weapon, let alone all of them. The fiercest opposition came, not from the US governmental experts (as might have been expected from recent US experiences with fire weapons), but from the West German ones. Possibly this reflected different modes of diplomacy; but it is also true that in preparing its defences against the forces of the Warsaw Treaty Organisation (WTO), the Federal Republic is in a somewhat different position from its NATO allies. Were there indeed to be an invasion of western Europe from the east, the initial onslaught would most likely be concentrated on West German territory. The nature of warfare demands that for success the attack must outnumber the defence, by perhaps as much as three to one. Should the WTO be able to muster and concentrate such a force, NATO would need to have some means at its disposal to compensate for localised numerical inferiorities. Area weapons are one possibility; but for the Federal Republic, possession of the most powerful of these – nuclear, biological, and chemical weapons – is forbidden under the 1954 revision of the Brussels Treaty. And in any event it would seem unlikely that a Federal government would willingly rest the Republic's security on an option to use on its own territory such cataclysmic categories of weapon. A definite stimulus therefore exists for the development and deployment of other area weapons; and in the Federal Republic there are indeed large investments in novel wide-area munitions, especially for indirect-fire weapons. Some of them, particularly fragmentation bomblet and minelet dispenser munitions, have been noted earlier; but there is also much West German activity in the development of novel flame and incendiary weapons.[71] In one part of the industrialised world, therefore, technological breakthroughs in incendiary warfare are likely to be exploited to the full.

These particularities are noted in order to illustrate a generality. Their purpose is not to assert that the controlled fireball concept is itself the most important new

[71] G. Buck, British patent no. 1 061 631, Incendiary compositions containing highly inflammable liquid carbon compounds and a thickener and a process for the production of such compositions (app. 12 Feb. 1964, pub. 15 Mar 1967); F. Hobart (ed.), Jane's Infantry Weapons 1975 (London, 1974); R. T. Pretty and D. H. Archer, Jane's Weapon Systems 1974–75 (London, 1974).

development in incendiary warfare, though it may well be. Neither are they intended to suggest that it is in the United States that such new developments are being sought the most avidly nor that it is in West Germany that they are being applied the most enthusiastically. The point is rather that, to an outside observer, there seems at present to be a ferment in the technology of incendiary warfare, and that real pressures exist for the rapid exploitation of whatever technical refinements may result. Were the open literature more extensive, one could probably provide additional illustrations from countries outside the Atlantic alliance. If true, one implication is obvious: by the time international agreement is reached, if it ever is, on specific incendiary-warfare constraints, the technology may well have developed far beyond the point where the constraints have much practical significance. ·

The most contentious aspect of incendiary warfare today is the tactical use of air-dropped napalm. Many people consider that the principal weapon used for this purpose, the napalm firebomb, is one of the most inhumane weapons ever invented, and one which by its very nature must act to subvert the ancient principle of the immunity of the non-combatant in time of war. This opinion has stemmed from three main considerations: first, the effects of napalm on the human body are judged to be extraordinarily horrible; second, the area of ground over which a single napalm firebomb may exert its effects is judged to be too great to permit discriminate use against enemy combatants in a region where there are also non-combatants, having regard also to the inaccuracy with which firebombs are commonly dropped; third, the inherent indiscriminacy of a weapon so readily available as the napalm firebomb is judged likely to encourage resort to indiscriminate modes of warfare. On each of these three matters, however, the opposite opinion has also been expressed strongly, particularly by military experts: first, all weapons can produce terrible injuries, and there is no hard evidence to show that napalm is any worse than, say, a fragmentation bomb; second, the latest types of napalm firebomb, if used with the latest techniques of fire control, are at least as accurate as any other form of indirect fire, and can actually be used with greater discrimination; third, except where demanded by military necessity, no principled military staff would ever resort to indiscriminate warfare.

These arguments *pro* and *contra* are set out in greater detail in the reports from the Lucerne and Lugano conferences, and it is not the intention here to traverse the same ground. Instead the purpose is to relate this flow of argument to new developments in incendiary-warfare technology, in particular the controlled fireball concept.

The firebomb originated with the US Army Air Force during the Pacific campaigns of the Second World War, subsequently being introduced into the European theatre as well.[72] At first no more than a field expedient comprising an auxiliary wing-mounted fuel drop-tank charged with some hundreds of litres of thickened aviation fuel, it was later developed into a regular inventory item, and is now a munition much like any other aircraft bomb. The results obtained with it during the Korean War, and during French operations against the Viet-Minh in Indochina, were sufficiently impressive for the firebomb subsequently to be adopted as standard air force armament by

[72] B. E. Kleber and D. E. Birdsell, *The Chemical Warfare Service: Chemicals in combat*, US Dept of the Army, Office of the Chief of Military History (Washington, D.C., 1966).

virtually all countries that possessed a significant air force. As such, the firebomb has been repeatedly used throughout the world: against guerrilla fighters. in several South American, southern African, and Middle Eastern countries; in the border wars that have been, and are being fought in the horn of Africa; in the Arab-Israeli wars; in the Algerian war of independence; in Cyprus; in at least one of the Indo-Pakistan wars; in the Ussuri fighting of 1969; and so on. The most extensive, and certainly the most publicised, use has been during the last Indochina war, where it seems (from the US firebomb procurement data) that something in the order of a million and a half firebombs were dropped.

The popularity of the napalm firebomb in military circles has mainly resided in its flexibility: its apparent efficacy against many of the widely differing targets that strike aircraft may be called upon to attack, whether in ground-support or in interdiction. The French in Indochina and UN forces in Korea made much of its area anti-personnel role against concentrations of enemy troops; and for many years the firebomb continued to be one of the best weapons available for destroying tanks and weapon emplacements, and for reducing the simpler types of fortification. More recently, however, it is clear that some of the military attractions of the firebomb have declined. Techniques have been developed for lessening the vulnerability of important materiel, particularly tanks, towards napalm, and, at the same time, other more effective anti-materiel weapons have become available. Anti-personnel applications have thus come, increasingly, to dominate the uses to which the firebomb is put.

Such applications capitalise on the terror effects of napalm. The military manuals are quite explicit on this point. 'The psychological impact is probably one of the greatest effects of flame', says one, and explains how flame weapons may be 'used to demoralise troops and reduce positions that have resisted other forms of attack'.[73] On the regular battlefield, terror may thus secure immediate tactical gains; in irregular operations terror tactics can contribute and have contributed in several ways to counter-insurgency objectives, to such an extent that napalm has become a pre-eminent symbol of coercive warfare. A firebomb drop is an awesome sight. An area little smaller than a football pitch is suddenly smothered in a tide of orange fire, the flames reaching above treetop level, burning with enormous ferocity and billowing out dense black smoke. It seems impossible that anyone or anything could survive within the fireball, and those faced by it must feel that death of a most painful kind is upon them. Such fears can only be heightened in those who have seen the bodies of napalm victims.

Yet there is evidence to suggest that, setting aside their demoralising effects, napalm firebombs are not in fact very effective anti-personnel weapons. There is also evidence which suggests, to the contrary, that firebombs cause singularly high casualty rates, and that the mortality rate among the casualties is also high. It is perhaps surprising that, for a weapon so widely used, such fundamental disagreement should persist. But it has to be taken into account that weapon-effects data constitute valuable intelligence, and that it is in the nature of things that the side attacked with firebombs is generally in a position to deny the attacker quantitative data about the

[73] US Dept of the Army, *Combat Flame operations*, Field manual FM 20–33 (1970).

results of the attack. Under these circumstances, and in view of the sensitivity and responsiveness of public opinion regarding napalm, it could well be that the public literature on the subject has tended to over-estimate rather than under-estimate, the lethality of napalm.

As to case-fatality rates, estimates have been published – all of them during, or very soon after, the commitment of US forces to the Vietnam war – by Vietnamese, East German, Czech, and Soviet authorities.[74] The first three of these related to Vietnamese casualties, stating an average mortality rate of about one in three among people burnt by napalm. This referred to immediate mortality, i.e. within about half an hour. An additional, similarly large, proportion of casualties were said to have died subsequently. The Soviet authority, referring to Korean War cases, also wrote of a one-in-three immediate mortality rate but neither he, nor the other three authorities, provided estimates of the casualty rates within which this mortality was displayed. The Czech authority reported that more than one in four of the people struck by napalm during the types of napalm attack then current were likely to suffer burns over more than 25 percent of their body surface, half of these burns being of the third degree. None of the four sources provide detailed substantiating data for the estimates, which, as regards evidence, are therefore left to be taken on trust or rejected on mistrust.

Conflicting data were presented by a US expert at the Lucerne conference, the data subsequently being expanded and amended at the Lugano conference, with the purpose of refuting the foregoing estimates. The new data related to seven accidental firebomb drops on US units in South Vietnam during 1968–9. Of the 51 people who had been caught within the fireballs, 48 suffered burns. There were no immediate deaths, but three later deaths, the first of them after seven days. One in four of those burnt had burns over more than 20 percent of their body surface. Data were not presented on the depths of these particular burns, though one in five of the casualties was reported to have suffered third-degree burns over more than 10 percent of their body surface.

These two contrasting sets of data are clearly incompatible as regards immediate mortality. As regards late mortality they are not necessarily incompatible, since it is to be supposed that the US casualties received medical aid far more rapidly than did the other casualties, and that medical aid was of a superior quality. Neither are the figures given on the extent of the individual burn injuries necessarily incompatible; and in the event of medical aid having been unavailable, high late-mortality rates would surely have been likely for burns of such severity. It must be said, however, that in both cases the information available is inadequate for reaching any firm,

---

[74] Do Xuan Hop 'Napalmoch fosforbomber', in J. Takman (ed.), *Napalm* (Stockholm, 1967); H. R. Gestewitz, 'Der Einsatz von Brandbomben (-granaten) der US-Fliegerkräfte gegen die Demokratische Republik Vietnam und ihre Wirkung unter besonderer Berücksichtigung von Napalm', *Zeitschrift für Militärmedizin* (1968) no. 5, pp. 275–80; J. Franek cited as a private communication in United Nations Secretary General, *Napalm and other incendiary weapons and all aspects of their possible use*, UN document A/8803/rev. 1 (New York, 1973) p. 33; V. A. Dolinin, 'Clinical picture, organisation and level of medical care for napalm casualties', *Voenno-Meditsinsky Zhurnal* (1975) no. 12, pp. 33–7.

general conclusions about the quantitative casualty effects of napalm in comparison with better known weapons. The US figures describe only a small sample, while for the other figures the size and circumstances of the samples on which, presumably, the estimates were based have not been given at all, despite the opportunity for doing so afforded by the Lucerne and Lugano meetings.

Subsequent studies may have clarified the situation, but, if they exist, they are to be found only in the secret literature. It may be noted that, by 1973, the US armed services had a 'Tri-Service Comprehensive Anti-Personnel Thermal Evaluation Program' under way; the 'Human Skin Burn and Pain Model' under development by the Radiation Engineering Branch of the US Army's Ballistic Research Laboratories as an 'effectiveness measurement tool' presumably contributed to the programme.[75]

Among those who were privy to it, the US accident-data study, which was not completed until some four years after the accidents happened, is said to have had a profound effect on appraisals of the efficacy of napalm. For years it had been supposed that everyone caught within a napalm fireball was likely to be incinerated: an exceptionally high casualty and mortality rate was accepted without question: and who seeing the inferno created by a firebomb might think to doubt it? Yet it now seemed that the lethality was much the same as that of any other weapon. But more important from a military standpoint was the finding that every single one of the burn victims had remained capable of performing whatever physical activity was attempted during the period immediately after the napalm drop. Thus, not only did napalm fail to kill many of its victims but also, apparently, it did not even immediately incapacitate them. For US armed forces, the chief function remaining for the firebomb by that time was in close air support; and the primary requirement made of any close support weapon is that it should disable enemy troops fast. For the US, and for other industrialised countries which could afford better, napalm firebombs were already becoming outmoded as anti-materiel weapons; now there seemed to be little point in retaining them as anti-personnel weapons. With hindsight, therefore, one may cease to be surprised at the number of governments which, in a succession of public statements since 1973 aimed largely at humanitarian audiences, have announced that their armed forces are giving up napalm firebombs.[76]

It would undoubtedly be wrong to suppose that there is now a universal disparagement of napalm firebombs in military circles. This may be so for the advanced armed forces of the West, though even here napalm enthusiasts remain; and no doubt also for the Warsaw Pact. But in other parts of the world it is not the case that the chief remaining function of the firebomb is in close air support (a function which is in any case ruled out for all but the most sophisticated air forces). The armed forces of many countries are equipped with obsolete hand-me-downs from richer countries:

---

[75] US Army Armament Command, *Laboratory posture report FY 74*, RCS AMCDL-lOl (1974) and *FY 75*, RCS AMCDL-101 (1975).

[76] United Nations Secretary General, *Napalm and other incendiary weapons and all aspects of their possible use*, UN document A/9207 with Corr. 1 and Add. 1 (1973) and UN document A/10223 (1975).

tanks, for example, of a vintage still highly vulnerable to napalm. And the firebomb is still a terror weapon, valuable as such to military or counter-insurgency forces that see a need for such weapons.

Neither is it the case that the decline in anti-materiel utility of firebombs has been swift and clear-cut. Over the years the weapons technologists have contrived a succession of improvements in the composition of firebomb fills whereby anti-materiel efficacy has in some – though not all – cases kept pace with developing anti-flame counter-measures. For example, several 'loaded' napalms have been introduced in which special ingredients are incorporated in order to enhance the aggressive properties of the burning napalm against particular types of materiel.[77] This has of course meant an overall diminution of the variety of targets against which firebombs continue to remain cost-effective, but in some cases – and the special predicament of West Germany referred to earlier may be one such case – the loss of flexibility may be considered militarily acceptable. So far as may be judged from the limited information available in the open literature, the gains in anti-materiel efficiency obtained by 'loading' napalm are at the expense of anti-personnel efficiency: slag-forming additives, for example, decrease the area of ground over which the firebomb fill is spattered and therefore reduce the number of people placed at risk by each firebomb. Not all developments in napalm have been so specialised, however. The prime example of one which has not been is the development of 'Alectogel' or 'Napalm-B' in the mid-1960s. In composition, this new agent was a radical departure from earlier napalm, a viscous 50 percent solution of a polystyrene in equal parts of benzene and gasoline, rather than gasoline or aviation fuel gelled with a few percent of special soap. In performance, firebombs loaded with Napalm-B combined a doubled area-coverage with a substantially greater destructiveness of materiel, such as armoured personnel carriers. The latter property it owed to its increased propensity for starting secondary fires, stemming from its increased burning time. The results obtained with the new agent in Vietnam from 1966 onwards stimulated the development of new anti-materiel flame weapons, in particular napalm-B cluster munitions, as well as improved firebombs of the more traditional kind. Whether the new firebombs were more powerful than the old ones as anti-personnel weapons is not disclosed in the open literature, but such technical data as are available certainly suggest some increase in performance. It is not known whether the US accident data quoted above referred to napalm-B.

On the face of it, then, the introduction of napalm-B served to prolong military interest in air-delivered flame. This illustrates one characteristic feature of the

---

[77] *Napalm and other, incendiary weapons* (*supra*, n. 41) Editors' note. This corresponds to Do Xuan Hop 'Napalm-och fosforbomber', in J. Takman (ed.), *Napalm* (Stockholm, 1967); H. R. Gestewitz, 'Der Einsatz von Brandbomben (-granaten) der US-Fliegerkräfte gegen die Demokratische Republik Vietnam und ihre Wirkung unter besonderer Berücksichtigung von Napalm', *Zeitschrift für Militärmedizin* (1968) no. 5, pp. 275–80; J. Franek cited as a private communication in United Nations Secretary General, *Napalm and other incendiary weapons and all aspects of their possible use*, UN document A/8803/rev. 1 (New York, 1973) p. 33; V. A. Dolinin, 'Clinical picture, organisation and level of medical care for napalm casualties', *Voenno-Meditsinsky Zhurnal* (1975) no. 12, pp. 33–7.

history of this particular category of munition: that, over time, the properties for which the firebomb has been most valued have oscillated between anti-materiel and anti-personnel applications, the net result being the retention of firebombs in the operational inventory and hence their availability for use in either role. The process of technological evolution has, with each new technical change, carried the firebomb through the successive inflexions of its operational history. Thus, although the US accident-data report may at present seem to be ejecting the firebomb from the arsenals of the rich countries, technical change may still bring about a new upturn of military interest.

Application of the controlled fireball concept may very well prove to be that critical technical change. Not much has been openly published about the concept. It seems to have originated in a new weapon which the US Army was developing for use in Vietnam in place of the portable flamethrower. Most of the operational limitations of flamethrowers stem from the necessity of igniting the jet of fuel at the nozzle of the weapon, for much of the payload is thereby wasted en route to the target and the operator becomes highly conspicuous to enemy counterfire. Hence the idea of using a flame agent which would ignite spontaneously in the air and of projecting it on to the target in frangible, rocket-propelled capsules rather than in a continuous jet. Many chemicals existed that had the necessary pyrophoricity, and some of them, notably the metal alkyls, had a heat of combustion per unit weight which exceeded even that of gasoline. They are extremely dangerous substances to manufacture and handle in bulk, since they can ignite with explosive violence in contact with air or water; but this was no great problem in a chemical industry capable of producing, for example, hypergolic rocket fuels. The necessary production and containment techniques were worked out, and a thickener was chosen which could hold the agent payload together so as to give a sufficiently dense and adhesive impact pattern on target. The new category of 'thickened pyrophoric agents' (TPA) thus came into being alongside the several other categories of flame and incendiary agent. The preferred TPA at that time was triethylaluminium (TEA) thickened with some six percent of a polyisobutene (PIB). The new weapon was standardised in 1972 as the M74 Rocket-clip/M202 launcher, a shoulder-fired, four-tubed system.[78]

During 1969–70, prior to standardisation, the US Army and the US Marine Corps between them procured nearly 600,000 of the rockets, together with 3200 of the reusable launchers for them.[79] Despite the difficult manufacturing processes associated with the rockets, the price paid for them averaged out at \$0.87 per kilogram – only about double that of the napalm firebomb, the cheapest of all munitions per unit weight. Much of the procurement was presumably used in Indochina.

It soon became apparent that the new weapon was remarkably powerful, even though each rocket is of a mere 66 mm calibre weighing no more than a kilogram

[78] US Army Munitions Command, *Laboratory posture report FY-72* (1972).

[79] US Dept of Defense, Reports of programs for fiscal years 1964–73: 'Procurement of equipment and missiles, Army: Ammunition, P-4800', and 'Procurement, Marine Corps: Budget activity 1: Ammunition' (*supra*, n; 6). Editors' note. The latter corresponds to US Dept of Defense, 'Procurement, Marine Corps: Budget activity 1: Ammunition'; (30 June 1966–31 Dec 1972).

and a half, and the total payload of flame agent in a four-rocket clip is about one-sixth of the napalm in current US portable flamethrowers. In May 1972, two North Vietnamese T54 tanks were destroyed with M74 rockets,[80] and tanks are not the sort of target against which portable flamethrowers are normally recommended for use.

Thickened TEA is inherently more destructive than napalm. It releases more heat per unit weight when it burns, and it tends to burn at a higher temperature, which means that a higher proportion of the heat which it gives out appears as thermal radiation. When napalm burns, some thermal radiation is of course emitted, but the dominant modes of heat transfer are conduction, through whatever material happens to be in contact with the burning agent, and convection, the creation of hot air currents. The hotter the air the faster it rises; this means, in effect, that the most destructive region of a napalm fireball is its outer surface: inside the fireball, except where there is direct contact with gobbets of burning napalm, the environment is not as hostile as it may appear to the onlooker. This came out clearly in the statements recorded by survivors in the US accident study. Radiant heat, in contrast, is omnidirectional (unless the emitter is shielded in a particular direction) and transmitted, so that radiation damage may be suffered even at a distance from the emitter without direct contact with it. The stronger the thermal radiation, the greater is the distance over which it may damage. Thus, in contrast to a napalm fireball, the internal environment of a TEA fireball is intensely destructive, the zone of destructiveness reaching out, moreover, to a distance considerably beyond the periphery of the fireball.

Further laboratory studies showed that TPA munitions could be made more powerful stilL By reducing the proportion of PIB thickener, the heat pulse became so great that people inside a bunker would, it was calculated, receive third-degree burns over most of their body surface within a few seconds.[81] Experiments with pigs showed that this was indeed the case.[82] Materiel targets of various types were also found to be vulnerable if the composition of the formulation, and hence its combustion rate, was appropriately adjusted. From this it was but a short step to the idea of 'controlled' fireball: formulating the mixture so as to provide particular combustion rates from specific munitions, which would then be optimal for the destruction of particular types of target. The technologists involved reported, 'Predictive mathematical models indicate that this controlled fireball technique can be weaponised in an unlimited number of concepts since the flame agent can be tailored to specific operational requirements'.[83]

Orders were soon placed for other TPA munitions. The army wanted an 'encapsulated flame round' for its 152 mm tank armament, and also started work on a 'Directional Ground Flame Munition System', an emplaceable TPA device for perimeter

[80] US Army Munitions Command (*supra*, n. 45). Editors' note. This corresponds to US Army Munitions Command, *Laboratory posture report FY-72* (1972).

[81] Ibid.

[82] US Army Armament Command, FY 74 (*supra*, n. 42). Editors' note. This corresponds to US Army Armament Command, *Laboratory posture report FY 74*, RCS AMCDL-1O1 (1974) and *FY 75*, RCS AMCDL-101 (1975).

[83] US Army Munitions Command (*supra*, n. 35). Editors' note. This corresponds to US Army Munitions Command, *Laboratory posture report FY-73* RCS AMCDI-101 (1973).

defence and terrain denial.[84] The marine corps, already the chief customer for the M74/M202 system, wanted a 105 mm round. The air force wanted TEA bomblets for their aircraft dispenser munitions, and they started work also on TEA air-to-ground rocket warheads and TEA firebombs.[85] Stimulating results from field trials continued to be reported, particularly test firings against materiel targets including armoured vehicles. For the developmental 152 mm round, the results 'indicated that a near miss, 50 feet short in range, would still permit defeat of hard targets'.[86] A 1000-lb general-purpose HE bomb must fall within 10–15 m of a tank even to stop it, let alone destroy it.[87] The 152 mm round could create a fireball 25 m in diameter within which the temperature could reach above 1200°C and the peak flux 6 kilocal-sec/cm^2.[88] Here indeed was a fireball within which everyone caught unprotected was likely to be incinerated.

So far as public sources reveal, none of these new TPA munitions have yet emerged from the development laboratories; and it may be that technical problems have been encountered. One can certainly see grounds for hesitation on the part of the air force since there would be exceptional danger in carrying TEA inside or underneath an aircraft. But it seems that there is a way round this problem, for the army is now working on what it calls 'binary flame agents', a concept whose feasibility was, in 1974, said to have been confirmed, thereby making TEA 'a candidate agent for air-delivered flame weapons'.[89] The scheme is to mix the TEA with a volatile diluent, such as a hexane or heptane, so that spontaneous combustibility is lost, returning only when the diluent evaporates away after rupture of the munition on target. It seems that a large quantity of diluent is needed for this: the payload in the developmental 'Directional Ground Flame Munition System' noted above contains 35 percent of heptane. But one would apparently be wrong to suppose that the need to devote limited payload space to diluent must substantially weaken the power of the munition: the army has found that if things are arranged properly the evaporating diluent will form itself into a cloud of vapour which explodes when the TEA ignites. As a result, a totally new type of munition seems to be emerging, one which combines

[84] US Army Armament Command (*supra*, n. 42). Editors' note. This corresponds US Army Armament Command, *Laboratory posture report FY 74*, RCS AMCDL-lOl (1974) and *FY 75*, RCS AMCDL-101 (1975).

[85] 'Ballistics unit tests research concepts', *Aviation Week and Space Technology* (15 July 1974) p. 275.

[86] US Army Munitions Command (*supra*, n. 35). Editors' note. This corresponds to US Army Munitions Command, *Laboratory posture report FY-73* RCS AMCDI-101 (1973).

[87] UK Royal Air Force, statement by A. W. Fraser at the First Session of the Conference of Government Experts on the use of Certain Conventional Weapons, Lucerne (9 Oct 1974).

[88] US Army Armament Command, FY 74 (*supra*, n. 42). Editors' note. This corresponds to US Army Armament Command, *Laboratory posture report FY 74*, RCS AMCDL-lOl (1974) and *FY 75*, RCS AMCDL-101 (1975).

[89] Ibid.

intense wide-area incendiary effects with the area blast effects typical of the fuel-air explosive munitions described below. There is now talk of 'flameblast' munitions.[90]

Until a good deal more is published about flameblast experiments, one must remain sceptical both about the feasibility of flameblast munitions and about their military utility. Here is yet another instance, surely, of the internal momentum of weapons technology carrying it far beyond military requirements, though in time the requirements will no doubt adapt themselves so as to demand, as usual, whatever happens to be available.

Be that as it may, it is certainly remarkable how slight has been the attention given to TEA and other such late-model incendiary weapons at the conferences where possible incendiary-warfare constraints have been discussed. The talk has been primarily about napalm, as though napalm was, from the humanitarian point of view, the worst of the incendiaries. Indeed, the neglect of TEA – and it is but one of several varieties of novel incendiary agent under examination – appears very much to be a studied neglect.

8.2.2.3 Fuel-Air Explosives

The destructiveness of a munition is limited by its payload capacity: only so much high explosive can be packed into a shell that is to be fired from a particular gun or into a bomb that is to be dropped from a particular bomb-rack. A search for increased destructiveness is thus in part a search for more efficient ways of using payload capacity. In this respect incendiary munitions offer the decided attraction that they need carry less than half the weight of what is needed to do damage: the remainder is drawn from the environment, in the form of the atmospheric oxygen that reacts with the incendiary agent to create fire, and also, in the case of some anti-materiel applications, from the target itself, in the form of additional fuel for the fire. It has long been recognised that a similar saving of payload capacity is in principle also possible with explosive munitions, since a chemical (as opposed to nuclear) explosion is nothing more than an extremely rapid combustion reaction, no less dependent upon an adequate supply of oxygen. Conventional chemical explosives carry the oxygen with them – in the nitro groups of TNT, for example. But there are many fuels which, if intimately mixed with air, can be made to explode rather than deflagrate. This is what happens in an internal-combustion engine. The phenomenon is now being exploited to an increasing extent in the novel class of munitions known as 'fuel-air explosives' (FAEs).

The basic problem in building an FAE munition is the mixing of the fuel with air: if there is too much or too little air, the mixture will not explode.[91] Volatile liquid hydrocarbons are the most amenable fuels for it is relatively easy to suspend them in

---

[90] US Army Munitions Command (*supra*, n. 35). Editors' note. This corresponds to US Army Munitions Command, *Laboratory posture report FY-73* RCS AMCDI-101 (1973).

[91] A. G. Bilek, assignor to the Secretary of the US Air Force, US patent no. 3 999 482, *High explosive launcher system* (app. 9 July 1975, iss. 28 Dec 1976); G. A. Carlson, assignor to the US Energy

air as a vapour or aerosol cloud of the requisite dilution. Finely-powdered solids, such as coal dust, can also explode in air, as innumerable accidents have demonstrated; and there has been talk of using a metal-powder FAE in the propellant system of rapid-fire guns. The latter scheme would certainly present engineering problems of great complexity; even the problems in the design of a simple FAE bomb have only recently become manageable.

Antecedents of current FAE programmes can be found in Second World War work. Throughout the war the US Army Chemical Warfare Service was studying an FAE bomb that used butane as the fuel, but the technical difficulties proved insurmountable and the project was terminated.[92] Similar work was going on in Britain. So far as may be judged from the open literature, the Americans are now the leaders in this field. As for the Russians, open Western publications on Soviet FAE work are as schizophrenic as they are in reporting other areas of Soviet military technology: on the one hand there are claims that the Russians are extraordinarily interested in FAE and far advanced in exploiting it; on the other, there are claims that Soviet FAE devices are crude, their development running far behind current US programmes.[93] The reawakening of US interest in FAE is said to have begun in a navy ordnance laboratory around 1960. Concerted navy FAE R&D did not, however, commence until 1966, the air force following in 1967, building upon the lead technology established by the navy.

The outside world began to hear about US fuel-air explosives from odd rumours and guarded allusions,[94] including a reference in 1967 to a mysterious new navy weapon called 'FAX' that was capable, so it was said, of shearing the landing gear off

---

Research and Development Administration, US patent no. 3 955 509, *Fuel-air munition and device* (app. 21 Mar 1969, iss. 11 May 1976).

[92] L. P. Brophy, W. D. Miles, and R. C.Cochrane, *The Chemical Warfare Service: From laboratory to field*, US Dept of the Army; Office of the Chief of Military History (Washington, D.C., 1959).

[93] An appraisal of Soviet FAE work appears in a recent article by G. Johannsohn, a West German writer, who states that 'the USSR has probably had a very large arsenal of FAE weapons for several decades'. However, the article goes on to say that 'it can be concluded from the available test results that Soviet FAE weapon systems are comparable in type, size, and effect to those of the USA', a conclusion that is based on a 1965 paper in a Soviet journal reporting rather unremarkable findings from an experiment in combustion kinetics at the level of basic research. The author of the article refers to this experiment as 'the Kogarko tests' and then immediately quotes the Pentagon to the effect that 'the USSR continued testing in 1973 and thereafter.' He cites a published source for this; in fact it reads as follows: 'Soviet Union is in the early stages of testing fuel air explosives, but devices being used are somewhat crude in comparison to those in development in the US by navy and air force. In confirming the USSR's work in the area of FAE, Pentagon officials said the Soviets have made rapid progress during the past year (1972–3) but appear to be running far behind this country's development program.' It might be, of course, that the West German author had access to uncitable sources; but this is belied by the informedness of his article as regards other aspects of FAE, and by the extent to which it draws from recognisable prior publications. See G. Johannsohn, 'Fuel air explosives revolutionise conventional warfare', *International Defense Review* (Dec 1976), pp. 992–6. The published source which cites is C. A. Robinson, 'Soviets begin fuel air explosive tests', *Aviation Week and Space Technology* (22 Oct 1973).

[94] *The Reporter* (14 Sep 1961) p. 31.

parked aircraft without at the same time cratering air-field runways.[95] No connection was made in the public statements between FAX and the devices which the US Marine Corps were then just beginning to use in Vietnam for clearing helicopter landing-zones of mines and booby traps, which were in fact ground-functioned experimental versions of the BLU-73 FAE bomb, then under development as payload for the CBU-55 cluster bomb, the navy's first FAE munition.[96] In autumn 1968 the US Air Force began to employ its own early experimental FAEs – the big BLU-72 and BLU-76 bombs being developed in a programme code-named Pave Pat – for interdiction in Indochina, though with no great success and virtually no publicity.[97] By October 1970 the CBU-55 was in fullscale production for operational evaluation in Indochina, and in March 1972 the navy told Congress: 'FAE is one of our newest and most successful conventional munitions. It was a rapid development for the S.E. Asia conflict.'[98] US Navy, and later US Air Force, procurements through 1973 comprised some 30,000 FAE cluster bombs, a large proportion for supply to the South Vietnamese air force. Other potential foreign customers for US FAE munitions have been less successful. At least one NATO country – Canada – has been denied them, under the Ford administration; and Israel's order for some $7 million worth of CBU-72 clusters (probably about 3000–4000 of them), accepted in October 1976, was subsequently denied by President Carter.

The Pentagon had started to release detailed information about the new weapons by the end of 1970,[99] but even though this information described uses to which FAEs could be put, FAEs for a long time received almost no mention in the war-reporting from Indochina. Then quite suddenly, during the battle for Xuan Loc in April 1975, immediately preceding the fall of Saigon, there were press reports of the South Vietnamese air force using a devastating new anti-personnel weapon, said to be a bomb that caused instant asphyxiation over a wide area. Government spokesmen hastened to explain that the 'asphyxiation bomb' was not a poison-gas weapon, but merely the CBU-55FAE.[100]

It is surprising that use of so grossly spectacular an anti-personnel weapon should have remained unreported for as long as it did. Though the censorship and movement restrictions under which war correspondents have to operate may perhaps have contributed, this can surely not have been the only explanation. What seems more probable is that a weapon initially introduced for specialised anti-materiel

---

[95] Aviation Studies (International) Ltd, *Armament data sheets*, sect. 14, Novel weapons.

[96] C. A. Robinson, 'Special report: Fuel air explosives: Services ready joint development plan', *Aviation Week and Space Technology* (19 Feb 1973) pp. 42–6.

[97] Ibid., US Dept of the Air Force, Congressional testimony (*supra*, n. 28). Editors' note. This corresponds to US Dept of the Air Force, Congressional testimony, R. D. Anderson, statement before a subcommittee of the Senate Armed Services Committee (19 Nov 1970).

[98] US Dept of the Navy, Congressional testimony, FY1973 RDT&E budget line-item submitted for the record of the Senate Defense Appropriations Subcommittee (8 Mar 1972).

[99] US Dept of the Air Force, Congressional testimony (*supra*, n. 28). Editors' note. This corresponds to US Dept of the Air Force, Congressional testimony, R .. D. Anderson, statement before a subcommittee of the Senate Armed Services Committee (19 Nov 1970).

[100] 'Vietcong protest at new type of bomb', *The Times* (24 Apr 1975).

applications finally came to be used, under desperate circumstances, against troop concentrations in the open.

An FAE munition is a pure blast weapon, lacking the fragmentation effect of conventional HE munitions. Most of the people killed by HE munitions die from fragments (shrapnel), though blast may contribute, and fragment injuries are immediately obvious for what they are. Blast causes internal damage to the human body, and the cause of death in blast victims may not at once be apparent. There were instances during the city air-raids of the Second World War in which underground air-raid shelters buried in bomb debris were later found to be occupied by rows of people seemingly frozen, and with no visible injury, They were dead from blast. Blast injuries are not, of course, always fatal, neither are blast deaths always quick. The effects of shock waves on the human body are complex and still not fully understood. The greatest harm seems to arise from the passage of the shock wave through the gas-filled cavities of the body.[101] Pulmonary damage is typical: the lungs are squeezed fiercely as the shock wave decelerates into the thoracic cavity. Bubbles of air may then be forced into the blood circulation by the collapsing lungs, causing that brain damage or rapid death characteristic of air embolism. There will usually also be damage to the abdominal organs. Swedish experiments with 30-kilogram ethylene oxide FAEs, reported at the Lucerne and Lugano conferences, indicate that the mortality rate among FAE blast casualties is likely to be about one in two.[102] The mortality rate would probably rise to 85–90 per cent for a single cluster of 30-kg ethylene oxide bomblets.[103] If this were so, FAE munitions would indeed be exceptionally lethal weapons, as the reports from Xuan Loc suggested. The US team of experts at Lugano denied this, however, and asserted (without supporting data) that the killed-to-wounded ratio among FAE casualties was not significantly greater than that among casualties from more conventional explosives.[104]

Wherever the truth may ultimately be found to lie in this matter, the salient operational feature of an FAE munition is that it can subject a wide area of ground to a blast impulse that is probably stronger, and certainly more uniform, than that from a comparable HE munition. This unique attribute has found several military applications. The most highly valued of these so far, and the one which has probably done most to stimulate FAE R&D, is the clearing of minefields, for an FAE shock wave can detonate all types of pressure-sensitive mine and damage other types of mine sufficiently to render them harmless; it can do this in rough terrain where some mines might be shielded from the blast from ground-burst HE

US experiments with FAE warheads charged with 38.6 kg of propylene oxide fuel have shown a 100 percent mine-killing area of effectiveness of 246 m^2 in the case of

---

[101] US Dept of the Army, Office of the Surgeon General, *Wound Ballistics* (Washington, D.C., 1962) esp. R. W: French and G. R. Callender, 'Ballistic characteristics of wounding agents', pp. 91–141.

[102] Conference of Government experts on the use of certain conventional weapons, 2nd session, *Report* (Geneva, 1976).

[103] Ibid.

[104] Ibid.

pressure-fused mines and of 2110 m^2 in the case of pull-fused tripwire mines.[105] The increasing prominence of mine warfare in military doctrine has been noted earlier; and FAE munitions represent the only major new mine-neutralising device since the Second World War. This factor alone will probably guarantee a place for FAE munitions in more and more military arsenals around the world; and the munitions will at the same time be available for any other applications to which they may lend themselves.

As to these other applications, the experience of the Vietnam war has already indicated some of the possibilities. Others are suggested by the different FAE R&D projects that have been under way in the United States: some of these are detailed in Table 8.9. In particular, US Army proving-ground test-firings of FAE devices are reported to have 'demonstrated the combat applications of FAE blast effects for ... inflicting casualties to personnel in foxholes, producing mobility kills on wheeled vehicles, and destroying artificial and natural camouflage.'[106] In one demonstration, a single FAE warhead denuded a 90-foot diameter area 590 m^2 of thick heavy underbrush, trees, and a tactically emplaced army camouflage net. In another demonstration, a 2½-ton cargo truck was destroyed by an FAE warhead, being instantly set on fire, its engine and chassis severely damaged, and major parts blasted away; wooden targets in nearby foxholes were shattered. The effectiveness of FAE blast in killing or injuring people both in the open and inside structures into which the blast can penetrate is remarked in several sources.[107] Other US Army FAE developments noted in Table 8.9 include an FAE weapon intended for urban-warfare use and an FAE anti-ballistic missile warhead. The US Navy is reported to have a keen interest in FAE as a kill mechanism for anti-shipping weapons, including ship-launched cruise missiles.[108] In one trial, a simulated near miss with a prototype FAE weapon against a destroyer caused such damage that the ship subsequently sank.[109]

Relatively small blast overpressures – on the order of 40 kPa – can have great effects on ships, equipment such as masts, radar antennae, electronic-warfare devices, on-deck aircraft, and so forth being especially vulnerable. In these respects, areas of effectiveness on the order of 10–15 ha are expected from the 1500-lb class of FAE antishipping warhead currently under study.

---

[105] US Dept of the Army, congressional testimony . Office of the Assistant Secretary of the. Army (R&D) written statement for the record of the House Defense Appropriations Subcommittee (Apr 1974).

[106] J. R. Dennis, 'MERDC demonstrates fuel air explosive mine neutralization capabilities', *Army Research and Development Newsmagazine* (Jan–Feb 1975) pp. 12–13.

[107] US Dept of Defense, Congressional testimony, M. Currie, prepared statement for the Senate Armed Services Committee (Mar 1975); 'Armament lab verifies new technology', *Aviation Week and Space Technology* (15 July 1974) pp. 270–2; 'Jungle flatteners', *Ordnance* (Sep–Oct 1971) p. 157; C. A Robinson (*supra*, n. 60). Editors' note. The latter refers to note 387 in this volume.

[108] G. Johannsohn (*supra*, n. 60). Editors' note. This corresponds to G. Johannsohn, 'Fuel air explosives revolutionise conventional warfare', *International Defense Review* (Dec 1976), pp. 992–6.

[109] C. A. Robinson (*supra*, n. 63). Editors' note. This corresponds to C. A. Robinson, 'Special report: Fuel air explosives: Services ready joint development plan', *Aviation Week and Space Technology* (19 Feb 1973) pp. 42–6.

Table 8.9 Some US fuel-air explosive munitions and development projects[110]

Munition or project	Particulars
750-lb FAE bomb, BLU-64	200 kg of hydrocarbon fuel. An early US Air Force development project, soon cancelled. [a]
2500-lb FAE bomb, BLU-72 (Pave Pat I)	US Air Force development, designated as a 'general purpose' bomb.[a] Basically a drogue-chute retarded commercial propane tank (450 kg propane) fitted with a burster and a 3–4 second time-delay cloud detonator. Used experimentally in air interdiction operations over Indochina from Autumn 1968, carried on low-performance aircraft (e.g. A-1E), but reported 'not very successful'. [b]
130-lb FAE bomb, BLU-73	US Navy development as payload for the CBU-55 FAE cluster bomb. Fuel is ethylene oxide (about 33 kg). Weighs 60 kg. Drogue-chute retardation; 4-foot probe for activating burster charge. Two 125-millisec time-delay cloud detonators ejected from wall. [c] Combat use from 1967 onwards. [d] Said to be capable of completely defoliating a 700 m^2 area even in thick forest. [e]
2500-lb FAE bomb, BLU-76 (Pave Pat II)	A streamlined version of the BLU-72, for high speed carriage (e.g. three per F-4) though not high speed delivery. Not used in combat before 1971.[b] Unit cost in 1972 was $5000. Two BLU-76 bombs are capable of stripping vegetation and blowing down trees over an area of 0.84 hectares. [f]
500-lb FAE cluster bomb, CBU-55	Navy/Honeywell development, entering full-scale production in October 1970. The first successful FAE weapon. For low-performance aircraft (e.g. 14 per A-1E or 2 per UH-1). Comprises an SUU-49 rear-ejection free-fall dispenser with three BLU-73 bombs.[c, d] Unit cost in 1972 was $1520
500-lb FAE cluster bomb, CBU-72	Streamlined version of the CBU-55, for high speed (up to 450 knots) carriage; qualified for A-4 and A-7 aircraft.[d] Developed by 1971.[e] 'Designed to be an effective blast weapon against protected or concealed troops, land mines, booby traps and light material'. Unit cost in 1974 was $1770. [g]
Air-to-ground FAE rocket	For 4-tube launcher; studied in the early Air Force FAE work. [e]

(continued)

[110] Editors' note. This table was 'Table 3.8' in the original.

Table 8.9 (continued)

Munition or project	Particulars
500-lb high-speed FAE (? cluster) bomb	Being developed by the navy as a second-generation FAE weapon that can be delivered on target at high speeds, thus eliminating the delivery restrictions on the CBU-55/72. [d] In engineering development in 1972, but subsequently moved back into advanced development for further technology investigation rather than immediate weaponisation. [h] In July 1975 was scheduled to enter a joint navy/air force FAE-2 engineering development programme. Seems to use propylene oxide as the fuel
500-lb high-speed FAE bomb	In advanced development in 1973. [j] The air force contender for FAE-2. A version field-tested during 1974-5 had a 12-foot probe for the burster and trailed a cloud detonator on a 65-foot line, intended to function in the centre of the fuel-air cloud.[j] a 1975 air force patent application suggests that it uses propylene oxide (148 kg) as the fuel.[k] Other fuels under study by the air force include MAFP – a mixture of propyne, propadiene, propane and a trace of butane. [d]
2000-lb high-speed FAE bomb	Also part of the air force FAE-2 project. In concept-formulation stage in mid-1974. May use a proximity fuse rather than a probe for the burster.[i] A 1975 patent application suggests a 725-kg propylene oxide payload. [k]
FAESHED (helicopter-delivered FAE system)	An army mine-clearing development undergoing DT-III testing in 1974-5. Comprises a CBU-55 munition attached to each side of a UH-1H helicopter plus aiming and release mechanisms. Type-classification as standard was scheduled for early 1976. [m] Unit procurement cost was estimated in 1974 to be $8000-$9000. [n]
SLUFAE (surface-launched FAE unit)	Another army mine-clearing system in development at least since 1972 with completion scheduled for 1980. [j] Comprises a 30-tube armoured launcher mounted on an M548 5-ton tracked prime mover for single or ripple firing of modified navy Zuni 5-inch rockets, each with an oversize warhead containing 39 kg of propylene oxide and fitted with a foliage discriminating probe fuse. Maximum range approaches 1000 metres; can clear 8 × 100 m of minefield per loading, either by detonating pressure-sensitive mines or by damaging them so that they can no longer function. [m] Successfully tested against mines fitted with complex, long-impulse and double-impulse fuses, including hydraulic long-impulse, seismic infra-red, electronic, and magnetic influence fuses. [o] The 1974 estimate of eventual procurement cost was $75,000–$100,000 each, including the launch vehicle; total R&D cost to completion estimated at $12.677 millions. [n]

(continued)

Table 8.9 (continued)

Munition or project	Particulars
MADFAE (mass air-delivery FAE system)	A marine corps development under way in 1973. Comprises a downward-ejection dispenser, weighing just under a tonne fully loaded, with twelve 60-kg FAE bombs, for sling-loading from helicopter cargo hooks, two per helicopter. [j]
LANDFAE (large-area nozzle delivery of fuel-air explosive)	Demonstrated feasibility reported in Spring 1976. [o] An army mine-clearing concept in which a M67A2 mechanized flame thrower with a modified (high pressure) nozzle is used to spray fuel over a minefield, the fuel-air cloud being detonatable. Tests have shown that a traversible path 85 feet long can be cleared of mines with one spray; repeated sprayings could be done within seconds
Improved LANDFAE	In development in 1976: a high-pressure nozzle kit attached to an M60 tank enabling use of FAE 'in urban warfare and against other blast-sensitive targets requiring close-in attack by Army ground forces'. [p]
FAE warheads for space weapons or anti-ballistic missiles	In February 1973 it was reported that the navy was conducting detonation tests in vacuo of clouds from FAE munitions and that the results were 'expected to provide information leading toward perfecting a non-nuclear warhead for use in space'. [q] It seems that even under oxygen-sparse simulated high-altitude (10,000 metres) conditions FAE blast effects were significantly greater than those from comparable HE charges; this observation is said to have stimulated the army into evaluating the application of FAE to ABM systems. [e]
FAD/FAE follow-through hollow-charge warheads for 66 mm rockets	One of several SCIFT (shaped-charge incendiary follow-through) anti-armour munitions being developed by the army, this one for the M202A1 four-tube portable rocket launcher. Uses about 0.5 kg of liquid hydrocarbon fuel as follow-through for the hollow-charge jet: pentane, pentene, and propylene oxide have been studied. Larger calibre (80–90 mm) may be preferred. Stimulated by marine corps interest in more powerful portable weapons for use against bunker-type targets. In advanced development by 1975. [q]
FAE warheads for ship-launched cruise missiles	Navy development project, said to be under way in 1976. [e]
Shipborne FAE antimissile system	Early phase of a navy development project said to be under way in 1976. [e]
FAE warheads for antisubmarine torpedoes	Feasibility studies by the navy said to be under way in 1976. [e]

(continued)

Table 8.9 (continued)

Munition or project	Particulars
FAE warhead for Lance battlefield support guided missile	An army project apparently in concept-formulation stage in October 1976. [e]

Notes

[a] US Air Force Armament Laboratory, Guide to non-nuclear munitions and associated equipment under the cognizance of AFATL (1970)
[b] R. D. Anderson (Tactical Division, US Air Force Directorate of Operations), statement before a subcommittee of the Senate Armed Services Committee (19 Nov 1970). Investigation into electronic battlefield program: Hearings (Washington, D.C.: committee print, 1971), pp. 135–6
[c] US Naval Air Systems Command, Technical Manual NAVAIR 11-5A-26, Fuel-air explosive bomb cluster CBU-55/B
[d] C. A. Robinson, 'Fuel air explosives: Services ready joint development plan', Aviation Week and Space Technology (19 Feb 1973), pp. 42–6
[e] G. Johannsohn, 'Fuel air explosives revolutionise conventional warfare', International Defense Review 9(6) (Dec 1976), pp. 992–6
[f] Engineer Strategic Studies Group (Office of the Chief of Engineers, US Army), Herbicides and military operations, Vol 3 (Feb 1972)
[g] Admiral Gaddis, written response to the Chairman of the Senate Defense Appropriations Subcommittee, spring 1974: Department of Defense appropriations for fiscal year 1975: Hearings (Washington, D.C.: committee print, 1974)
[h] US Navy, justification statement in support of the 'Air-to-ground weapon technology (engineering)' FY 1973 RDT&E budget line-item submitted for the record of the Senate Defense Appropriations Subcommittee: Department of Defense appropriations for fiscal year 1973, Hearings (Washington, D.C.: committee print, 1972), Vol. 3, p. 1168
[i] 'Armament lab verifies new technology', Aviation Week and Space Technology (15 July 1974), pp. 270–2
[j] C. A. Robinson, 'Soviets begin fuel-air explosive tests', Aviation Week and Space Technology (22 Oct 1973)
[k] A. G. Bilek, assignor to the Secretary of the US Air Force, US patent no. 3,999,482 (app. 9 July 1975, iss. 28 Dec 1976), High explosive launcher system
[l] G.A. Carlson, assignor to the US Energy Research & Development Administration, US patent no. 3,955,509 (app. 21 Mar 1969, iss. 11 May 1976), Fuel-air munition and device.
[m] J. A. Dennis, 'MERDC demonstrates fuel air explosive mine neutralization capabilities', Army R&D Newsmagazine 16(1) (Jan–Feb 1975), pp. 12–13
[n] Office of the Assistant Secretary of the Army (R&D), written statement for the record of the House Defense Appropriations Subcommittee, in Department of Defense Appropriations for 1975, Hearings (Washington, D.C.: committee print, 1974), Vol. 4, p. 1231
[o] Ibid. Vol. 4, p. 1233
[p] 'LANDFAE feasibility shown: 6 nations send representatives', Army R&D Newsmagazine (May–June 1976), p. 13
[q] US Army Munitions Command, Research, Development and Engineering Directorate, Laboratory Posture Report FY-72 (1972) and Laboratory posture report FY-73, RCS AMCDL-101 (1973); and US Army Armament Command. Research Development and Engineering, Laboratory posture report FY-74, RCS AMCDL-101 (1974) and Laboratory posture report FY-75 (1975); and R. E. Brown and G. E. Taylor, HEAT follow-through warheads for improved lethality, contract report from Shock Hydrodynamics Division, Whittaker Corporation, to US Army Edgewood Arsenal, report no. ED-CR-74001, Jan 1974
[r] 'Latest US Army equipment on show', International Defense Review 10(1) (Feb 1977), pp. 101–8

Further back in development are novel fuels and detonation mechanisms which promise an order-of-magnitude increase in FAE blast effects per unit weight of munitions. Thus peak overpressures exceeding 10 MPa are regarded as attainable within current programmes. Should such objectives in fact be attained, individual FAE weapons would then have the capacity for exerting blast damage over an area approaching that of a nuclear artillery shell. Of the many trends in munition design that are rooted in the Vietnam war, the fastening of interest on fuel-air explosives is perhaps the most threatening of all.

Chapter 9
Bibliography of Julian's Writing on Chemical and Biological Warfare

Julian Perry Robinson

Editors' note. This bibliography details Julian's writings on chemical and biological warfare (CBW). It is based on Julian's 'O' list – the system he devised to catalogue his CBW work. As well as showcasing the extent of his writing, the list gives an insight into the way Julian thought about and managed his work. Rather than compiling a conventional bibliography by simply listing separate publications, Julian assigned numbers to each of his writing projects on CBW issues, starting with his undergraduate dissertation (Item 1). Each item number then details how he developed his ideas and repurposed writings for different audiences.

The list is long and detailed, but it is not comprehensive – there are some gaps in the numbering where Julian had an idea for a paper but did not produce a written output, and the 'O' list only ever included his work on CBW. Moreover, he was not able to update his 'O' list before he died, and some entries are incomplete. We have worked to add extra information, but we are aware there are still gaps; the Harvard Sussex Program is continuing work to fill these.

We have lightly edited some aspects of Julian's original, including by specifying full bibliographic details such as journal titles. We have also formatted entries in such a way as to enable a clear distinction between drafts, unpublished papers and published outputs.[1]

1—'Some aspects of chemical warfare: An investigation into the influence of the Second World War on the development of chemistry, with particular reference to the war gases', March 1964, 264 pp. Dissertation, Oxford University.

2—'Chemical warfare', March 1965, 4 pp. For *Spectator* (unpublished).

3—'Preliminary report (regarding chemical warfare in the Yemen)', June 1966, 4 pp [private distribution].

4—'Chemical warfare today: a synopsis', November 1966, 10 pp. For Granada Television.

[1] This is the version updated by Richard Guthrie, 23 February 2024.

J. P. Robinson (✉)
Brighton, UK

H. Wilson et al. (eds.), *Julian Perry Robinson: Pioneer in the Prohibition of Chemical and Biological Warfare*, Pioneers in Arts, Humanities, Science, Engineering, Practice 32, https://doi.org/10.1007/978-3-032-04064-0_9

5—'Chemical Warfare', December 1966, 21 pp.

Published as—'Chemical Warfare', *Science Journal* (London), vol 3, no 4, April 1967, pp 33–40.

6—'Chemical warfare', April 1967, 25 pp.

Published as—'Chemistry in warfare', in N. Valery (ed): *Penguin Science Survey 1968: Physical Sciences* (London: Penguin Books, 1968): pp 187–205.

7—'Toxic agents in warfare', April 1967, 13 pp. Book synopsis for André Deutsch Ltd.

8—'Chemical Weapons', July 1967, 8 pp, Draft chapter for André Deutsch Ltd.

9—'Gas warfare in Yemen', Letter to Editor, *The Times*, (unpublished), 20 July 1967, 1p.

10—'Chemical weapons', February 1968, 28 pp. Paper presented at the Bernal Peace Library Conference on Chemical & Biological Warfare, London.

Published as—'Chemical weapons', in: S.P.R. Rose (ed), *CBW Chemical and Biological Warfare*, (London: Harrap, 1968): pp 19–34 and pp 191–94.

11—[with Robin Clarke] 'Research and policy on CBW in the United Kingdom', Paper presented at the Bernal Peace Library Conference on Chemical and Biological Warfare, London, February 1968. 19 pp.

Published as—'Research and policy on CBW in the United Kingdom', in: Rose op cit, pp 105–14 and pp 199–201.

12—'The capabilities of chemical weapons', Paper presented at the SIPRI Symposium on Chemical Warfare, Stockholm, August 1968, 23 pp [submitted as a background paper June 1968].

13—'The verification of compliance with chemical arms control and disarmament agreements', Paper presented at the SIPRI Symposium on Chemical Warfare, Stockholm, August 1968, 26 pp [submitted as a background paper May 1968].

14—'The SIPRI chemical warfare symposium', 4 pp. Paper presented at the 19th Pugwash Conference on Science and World Affairs, Nice, September 1968.

Published as—'The SIPRI chemical warfare symposium', *Proceedings of the Nineteenth Pugwash Conference on Science and World Affairs, September 1968*, pp 213–15.

15—'A procedure for investigating alleged uses of chemical weapons', 2 pp. Paper presented at the 19th Pugwash Conference on Science and World Affairs, Nice, September 1968.

Published as—'A procedure for investigating alleged uses of chemical weapons', *Proceedings of the Nineteenth Pugwash Conference on Science and World Affairs, September 1968*, pp 216–17.

16—'An analysis of British gas casualties in World War I'. 25 January 1969. Working paper for Item 18 and Item 38 below. 7 pp.

17—'Some properties of antipersonnel chemical weapons', 20 February 1969, Included in the submission by SIPRI to the UN Secretary-General's Expert Group on Chemical and Biological Weapons, 113 pp.

18—'Chemical warfare during World War I', 20 February 1969, 45 pp. Included in the submission by SIPRI to the UN Secretary-General's Expert Group on Chemical and Biological Weapons, Revised version in Item 42 below, pp. 26–58 and 125–141.

19—'Chemical warfare after the First World War', 20 February 1969. 21 pp, Included in the submission by SIPRI to the UN Secretary-General's Expert Group on Chemical and Biological Weapons. Revised into Item 31 below.

20—[with Matthew Meselson] 'The status of tear gas and other sensory irritants in chemical warfare', 23 May 1969, 5 pp. Submission by the Pugwash Chemical and Biological Warfare Study Group to the UN Secretary-General's Expert Group on Chemical and Biological Weapons.

Published as—'Escalation of chemical warfare', *New Scientist*, vol 43, 14 August 1969, pp. 339–41.

21—Contributions to the submission by the World Health Organisation to the UN Secretary-General's Expert Group on Chemical and Biological Weapons, May 1969.

Expanded and published as—*Health Aspects of Chemical and Biological Weapons: Report of a WHO Group of Consultants* (Geneva: WHO, 1970).

22—'Popular attitudes towards CBW, 1919-1939'.

Published as—'Popular attitudes towards CBW, 1919–1939' in Item 42 below, pp 231–67.

23—'National CBW programmes and policies, 1919–1939', 1969.

Published as—'National CBW programmes and policies, 1919 1939' in Item 42 below, pp 268–293.

24—'Developments in chemical and biological warfare', July 1969.

Published as—'Developments in chemical and biological warfare', in: *SIPRI Year-book of World Armaments and Disarmament 1968/69* (Stockholm: Almqvist & Wiksell; London: Duckworth; and New York: Humanities Press, 1969), pp. 112–34.

25—'Chemical warfare', July 1969.

Published as—'Gli aggressive chimici' in: *S & T 70: Annuario della EST* (Milan: Enciclopedia della Sctenza e della Tecnica Mondadorf, 1970), pp 397–404.

26—'CS overstatement' (letter to the Editor), *New Scientist* vol 44, 2 October 1969, pp 42–43.

27—'Quantitative estimates of the possible effects of CBW attacks on civilian targets', October 1969, 60 pp. Working paper for an appendix (unpublished) to Item 52 below.

28—'The Yemen case', February 1970, 25 pp. Incorporated into the 1970 provisional edition of Items 42 and 43 below.

Revised and published as—'The allegations of chemical warfare in the Yemen, 1963–67', Item 43 below, pp 225–238.

29—[with Robert Neild] 'Tear gas: a blunder', *Sunday Times* (London) 15 February 1970, p 12.

30—'Some notes on the British Government's decision that CS is beyond the scope of the Geneva Protocol', April 1970, 26 pp. Paper presented at a meeting of the British Pugwash Group, London, 29 May 1970.

31—[with Milton Leitenberg] 'Instances and allegations of CBW, 1914–1970', 1970.

Published as—'Instances and allegations of CBW, 1914–1970' in Item 42 below, pp 125–230.

32—'The developing technology of CBW', 1970.

Published as—'The developing technology of CBW' in Item 42 below, pp 25–124.

33—'The non-use of CB weapons during World War II', 1970.

Published as—'The non-use of CB weapons during World War II' in Item 42 below, pp 294–335.

34—'The manufacture of nerve gases', June 1970, 35 pp. Paper presented at a SIPRI Symposium intended to initiate an international CW inspection project, Stockholm, 25–27 July 1970.

35—'Military uses of CB weapons, their value, and their relation to national CBW armament or disarmament policy', September 1970, 78 pp [private discussion paper].

36—'Constraints on use of CB weapons', 3 December 1970, 9 pp. Incorporated into Item 38 below.

37—'Would CB disarmament be worthwhile?', 9 January 1971. 61 pp. Incorporated into Item 43 below, pp. 74–91.

38—[with Robert Neild] 'The claims that CB weapons are less inhumane than other weapons', February 1971, 19 pp.

Abbreviated version published as—Appendix 1 in Item 43 below, pp. 124–136.

39—[with Anders Boserup] 'Strengthening the existing prohibition of CBW', February 1971.

Published as—'Strengthening the existing prohibition of CBW' in Item 43 below, pp 21–73.

40—Verification of compliance with a ban on development on production of CW agents: The possibilities of direct inspection at the stages of development, production, transport and storage'. 2 April 1971. 16 pp. Incorporated into Item 43 below, pp 163–178.

41—'The CB weapons controls of the Western European Union', May 1971.

Published as—'The CB weapons controls of the Western European Union' in Item 43 below, pp 190–219.

42—The Rise of CB Weapons (Stockholm: Almqvist & Wiksell; and New York: Humanities Press, 1971), 395 pp. [Volume 1 of SIPRI's The Problem of Chemical and Biological Warfare].

43—[with Anders Boserup and Robert Neild] The Prevention of CBW (Stockholm: Almqvist & Wiksell and New York: Humanities Press, 1971), 287 pp. [Volume V of SIPRI's The Problem of Chemical and Biological Warfare]

44—'National CBW programme and policies: the USSR', June 1971, 23 pp. Incorporated into Item 49 below.

45—'Comment: Chemical weapons and inspection', *New Scientist*, vol 53, 9 March 1972, p 523.

46—[with Hans von Schreeb] 'The Utility of chemical and biological weapons', April 1972.

Published as—'The Utility of chemical and biological weapons' in Item 52 below, pp. 116–159.

47—[With Carl-Goran Heden] 'Research and Development: implications for the future of CBW', April 1972.

Published as—'Research and Development: implications for the future of CBW', in Item 52 below, pp. 260–332.

48—'Modern CB weapons and the defences against them', May 1972.

Published as—'Modern CB weapons and the defences against them' in Item 52 below, pp. 27–115.

49—'National CBW policies and programmes', June 1972.

Published as—'National CBW policies and programmes' in Item 52 below, pp. 160–259.

50—Review of J. B. Neilands *et al*, 'Harvest of Death: Chemical Warfare in Vietnam and Cambodia', in: *New Scientist*, 22 June 1972, p 702–03.

51—'Binary weapons – a mixed problem', *New Scientist*, 5 April 1973, pp 34–35.

52—[with Carl-Goran Heden and Hans von Schreeb] *CB Weapons Today* (Stockholm: Almqvist & Wiksell; New York: Humanities Press; and London: Paul Elek,

1973. Volume II of SIPRI's *The Problem of Chemical and Biological Warfare*). 420 pp.

53—'Notes on the development of dibenz[b,f][1,4]oxazepine (CR) as a chemical-warfare agent', 27 September 1973, 8pp [private distribution].

53a—'The magnitude of the US nerve-gas capability', October 1973. Incorporated into Item 55a below.

54—'Decision making with respect to chemical and biological warfare', 15 November 1973, 15 pp. Narrative report on a continuing research project.

55—'Chemical/biological warfare: An introduction and bibliography', 15 November 1973, *Political Issues Series* (Los Angeles: Center for the Study of Armament and Disarmament, California State University), vol 3, no 2, 1974, 34 pp.

55a—'The size of the US nerve-gas stockpiles'. February 1974. Included as an appendix in Item 56 below. 17 pp.

56—'Binary nerve-gas weapons: their economic, environmental and security implications for the United States, including their likely effects on the international chemical disarmament negotiations'. February 1974. Paper presented at the American Chemical Society Symposium on Chemical Weapons and US Policy, Los Angeles, April 1974, and at the 1st Pugwash Chemical Warfare Workshop, Helsinki, April 1974. 88 pp. Revised version printed in *US Chemical Warfare Policy*, Hearings before the Subcommittee on National Security Policy and Scientific Developments, Committee on Foreign Affairs, House of Representatives, US Congress, May 1974, pages 294–344. Further revised version as Item 68 below; and shortened and updated version as Item 81 below.

57—'A future for the chemical disarmament negotiations', February 1974, 5 pp. Paper presented at the 1st Pugwash Chemical Warfare Workshop, Helsinki, April 1974.

58—'Some implications of binary nerve gas weapons', 1 April 1974, In American Chemical Society, Chemical Weapons and U.S. Public Policy, a report of the Committee on Chemistry and Public Affairs (Washington, D.C. Am Chem Soc, 1977), pp. 11–12.

59—'Chemical disarmament, verification and security', 18 April 1974, 5 pp. Paper presented at the 1st Pugwash Chemical Warfare Workshop, Helsinki, April 1974.

60—'Nerve gas, national security and international security', April 1974, 18 pp. Included in Item 61 below.

61—Statement before the Subcommittee on National Security Policy and Scientific Developments, 2 May 1974, 19 pp.

Published as—Statement before the Subcommittee on National Security Policy and Scientific Developments, Committee on Foreign Affairs, House of Representatives,

US Congress, 2 May 1974. In Hearings, U.S. Chemical Warfare Policy (Washington, D.C: USGPO, 1974), pp 58–69, 91–92 and 96.

62—'The implications of procuring binary munitions', May 1974, 4 pp. Included in Item 63 below.

63—Statement before the Defense Subcommittee of the Committee on Appropriations, House of Representatives, US Congress, 21 May 1974. In Hearings, *Department of Defense Appropriations for 1975* (Washington, D.C.: USGPO, 1974) Part 9.

64—'Binary nerve gas', 2 June 1974, 3 pp. For Federation of American Scientists.

65—'Binary nerve gas and the limitation of chemical weapons', 30 June 1974, 3 pp.

Published as—'Binary nerve gas and the limitation of chemical weapons', *Arms Control Today*, vol 4, no 7/8, July/August 1974, pp 1–2 & 4.

66—'Decision making with respect to chemical and biological warfare', 28 July 1974, 15 pp. Narrative report on a continuing research project.

67—'Assimilation as a factor in armaments proliferation: the example of CB weapons, and its relevance to the present disarmament negotiations', 16 August 1974, 8 pp. Paper presented at the 24th Pugwash Conference on Science and World Affairs, Baden, August-September 1974.

Published as—'The special case of chemical and biological weapons', *Bulletin of the Atomic Scientists*, vol 31, no 5, May 1975, pp 17–23.

68—'Binary nerve-gas weapons' September 1974.

Published in—SIPRI, *Chemical Disarmament: New Weapons for Old* (1975), pp 21–99. [See also Item 56 above].

69—'Notes on the latest DoD CBW Obligations Report and related matters', 15 November 1974, 4 pp [private distribution]

70—Review of SIPRI, *The Effects of Developments in the Biological and Chemical Sciences on CW Disarmament Negotiations*, *New Scientist*, 21 November 1974, p 601.

71—'CW Stockpiles', 25 November 1974, 2pp.

Published in—SIPRI, *Chemical Disarmament: New Weapons for Old* (Stockholm: Almqvist & Wiksell and New York: Humanities Press, 1975), pp 100–01.

72—'The limitation of chemical weapons', November 1974, 18 pp. For *Scientific American* (unpublished).

73—'Comment: Behind the VX disclosure', *New Scientist*, 9 January 1975, p 50.

74—'The US chemical-agent stockpiles', March 1975. Summarized in Item 81 below, table 2.

75—'Controls on CW research and development', March 1975, 22 pp. Paper presented at the 2nd Pugwash Chemical Warfare Workshop, Stockholm, April 1975.

76—'US procurement of lethal/incapacitating chemical agents/munitions from fiscal year 1964 onwards', June 1975 [private working paper].

77—Review of *The Problem of Chemical and Biological Warfare, Volume VI*, SIPRI, *New Scientist*, vol 66, 12 June 1975, p 627.

78—'The Second Pugwash Chemical-Warfare Workshop', 20 June 1975, 6 pp [private distribution].

79—'The use of chemical irritants during the *Mayaguez* affair', 30 June 1975, 5 pp [private distribution].

80—'Temporarily disabling chemicals: their treatment in a chemical weapons convention', 20 August 1975, 10 pp [private distribution].

Revised version—'Incapacitants: a proposal for their treatment in a CW Convention', a paper for the 3rd Pugwash Chemical Warfare Workshop, London, April 1976, 9 pp [submitted March 1976].

81—'The United States binary nerve-gas programme: national and international implications', ISIO Monographs (University of Sussex), First series, No. 10 (1975), 51 pp. [See also Item 56 above].

82—'The former biological-weapons capability of the United States: its nature, and some residual questions', 12 October 1975, 13 pp [private distribution].

83—'Comment: US Army thwarted on binaries', *New Scientist*, vol 68, 16 October 1975, p 131.

84—'CW verification: a reappraisal in the light of the recent past', Paper presented at the 3rd Pugwash Chemical Warfare Workshop, London, April 1976, 9 pp [submitted March 1976].

85—Review of *The Law of War and Dubious Weapons*, and *Medical Protection Against Chemical-Warfare Agents*, both by SIPRI, *New Scientist* vol 71, 19 August 1976, p 404.

86—'Chemical warfare armament and disarmament', 10 November 1976, 4 pp.

Published as—'Chemical weapons: an urgent threat to world peace', *Sanity* (London: CND), 1976, pp 8–9.

87—'Chemical weapons for NATO? A framework for considering policy options', December 1976. Paper presented at the American Academy of Arts & Sciences Conference on Chemical Weapons and Chemical Arms Control, Boston, January 1977.

Published in—Matthew S Meselson (ed), *Chemical Weapons and Chemical Arms Control* (Washington D.C.: Carnegie Endowment, 1978): pp 21–49, 52–53 and 104–122.

Revised and published as—'Should NATO keep chemical weapons? A framework for considering policy alternatives', August 1977, SPRU Occasional Paper series (University of Sussex) no 4, 94 pp.

88—'Near-site verification technique: its place in a confidence-building strategy of verification', Paper presented at the 5th Pugwash Chemical Warfare Workshop, Cologne, August 1977, and at the 27th Pugwash Conference on Science and World Affairs, Munich, August 1977, 9 pp [submitted July 1977].

Published as—'Near-site verification technique: its place in a confidence-building strategy of verification', *Pugwash Newsletter*, pp 420–24.

89—'Comment: Progress on CW', *New Scientist*, 26 January 1978, p 202.

90—'Current state of chemical-warfare arms limitation', 12 pp. Paper presented at the 6th Pugwash Chemical Warfare Workshop, Kansas City, May 1978.

91—'The 6th Pugwash Chemical-Warfare Workshop: An amplified summary of the proceedings', 22 May 1978, 21 pp [private distribution].

92—'Porton participation in the Pugwash Chemical-Warfare Study Group', 14 June 1978, 2 pp [private distribution].

93—'Controls on chemical-warfare research and development', July 1978, 29 pp [private distribution]. This is a revised version of Item 75.

94—'Estimates for the US lethal CW stockpile', January 1979. Working paper for Item 96 and Item 102 below.

95—'CBW talks', 4 June 1979, 1 p.

Published in—*ADIU Report*, vol 1, no 1, 5 June 1979.

96—'Chemical warfare capabilities of the Warsaw and North Atlantic Treaty Organisations: An Overview from open sources', June 1979. Paper presented at the SIPRI Symposium on Chemical Weapons, Stockholm, June 1979.

Published as—Printed in SIPRI, *Chemical Weapons: Destruction and Conversion* (London: Taylor & Francis, 1980), pp 9–56.

97—'Chemical warfare, chemical arms limitation and confidence building: a review of the past year, with proposals', June 1979, 22 pp. Paper presented at the 7th Pugwash Chemical Warfare Workshop, Stockholm, June 1979.

98—'Negotiated chemical-warfare arms limitations: an analysis', August 1979, 20 pp [private distribution].

Expanded and published as—'The negotiations on chemical-warfare arms control' in *Arms Control* (London: Frank Cass), vol 1, no 1, May 1980, pp 30–52.

99—'Introduction' (August 1979) to SIPRI, *Chemical Weapons: Destruction and Conversion* (London: Taylor & Francis, 1980), pp 1–8.

100—'Vapour Chase', 20 October 1979.

Abbreviated version published as—'Vapour Chase on Salisbury Plain', *New Scientist*, vol 84, 25 October 1979, p 254.

101—'How many BM-21 launchers would be needed to achieve substantial casualty effects from hydrogen cyanide over a square-kilometre area?', 29 November 1979, 4 pp [private distribution].

101a—'US chemdemil operations since 1969', January 1980. Incorporated into Item 110 below.

102—[with Matthew Meselson] 'Chemical warfare and chemical disarmament', February 1980.

Published as—'Chemical warfare and chemical disarmament', *Scientific American*, vol 242, no 4, April 1980, pp 38–47.

103—'Chemical and biological warfare: Analysis of recent reports concerning the Soviet Union and Vietnam', 20 February 1980.

Published as—'Chemical and biological warfare: Analysis of recent reports concerning the Soviet Union and Vietnam', *ADIU Occasional Paper* no 1, March 1980. 42 pp.

104—'Chemical weapons', *ADIU Report* 2(1): 7 March 1980.

105—'Editorial: Chemical arms: a sea of troubles', *Nature*, 3 April 1980, p 387.

106—'East-west fencing at Geneva', *Nature*, 3 April 1980, p 393.

107—'Poison gas futures'.

Published as—'The invisible chemistry of death', *The Guardian*, 24 July 1980, p 13.

108—'Sibirskaya Yazva, binary nerve-gas and the future of arms control', 18 August 1980, 2 pp. Paper presented at the 30th Pugwash Conference on Science and World Affairs, Breuklen, August 1980.

109—'Retaliatory chemical ammunition requirements', September 1980, 9pp [private distribution].

110—'American stocks of lethal chemical-warfare agents and munititions', 30 September 1980, 37 pp [private distribution].

111—'Notes to accompany UK/END statement on chemical warfare', 20 November 1980, [private distribution].

Rev.1—published in mimeo form as Russell Committee Against Chemical Weapons, 'Oppose a new chemical arms race: Briefing notes on the Scientists' Campaign Against Chemical Rearmament', 26 February 1981, 38 pp.

112—'Chemical warfare intrigues', 4 January 1981.

Published as—'Giftigkorridorkrig' in *Forsvar* (Copenhagen) 1981(4): 7–8.

113—Letter to the Editor, *The Guardian* (London & Manchester), 10 January 1981, p. 8.

114—'Chemical warfare', February 1981, 17 pp. Paper presented at a private discussion meeting of the International Institute for Strategic Studies, London, 18 February 1981.

115—'Imminent chemical rearmament', 23 February 1981, 3 pp. Incorporated into: Steven Rose, 'The case against chemical weapons', *New Scientist*, 12 March 1981, pp 670–71.

116—Statement at a press conference in the House of Lords, London, 26 February 1981, 4 pp. Launch of the Scientists' Campaign Against Chemical Weapons.

117—'Chemical arms control and the assimilation of chemical weapons', *International Journal* (Toronto), 36(3), July 1981, pp 515–34 [submitted February 1981].

118—'The projected Chemical Weapons Convention: A review of recent salient events, with comments on the use of on-site inspection and a consultative committee for verification', Paper presented at the 8th Pugwash Chemical Warfare Workshop, Geneva, April 1981, 22 pp [submitted 29 March 1981].

119—'Environmental effects of chemical and biological warfare', 9 May 1981, 43pp. Text of a public lecture given at the invitation of the Minister of Agriculture, Stockholm, 25 May 1981, with appended 'Some characteristics of chemical and biological weapons'.

Published as—Printed in Environmental Advisory Council, *War and Environment* (Stockholm: LiberForlag, 1981), pp 73–117.

120—[with Karlheinz Lohs] 'Impressions of the Eighth Pugwash Workshop on Chemical Workshop'. *Pugwash Newsletter* 19(1): 3–5, July 1981.

121—'Tabular documented collation of data on US offensive CW capabilities', July 1981, 48 pp. Incorporated into: Item 155 below.

122—'Chemical weapons and Europe', August 1981.

Published as—*Survival*, vol 24, no 1, January/February 1982, pp 9–18.

123—'Chemical, biological and radiological warfare: futures from the past', September 1981, 25 pp. Submission to the Independent Commission on Disarmament and Security Issues.

124—'Haig's mycotoxins', 22 October 1981, 10 pp. A background paper for *Der Spiegel.*

125—'Haig's mycotoxins: A commentary on the new American charges of chemical/ biological warfare in Cambodia, Laos and Afghanistan', 26 October 1981, 21 pp [private distribution].

126—'Some recent developments concerning the American chemical rearmament programme', 6 November 1981, 2 pp [private distribution].

127—[with Sean Murphy and Alastair Hay] 'On chemical war' (letter), *Nature* vol 294, 19 November 1981, p 204.

128—Review of Sterling Seagrave, *Yellow Rain: A Journey through the Terror of Chemical Warfare*, 23 November 1981, 9pp. For *New York Review of Books* (unpublished).

129—[with Alastair Hay, Sean Murphy and Steven Rose] November 1981

Published as—*The Threat of Chemical Weapons*, Russell Committee against Chemical Weapons, (Nottingham: Spokesman Pamphlet No. 78, 1982), 38 pp.

130—'The changing status of chemical and biological warfare: recent technical, military and political developments', 6 December 1981.

Published as—'The changing status of chemical and biological warfare: recent technical, military and political developments', in *World Armaments and Disarmament: SIPRI Yearbook 1982* (London: Taylor & Francis, 1982), pp 317–61.

131—'Chemical warfare as a current disarmament and security issue: a summary analysis, with proposals', 6 January 1982, 8 pp. Note to the independent Commission on Disarmament and Security Issues.

132—'Chemical warfare as a current disarmament and security issue: a summary analysis, with proposals', 12 January 1982, 8 pp. Notes accompanying a lecture before the Arbetarrorelsens Fredsforum, Stockholm 13 January 1982.

133—[with Alastair Hay, Sean Murphy and Steven Rose] 'Death clouds over Europe' (excerpts from Item 129 above). *The Guardian.* 28 January 1982, p.13.

134—[with Alastair Hay, Sean Murphy and Steven Rose] 'The poison cloud hanging over Europe' (excerpts from Item 129 above). *New Scientist* 93: 630-635, 11 March 1982.

135—[with Alastair Hay, Sean Murphy and Steven Rose] 'Chemical warfare in the Eighties' (excerpts from Item 129 above). *Journal of the Medical Association for Prevention of War* 3(8): 348–358. Autumn 1982.

136—'Reagan's binaries: How should Europeans regard American poison-gas rearmament?', 24 February 1982, 17 pp [private distribution].

137—'Clarifying the numbers of stockpiled munitions', Letter to the Editor, *Armed Forces Journal International* 119(7): 8 March 1982, p 8.

138—'Chemical warfare: some events of the past year and their implications', 9 March 1982, 16 pp. Paper presented at the 9th Pugwash Chemical Warfare Workshop, Geneva, March 1982.

Rev.1—'Chemical warfare: some events of the past year and their implications', 16 pp.

Published as—'Chemical warfare: some events of the past year and their implications', *Pugwash Newsletter*, vol 19, no 4, April 1982, 157–64.

Published as—'Chemische Kriegführung – Pläne und Wirkungen' in H G Brauch and A Schrempf (eds), *Giftgas in der Bundesrepublik Deutschland* (Frankfurt: Fischer, 1982), pp 7–27 and 289–293.

139—'Narrative account of the Ninth Workshop of the Pugwash Chemical- Warfare Study Group', 25 March 1982, 16 pp [private distribution].

140—Presentation on the Sverdlovsk anthrax episode, 16 April 1982, 7 pp. CSACIS Symposium on Verification, Lancaster. Revised and expanded into Item 148 below.

141—[with Karlheinz Lohs] 'Impressions of the Ninth Workshop on Chemical Warfare'.

Published as—'Impressions of the Ninth Workshop on Chemical Warfare', *Pugwash Newsletter*, vol 19, no 4, April 1982, pp 152–57.

142—'The binary nerve gas program from a European standpoint', May 1982, 22 pp. Included in Item 143 below.

143—Statement before the Committee on Appropriations, US Senate, 5 May 1982.

Published as—'Binary Chemical Weapons', (Washington, DC: US Government Printing Office, 1983), pp 40–54 and passim.

144—'Estimates of current American holdings of toxic chemical munitions and agents', 28 May 1982, 24 pp. Working paper for Item 145 below.

145—'Recent developments in the field of chemical warfare', 30 May 1982, 16 pp.

Published as—'Recent developments in the field of chemical warfare', Royal United Services Institute for Defence Studies, *Defence Yearbook* 1983 (London: Brassey's, 1983), pp 166–82.

146—'Senate freezes U.S. poison-gas artillery stockpile', 3 June 1982, 3 pp.

Rev.1 [annotated version of above]—'Senate freezes U.S. poison-gas artillery stockpile', 16 June 1982, 3 pp [colour scan (much larger file)].

Published as—'Senate freezes U.S. poison-gas artillery stockpile', *ADIU Report*, vol 4, no 3, May/June 1982, pp 5–6.

147—'Estimates of current American holdings of toxic chemical munitions and agents', June 1982, 24 pp. Sussex paper no 2 for the Berghof Quasi Nuclear Weapons Project. Amended version of Item 144 above.

148—'Discussion of the Paper by Philip Towle, "The Soviet Union and the Biological Weapons Convention" and a guide to sources on the Sverdlovsk incident', 13 July 1982, 23 pp.

Published as—' "The Soviet Union and the Biological Weapons Convention" and a guide to sources on the Sverdlovsk incident', *Arms Control* (London), vol 3, no

3 [Special Issue on 'The Verification of Arms Control Agreements', edited by Ian Bellany and Coit D Blacker], December 1982, pp 41–56.

149—'Chemical warfare arms control from a European standpoint', July 1982. In US Congress, Committee on Foreign Affairs, Subcommittees on International Security & Scientific Affairs and Asian and Pacific Affairs, Hearings, *Foreign Policy and Arms Control Implications of Chemical weapons*, March 30 and July 13, 1982, committee print, pages 229–238.

150—[with Martin Kaplan] 'The Pugwash Chemical-Warfare Study Group: Progress and Prospects', 6 August 1982, 5 pp. Paper presented at the 32nd Pugwash Conference on Science and World Affairs, Warsaw, August 1982.

151—'Tabulated data on Soviet toxic chemical munition capabilities, with some derived estimates'. October 1982. Sussex paper no 3 for the Berghof Quasi Nuclear Weapons Project, second revision. 27 pp. Incorporated into Item 155 below.

152—'Bacteriological and chemical war'. 29 October 1982. Paper presented at the UNESCO Colloquium on Armament, Development, Human Rights and Disarmament, Paris, October 1982. 5 pp. Revised into Item 165 below.

153—'Ways open to individual Westerners for promoting chemical-warfare disarmament'. November 1982. Private distribution, CWDN/DP.2. 6 pp.

154—'Defining the work that a nongovernmental policy-research institute might best undertake in support of the present chemical-warfare arms control negotiations' plus annexe, 26 November 1982. Private distribution, SIPRI/CWACP/DP.1.

155—'Tabulated data on American and Soviet chemical weapons, with some derived estimates'. December 1982. Private distribution, CWDN/IP.2. 63 pp.

156—'Developments during 1982 in the field of chemical and biological warfare'. 25 December 1982. 36 pp.

Published as—As 'Chemical and biological warfare: developments in 1982' in *World Armaments and Disarmament: SIPRI Yearbook 1983* (London: Taylor & Francis, 1983): 391–423.

157—'A review of developments during 1982 bearing on chemical-warfare disarmament'. February 1983. Paper presented at the 10th Pugwash Chemical-Warfare Workshop, Geneva, 19–20 February 1983. 43 pp. (This is a slightly revised and updated version of Item 156.)

Abbreviated published version—Abbreviated version in *Pugwash Newsletter* 20(4): 136–145, April 1983.

158—[With Karlheinz Lohs] 'Report on the Tenth Workshop on Chemical Warfare'. 21 February 1983.

Published as—In *Pugwash Newsletter* 20(4): 133–136, April 1983.

158a—[with Matthew Meselson] Tabulated data on the environmental and biomedical samples from Southeast Asia and Afghanistan that have been analysed for trichothecenes. Presentation at the American Academy Conference on 'Yellow Rain', Cambridge, 15–16 April 1983. Successive subsequent revisions. Ca 12 pp.

159—'Draft report from the chemical warfare working group'. 25 April 1983. American Academy Conference on 'Yellow Rain', Cambridge, 15–16 April 1983. 8 pp.

160—'The concept of a chemical-weapons-free zone in Europe: some salient features of the current European chemical-warfare situation'. 3 May 1983. Paper presented at the LUISS Centro di Studi Strategici Symposium on Chemical Warfare, Rome, May 1983. 17 pp.

Published as—Printed in *Chemical Weapons and Arms Control: Views from Europe* (Rome: LUISS Centro di Studi Strategici, 1983): 42–56.

161—'Summary report of the LUISS Centro di Studi Strategici Symposium on Chemical Warfare, Rome, May 1983['. 10 May 1983. 6 pp. Printed in *Chemical Weapons and Arms Control: Views from Europe* (Rome: LUISS Centro di Studi Strategici, 1983). pp 5–8.

162—[with Matthew Meselson] 'Towards a unifying concept for chemical-warfare defence and arms-control policy'. 13 May 1983. Private distribution. 19 pp.

163—Review of Grant Evans: *The Yellow Rainmakers: Are Chemical Weapons being used in Southeast Asia?* 22 May 1983. 2 pp.

Published as—Abbreviated version in *New Society* 64: 358, 2 June 1983.

164—[with Peter Ashton, Matthew Meselson, Joan Nowicke and Thomas Seeley] excerpted in 'Comparison of Yellow Rain and bee excrement'. 31 May 1983. 5 pp. A paper made available at the 1983 Annual AAAS Meeting, Detroit.

165—'Five propositions relating bacteriological and chemical war to the theme armament-development-human rights-disarmament'. August 1983. 7 pp. Revised version of Item 152 above.

Published as—Published May 1985 in G Fischer (ed) *Armement-Développement - Droits de l'Homme - Désarmement.* (Paris: Publications de la Faculte^ de Droit, Paris V, 1985) pp 490–94.

166—[with Matthew Meselson] 'The pollen findings: A comparison of four hypotheses about the 'Yellow Rain' phenomenon', August 1983, 2 pp [Private distribution].

167—[with Peter Ashton, Matthew Meselson, Joan Nowicke and Thomas Seeley] 'Natural origin of Yellow Rain?' 1 September 1983. 4 pp. Submitted to *Science.*

Published as—(less Nowicke) 'Origin of Yellow Rain' in letters columns, *Science*, 28 October 1983, vol 222, pp 366 & 368.

168—'Approaches to chemical arms control', 30 September 1983, 26 pp. For Council for Arms Control.

169—'Chemical and biological warfare: developments in 1983'. 31 December 1983. 42 pp.

Published as—Printed in *World Armaments and Disarmament: SIPRI Yearbook 1984* (London: Taylor & Francis, 1984): 319–349.

170—'Contributions to chemical-warfare disarmament: Final report on the work during 1 January to 31 December 1983 supported by the Joseph Rowntree Charitable Trust'. End-of-grant report, CWDN/PR.1. 1 February 1984. 15 pp.

171—'Chemical/biological warfare (CBW)', 15 March 1984, 4 pp.

Published as—'Chemical/biological warfare (CBW)', *ADIU Factsheet* (University of Sussex: Armament & Disarmament Information Unit) No AFS.5, 15 March 1984, 4 pp.

172—'Biological weapons', 15 March 1984, 4 pp.

Published as—'Biological weapons', *ADIU Factsheet* No AFS.6, 15 March 1984, 4 pp.

173—'Chemical weapons', 15 March 1984, 4 pp.

Published as—'Chemical weapons', *ADIU Factsheet* No AFS.7, 15 March 1984, 4 pp.

174—'Technology and arms control: the case of chemical weapons', 28 February 1984, 8 pp. Notes for a lecture given at Imperial College, London.

175—'Chemical and biological warfare: how States behave', 29 February 1984, 11 pp. Notes for a Dunamis public lecture given at St James's Church, Piccadilly, 29 February 1984. Tape-recording of lecture distributed on cassette by Dunamis.

176—'Chemical warfare: status of technological developments and deployments', 15 March 1984, 19 pp. Background paper for Working Group 4 (Chemical Warfare) of the 34th Annual Pugwash Conference on Science and World Affairs, Bjõrkliden, Sweden, 9–15 July 1984.

Rev.1—'Chemical warfare: status of technological developments and deployments', 20 October 1984, 23pp.

Published as—'Chemical Warfare: Status of Technological Developments and Deployments', in J Rotblat & S Hellman (eds), *Nuclear Strategy and World Security* (Annals of Pugwash 1984), London: Macmillan, 1985, pp 155–75.

Published as—'Chemical warfare: status of technological developments and deployments', *1984 and Beyond: Science, Security and Public Opinion* (Proceedings of the 34th Pugwash Conference on Science and World Affairs), London: Pugwash, 1985, pp 336–50.

177—'Chemical warfare in the Iraq-Iran War'. 12 April 1984. 12 pp.

Rev.1—Revised for publication 24 May 1984.

Published as—*SIPRI Factsheet: Chemical Weapons 1*, May 1984. 11 pp. [An annotated version is also on file.]

178—'Conference in Ghent, Belgium, 21–24 May 1984: 'First World Congress: New Compounds in Biological and Chemical Warfare: Toxicological Evaluation' (Chairman: Professor Aubin Heyndrickx, Rijksuniversiteit- Ghent)', 31 May 1984, 13 pp. Confidential trip report.

179—'Workshop on Verification of the Destruction of Chemical Weapons, Munster, 12–14 June 1984 (organised by the FRG Foreign Office for the CD)', 15 June 1984, 6 pp. Confidential trip report.

180—'Prior Pugwash work on chemical warfare', 19 June 1984, 4 pp.

Published as—Digest of Pugwash Discussions Relevant to Working Group 4 Chemical Warfare', background document INF.4 provided by the British Pugwash Group for the 34th Annual Pugwash Conference, Björkliden, Sweden, 9–15 July 1984, 14 pp.

181—'Chemical-warfare arms control: a framework for considering policy alternatives', 28 September 1984, 115 pp. SIPRI/CWACP/1/Draft.

Published as—Revised version published (October 1985) as *SCBWS* no 2 (London: Taylor & Francis, 1985), 116 pp.

182—'An analysis of reports of Iraqi chemical warfare against Iran, 1980-84', 17 November 1984, 14 pp.

Rev.1—Revised for publication, 5 January 1985.

Published as—an appendix to Item 184 below.

183—'The value of chemical weapons for deterrence', 28 September 1984, 7 pp.

Published as—Included as an annexe in Item 181 above.

184—'Chemical and biological warfare: developments in 1984', 5 January 1985, 83 pp.

Rev.1—'Chemical and biological warfare: developments in 1984', 4 March 1985, 84 pp.

Published as—'Chemical and biological warfare: developments in 1984', in: *World Armament and Disarmament: SIPRI Yearbook 1985*, (London: Taylor & Francis, 1985), pp 159–219.

185—[with Peter Ashton, Matthew Meselson, Joan Nowicke and John Seeley] Letter to the Editor, *Nature*, 27 February 1985, 4 pp.

Published as—'Chemical weapons evidence unconvincing' in 'Scientific Correspondence' columns, *Nature*, vol 315, 23 May 1985, p 284.

186—'Chemical-warfare weapons and relevant aspects of their manufacture', 20 May 1985, 36 pp. CIPCWC/BP.1 DRAFT. Revised version is Item 191 below.

187—'Some background material for a discussion of chemical weapons in Soviet military doctrine', 5 June 1985, 11 pp. Discussion paper for the LUISS Centro di Studi Strategici panel meeting in Rome, 7–8 June 1985. Revised and expanded version is Item 189 below.

188—'A summary overview of the proceedings of the LUISS Centro di Studi Strategici panel meeting in Rome, 7–8 June 1985, 'Chemical warfare in Soviet military doctrine'. 11 June 1985, 4 pp.

Published as—Published in E Jacchia (ed), Item 189 below, pp 7–9.

189—'Views of Soviet chemical-warfare armament', 22 June 1985, 24 pp. Expanded version of Item 187 above.

Published as—'Views of Soviet chemical-warfare armament', in: E Jacchia (ed) *Chemical warfare in Soviet military doctrine* (Rome: LUISS, 1985) 51–68.

190—'An assessment of the chemical and biological warfare threat to London', 21 July 1985, 32 pp. Draft Consultant's paper for the Greater London Area War Risk Study (GLAWARS).

191—'Chemical-warfare weapons and relevant aspects of their manufacture', August 1985, 36 pp. Background paper CIPCWC/BP.1 for the SIPRI/ Pugwash conference on 'The Chemical Industry and the Projected Chemical Weapons Convention', Stockholm, 24–26 October 1985.

192—'British poison-gas weapons?', 8 September 1985, 16 pp.

Published as—'New poison gas weapons ahead?', *ADIU Report* vol 7, no 5, Sept/ Oct 1985, pp 1–2, 4–5.

193—'CW possessor states (according to pre-1985 literature)', 5 March 1985, 1 p.

194—'Preface', 2 October 1985, 2 pp.

Published as—'Preface', in: Ralf Trapp (ed.), 'The Detoxification and Natural Degradation of Chemical Warfare Agents', *SIPRI Chemical & Biological Warfare Studies* no 3 (Oxford: Oxford University Press), 1985, pp ix–x.

195—'Correlating categories of chemical with categories of control under the projected Chemical Weapons Convention', 17 October 1985, 15 pp. Working paper CIPCWC/WP.9 for the SIPRI/Pugwash Conference, Item 191 above.

196—'The chemical industry and chemical-warfare disarmament: categorizing chemicals for the purposes of the projected Chemical Weapons Convention', 4 November 1985, 50 pp. [Integration of Items 191 and 195 above.]

Published as—'The chemical industry and chemical-warfare disarmament: categorizing chemicals for the purposes of the projected Chemical Weapons Convention', in: 'The Chemical Industry and the Projected Chemical Weapons Convention: Proceedings of a SIPRI/Pugwash Conference. Volume 1: Summary Report and Background Papers', *SIPRI Chemical & Biological Warfare Studies* no 4 (Oxford: Oxford University Press), 1986, pp 55–104.

197—'US binary nerve gas: Congress still undecided', 11 November 1985, 1 p.

Published as—For *ADIU Report.*

198—(Lecture at Southampton University)

199—'Chemical weapons: the technical and military aspect', 24 November 1985, 11 pp. Text of a lecture given at the University of Florence on 30 November 1985 (during the Day of Study on Chemical Weapons organised by the Forum per i Problemi della Pace e della Guerra).

Rev.1—'Chemical weapons: the technical and military aspect', 30 November 1985, 17 pp.

Published as—L'evoluzione della guerra chimica, in S Aiello and G Gaja (eds), *Le Arme Chimiche*, (La Nuova Italia Scientifica Giunta regionale toscana) 1987, pp 13–26.

200—'Summary of a presentation made during the Day of Study on Chemical Weapons organised in the University of Florence by the Forum per i Problemi della Pace e della Guerra on 30 November 1985', 28 November 1985, 2 pp [summary of item 199].

201—[with Ralf Trapp, assisted by K Martensson] 'Summary report of the proceedings [of the SIPRI/Pugwash CIPCWC conference, Stockholm, 24–26 October 1985]'. 17 December 1985, 44 pp.

Rev.1—'Summary report of the proceedings [of the SIPRI/Pugwash CIPCWC conference, Stockholm, 24–26 October 1985]', 16 January 1986, 34 pp.

Published as—'Summary report of the proceedings', in: 'The Chemical Industry and the Projected Chemical Weapons Convention: Proceedings of a SIPRI/Pugwash Conference. Volume 1: Summary Report and Background Papers', *SIPRI Chemical & Biological Warfare Studies* no 4 (Oxford: Oxford University Press), 1986, pp 55–104.

202—'Preface to *SCBWS* 4/5, the proceedings of the SIPRI/Pugwash CIPCWC conference, Stockholm, 24–26 October 1985', 16 January 1986, 7 pp.

Published as—'Preface', in: 'The Chemical Industry and the Projected Chemical Weapons Convention: Proceedings of a SIPRI/Pugwash Conference. Volume 1: Summary Report and Background Papers', *SIPRI Chemical & Biological Warfare Studies* no 4 (Oxford: Oxford University Press), 1986, pp v–xi.

203—'Chemical and Biological Warfare: Developments in 1985', 22 January 1986, 41 pp.

Published in shortened form as—'Chemical and biological warfare: developments in 1985', *SIPRI Yearbook 1986*, pp 159–79. [Longer version published as item 205 below]

204—'US binary nerve gas', 24 January 1986, 4 pp,

Published as—'US binary nerve gas', *ADIU Report*, vol 8, no 1, January-February 1986, p 6.

205—'Chemical and biological warfare developments: 1985'. First draft. 12 March 1986, 110 pp.

Published as—Published as SCBWS no 6.

206—'Institute for East-West Security Studies: possible contributions to work on chemical-warfare weapons', 19 March 1986, 19 pp.

207—'Prospects for the international control of chemical-warfare weapons', 23 April 1986, 13 pp. Lecture at Royal Society of Medicine during the MAPW conference on 'Chemical and Biological Weapons: A Matter of Medical Concern', 26 April 1986.

Rev.1—'Prospects for the international control of chemical-warfare weapons', 28 April 1986, 16 pp.

208—'Approaches to general agreements on chemical-weapons limitations'. 30 May 1986, 25 pp. Draft of a paper for the Aspen Institute Berlin Conference on Chemical Weapons and European Security.

Rev.1—'Approaches to general agreements on chemical-weapons limitations', 7 June 1986, 30 pp. Adapted into Item 209 and Item 211 below.

Rev.2—Revised version for Conference proceedings, 5 October 1986, ca 40 pp.

209—'Disarmament and other options for western policy-making on chemical warfare', 23 June 1986, 32 pp.

Rev.1—'Disarmament and other options for western policy-making on chemical warfare', 15 September 1986.

Published as—'Disarmament and other options for western policy-making on chemical warfare', *International Affairs* (Chatham House), vol 63, no 1, (Winter 1986/87), pp. 65–80.

210—'NATO chemical weapons policy and posture', 28 July 1986, 91 pp including appendices.

Published as—'NATO Chemical Weapons Policy and Posture', *ADIU Occasional Paper* 4, September 1986, 91 pp.

Published as—*L'OTAN face aux armes chimiques* in GRIP Dossier 'notes et documents' No 112–113, Août-Septembre 1987

Published as—and as *La doctrine des deux blocs* in GRIP Informations (Dossier Trimestriel No 13 - Hiver 1987) pp 30–32.

211—'Alternatives for western policy on CW armament', 4 July 1986, 41 pp. For CCADD International Conference, FRG, 28 August–1 September 1986.

212—'An historical context for European chemical weapon-free zone concept, with an account of current European chemical-warfare forces', 21 September 1986, 28 pp.

Published as—SCBWS no 7 (1987), pp 1–29.

213—'Preface to *SCBWS* No 7 - Chemical Weapon Free Zones?', 21 October 1986.

Published as—Published in SCBWS No 7 (1987), pp v–viii.

214—'Chemical and biological warfare: developments in 1986'. Initial draft 15 January 1986, pp 23 and pp 13 of tables. Published in *SIPRI Yearbook 1987: World Armaments and Disarmament*, Oxford University Press, 1987, pp 13 and pp 7 of tables.

Rev.2—submitted as a background paper to the Pugwash Chemical Warfare Study Group 12th Workshop, Berlin, GDR, 5–8 March 1987, 'Some Recent Developments in the Field of Chemical Warfare', 21 February 1987, 31 pp.

215—'Supply, demand and assimilation in chemical-warfare armament', 27 February 1987, 18 pp.

Published as—Published in *Military Technology, Armaments Dynamics and Disarmament*, edited by Hans Günter Brauch, Macmillan Press, 1989.

216—'Some recent developments in the field of chemical warfare' for 12th Pugwash Chemical Warfare Workshop. 21 February 1987, 30 pp.

217—'Some categories of chemical—or what use are toxicity criteria?' n.d. [3 March 1987], 4 pp. Prepared for but not in fact presented at the 12th Pugwash Chemical Warfare Workshop.

218—'Yellow Rain chronology, 1975 through 1986'. 27 February 1987, pp 147. This is the CONFCHYR version. There were several earlier drafts, going back to 27 October 86.

219—[with Matthew Meselson and Jeanne Guillemin] 'Yellow rain' and treaty compliance, 6 April 1987, 25 pp. There were several subsequent (and earlier) drafts.

Published as—'Yellow Rain: the Story Collapses', in *Foreign Policy*, No 68 (Fall 1987), pp 100–17.

Published as—'Yellow Rain in Southeast Asia: the Story Collapses', (version with footnotes), in: Susan Wright (ed.), *Preventing a Biological Arms Race*, MIT Press, Cambridge, Massachusetts, and London, 1990.

220—'The view of the NUSNAT allies in past and current multilateral negotiations on global arms control: the case of chemical and biological weapons', for RIIA/CIIPS Workshop, extended draft outline, 30 April 1987, 9 pp.

221—[with Karlheinz Lohs] '12th Workshop of the Pugwash Chemical Warfare Study Group', including 'Personal impression of the workshop', [undated], 10 pp.

Published as—12th Workshop of the Pugwash Chemical Warfare Study Group, *Pugwash Newsletter*, April 1997, pp 106–16.

222—'Summary record of the 12th PCWSG Workshop'. CONFIDENTIAL. n.d. [ca 23 March 1987], pp 15 and pp 4 attachments.

223—'Current developments and prospective developments of chemical weapons', 27 May 1987, pp 21. For 37th Pugwash Conference, Gmunden, 1–6 September 1987.

Published as—'Current developments and prospective developments of chemical weapons', in: *Proceedings of the Thirty-Seventh Pugwash Conference on Science and World Affairs*, pp 417–32.

224—[with Martin Kaplan] 'Verification of chemical disarmament', 8 June 1987, 20 pp. Initial draft for joint paper for 37th Pugwash Conference, Gmunden, 1–6 September 1987.

Published as—'Verification of chemical disarmament', in: *Proceedings of the Thirty-Seventh Pugwash Conference on Science and World Affairs*, pp 326–39.

225—'Grant no 84-242: Narrative account no2', 21 June 1987, pp 16. Report to the MacArthur Foundation on work done on the Harvard-Sussex CBW programme, August 1985 – December 1986.

226—'Models for chemical weapons control and their implications for Switzerland from a military, security policy of economic perspective', 16 December 1987, 82 + ix pp plus appendices. For CASIN, Geneva.

227—'The scope of the projected Chemical Weapons Convention', 6 October 1987, 9pp. Speaking notes for Ottawa Conference on Implementing a Global Chemical Weapons Convention, 7–9 October 1987.

Rev.1—'The scope of the projected Chemical Weapons Convention', 10 October 1987, 9pp.

Published as—'The scope of the projected Chemical Weapons Convention', in: Jane Boulden (ed.) 'Towards a Chemical Weapons Convention: Proceedings of a Conference entitled "Implementing a Global Chemical Weapons Convention" held in Ottawa, October 7–9, 1987', *Aurora Papers 9*, January 1989 pp 33–36.

228—'The Harvard/Sussex Programme on Chemical and Biological Warfare: Plans for the 1988–90 Trilennium', 27 November 1987, 5pp. For the MacArthur Foundation.

229—'Chemical and biological warfare: some developments in 1987', 16 January 1988, 26pp.

Published as—A contribution to: SJ Lundin, JP Perry Robinson and Ralf Trapp, 'Chemical and biological warfare: developments in 1987', *SIPRI Yearbook 1988*, pp 101–25.

230—'The present situation in the negotiations for a Chemical Weapons Convention', for 13th Workshop of the Pugwash Study Group on Chemical Warfare, monitoring a Chemical Weapons Treaty, Geneva, Switzerland, 23–24 January 1988, 33 pp, including appendix 'Some developments over the past year in the field of chemical-warfare armament', see Item 229, above.

231—'Summary record of the 13th Pugwash Study Group on Chemical Warfare, Geneva, Switzerland 23–24 January 1988, 12pp plus 3pp attachment [private distribution].

232—[with Karlheinz Lohs], 'A Summary Report of the Workshop', 13th Workshop of the Pugwash Study Group on Chemical Warfare, Geneva, Switzerland, 23–24 January 1988, 4pp.

233—'The Verification of a New treaty on Chemical-Warfare Weapons', Background Paper, CAC Seminar on verification, London, 24 February 1988, 29pp.

Published as—'Can a Chemical Warfare Treaty be Verified?', *Common Security in Europe 1988* papers, Foundation for International Security June 1988, 22 pp.

234—'Verifying a Ban on Chemical-Warfare Weapons', 15 March 1988, 36pp + tables.

Published as—'Verifying a Ban on Chemical-Warfare Weapons', *Faraday Discussion Paper No 12*, July 1988, 28 pp.

235—'Chemical Warfare – the Moral Dilemma', lecture given at RSC Professional Symposium, Canterbury, 13 April 1988. 15pp.

236—'The Negotiations on Chemical Weapons', 2 May 1988, 15pp. Speaking notes for Course 72 special lecture prepared for NATO Defense College (Rome), 2 May 1988.

Rev.1—The Negotiations on Chemical Weapons', 1 May 1988, 27pp.

237—'Adequacy vs feasibility in the scope of the projected chemical weapons convention', 9 May 1988, 10pp. Paper commissioned for agendum 2–4 of the 38th Pugwash Conference, Dugomys, USSR, 29 August–3rd September 1988.

Rev.1—'Adequacy vs feasibility in the scope of the projected chemical weapons convention', 25 May 1988, 10pp.

Conference version—'Adequacy vs feasibility in the scope of the projected chemical weapons convention', 10pp.

Published as—'Adequacy vs feasibility in the scope of the projected chemical weapons convention', in *Proceedings of the Thirty-Eigth Pugwash Conference on Science and World Affairs: 'Global Problems and Security'*, (1989) pp 333–39.

Published as—'Adequacy vs feasibility in the scope of the projected chemical weapons convention', in *Annals of Pugwash 1988:* 'Global Problems and Common Security' (1989), pp 105–11.

238—'Skikhany to the Moscow Summit: A chronology of CBW news: October 1987 to June 1988'. For *Chemical Weapons Convention Bulletin* No 1, 22 June 1988, 19pp.

239—'The Moscow Summit to the Mitterrand Proposals: A chronology of CBW news, June through September 1988', 10 October 1988, 29pp. version proposal for *Chemical Weapons Convention Bulletin* no 2.

240—'World CW Armament, Part One: The United States'. 8 October 1988, 15pp.

Published version—For *Chemical Weapons Convention Bulletin* no 2.

241—'Since the Kurdish exodus from Iraq : A chronology of CBW news, September through December 1988', 1 November 1988, 16pp.

242—'General Introduction: Information Processes and the Projected Chemical Weapons Convention'. Presentation at the WPCBW conference, London, 15–16 Sept 1988.

Rev.1—'General Introduction: Information Processes and the Projected Chemical Weapons Convention', 15 January 1989. 13pp.

Published as—[translated by Joachim Badelt] 'Die Verhandlungen über ein vollständiges Verbot Chemischer Waffen: Rolle und Bedeutung eined umfassenden Informationsprozesses', *CBW Infodienst*, Nr 1, February 1990, pp 2–4.

243—'Verification Procedures for a Chemical Weapons Treaty' for PRIF Verification Workshop, London, 30 Nov–2 Dec 1988. 11pp.

Published version—'Nuclear, Conventional and Chemical', *Verification of Arms Reductions*, 1989, pp 201–09.

244—'Review: World CW Armament', 2 pp

Published as—for *Chemical Weapons Convention Bulletin* no 3, 4 Feb 1989. 2pp

245—[with Matthew Meselson] 'Chemical and biological weapons', 1 March 1989, 6pp.

Published as—'Chemical and biological weapons', *Encyclopaedia Britannica*, 15th Edition, 1990 printing, vol 29, pp 580–1.

Published online as—'Chemical and biological weapons', *Encyclopaedia Britannica.*

246—'World CW Armament Part Two: Soviet Union', 7 April 1989, 24pp.

Updated—'World CW Armament Part Two: Soviet Union', 17 April 1989, 20 pp.

Annotated—'World CW Armament Part Two: Soviet Union', 17 April 1989, 20 pp.

Rev.1—'World CW Armament Part Two: Soviet Union', 2 May 1989, 20 pp.

Published as—*Chemical Weapons Convention Bulletin* no 4.

247—'Benefits and Costs of a Chemical Warfare Disarmament Regime', 11 April 1989, 6pp. Speaking notes for 'Towards a Universal Chemical Weapons Convention', Les Avants, 13–16 April 1989.

248—[with Karlheinz Lohs and Nikita P Smidovich] 'Verification and chemical-warfare weapons', May 1989, 23 pp. For the Pugwash Study on Verification Issues: (based on the New York, Moscow, London and Woods Hole discussions).

Rev.1—'Verification and chemical-warfare weapons', 24 September 1989, 24 pp;

Rev.2—'Verification and chemical-warfare weapons', 5 October 1989, 24 pp.

Published as—Karlheinz Lohs, Julian P Perry-Robinson and Nikita P Smidovich, 'Verification and Chemical-Warfare Weapons', in F Calogero, M L Goldberger and S P Kapitza (eds), *Verification—Monitoring Disarmament* (Boulder, Colorado: Westview Press, 1991), pp 123–48.

249—'39th Pugwash Conference (Cambridge, Mass., 23–28 July 1989): Report from Working Group 3', 28 July 1989, 5pp.

Published as—'Report on Working Group 3 – Chemical Weapons and Their Control', *Pugwash Newsletter*, vol 27, no 2, October 1989, pp 78–81.

250—'Implementing the Chemical Weapons Convention: Actions of Governments', 19 September, 8 pp. Statement to the Government-Industry Conference against Chemical Weapons, Canberra, 18–22 September 1989.

Published as—'Implementing the Chemical Weapons Convention: Actions of Governments', *Final Record of the Government-Industry Conference against Chemical Weapons, Canberra, 18–22 September 1989*, (Department of Foreign Affairs and Trade, Canberra) pp 187–99.

251—'Impressions of the Canberra Conference', 31 October 1989, 13 pp.

Published as—'Review: the Canberra Conference', *Chemical Weapons Convention Bulletin*, no 6, (November 1989), pp 16–22.

252—Review of *Chemical Weaponry: a continuing challenge* by Edward M Spiers.

Published as—'Chemical Weaponry: a continuing challenge by Edward M Spiers', book review, *International Affairs*, vol 66, no 1, January 1990, pp 149–150.

253—'Chemical Weapons Proliferation in the Middle East', for IEWSS Committee on Regional Arms Transfers and Arms Control in the Middle East, Wiston House, Sussex, 7–9 February 1990. 21pp.

254—Review of 'Chemical warfare, chemical disarmament: beyond Gethsemane', by Valerie Adams, 5 February 1990, 2 pp.

Published as—'Chemical warfare, chemical disarmament: beyond Gethsemane by Valerie Adams', book review, *International Affairs*, vol 66, no 3, July 1990, p 582.

255—[with Ralf Trapp] '15th Workshop of the Pugwash Study Group on Chemical Warfare', Geneva, 27–28 January, 1990, 6pp.

Rev.1—'15th Workshop of the Pugwash Study Group on Chemical Warfare', Geneva, 27–28 January, 1990, 6pp.

Published as—'Report', *Pugwash Newsletter*, vol 27, no 4, April 1990, pp 166–76.

256—'Chemical weapons proliferation: the problem in perspective', 22 March 1990, 21 pp. Revised text of presentation given as the keynote address at the ANU/RSPacS/PRC workshop *Chemical Weapon and Missile Proliferation – Implications for Asia/Pacific*, Canberra, Australia, 8–9 March 1990.

Published as—'Chemical weapons proliferation: the problem in perspective', in: Trevor Findlay (ed.), *Chemical Weapons & Missile Proliferation With Implications for the Asia/Pacific Region*, (Boulder and London: Lynne Rienner, 1991), pp 19–35.

257—'Chemical weapons proliferation: security risks', 16 March 1990, 23 pp. For VUB 2nd Annual Conference on Chemical Warfare, *Chemical Weapons Proliferation: Policy Issues Pending an International Treaty, Brussels*, 16 March 1990.

Published as—'Chemical weapons proliferation: security risks', Jean Pascal Zanders and Eric Remacle (eds.), *Chemical Weapons Proliferation: Policy Issues Pending an International Treaty*, (Brussels: VUB Centrum voor Polemologie, 1991), pp 69–92.

258—'A Chemical Disarmament Regime: Benefits and Costs', 7 May 1990, 5pp. Speaking notes for conference 'Towards a Universal Chemical Weapons Convention', Ma'in, Jordan, 7–9 May 1990, 5pp.

259—'Prevention of Use of Biological & Chemical Weapons vs Total Bans on Production', speaking notes for Wilton Park conference 351, 'New Threats to Global Security: Proliferation of Nuclear, Chemical and Biological Weapons', Wiston House, Sussex, 29 May–2 June 1990, 8 pp.

260—[with Ralf Trapp] 'Production and chemistry of mustard gas', 21 January 1991, 12pp.

Published as—Production and chemistry of mustard gas', in S J Lundin (ed.), *Verification of Dual-use Chemicals under the Chemical Weapons Convention: The Case of Thiodiglycol* (Oxford University Press, 1991: *SIPRI Chemical & Biological Warfare Studies* no 13) pp 4–23.

261—'Is the illegitimacy of chemical warfare surviving?', 17 September 1990, 5 pp. Presentation at a plenary session of the 40th Pugwash Conference, Egham, UK, 15–20 September 1990.

Published as—'Banning chemical weapons', *Proceedings of the Fortieth Pugwash Conference on Science and World Affairs* (London: Pugwash, 1991), pp 135–8.

262—'The Prohibition of Chemical and Biological Warfare in International Law', 9 November 1990, 16pp.

Published as—'The Prohibition of Chemical and Biological Warfare in International Law', *Genocide: The Winds of Death in Kurdistan*, (London: Kurdish Medical and Scientific Association).

263—'The 1972 Biological Weapons Convention: aspects of adherence and effectiveness', 25 February 1991, 10pp. Extended speaking notes for the Dutch Government's *Seminar on the Biological Weapons Convention in the Perspective of the Forthcoming BW Convention Third Review Conference*, Noordwijk, 6–8 February 1991.

264—[with Nicolai Platé] 'Report [on the 16th Workshop of the Pugwash Chemical Warfare Study Group, Geneva, 26-27 January 1991]', 4 pp.

Published as—'Report [on the 16th Workshop of the Pugwash Chemical Warfare Study Group, Geneva, 26–27 January 1991]', *Pugwash Newsletter*, April 1991, pp 244–47.

265—'Chemical and Biological Arms Control Since World War II', 12 November 1990, 5pp. Thematic outline for essay for Scribner's Encyclopedia of Arms Control and Disarmament.

266—'The problem of chemical-weapon proliferation in the 1990s: issues and perspectives', May 1991, 143 pp [private distribution].

267—'Adherence to the 1972 Biological Weapons Convention', 20 June 1991, 10 pp. Speaking notes and tabulated data for the Quaker Residential Conference *Strengthening the Biological Weapons Convention*, Château de Bossey, Switzerland, 31 May–2 June 1991.

Rev.1—'Adherence to the 1972 Biological Weapons Convention', 21 June, 9 pp.

268—'The control of chemical and biological weaponry', 2 July 1991, 8 pp. Extended speaking notes for the SPRU 25th Anniversary Conference parallel session on *Regulating the Diffusion of Military Technology*, University of Sussex, 3 July 1991.

269—'Supply-side control of the spread of chemical weapons', 22 October 1991, 13 pp. Revised text of presentation at the CIIPS conference *Supply-Side Control of the Spread of Weapons*, Ottawa, Canada, 18–21 June 1991.

Published as—'The Supply-Side Control of the Spread of Chemical Weapons', in Jean-François Rioux (ed.), *Limiting the Proliferation of Weapons: The Role of Supply-Side Strategies*, (Ottawa: Carleton University Press), 1992 pp 57–73.

270—'The Australia Group: a description and assessment', 18 October 1991, 21 pp; plus 'Abstract', 21 October 1991, 2 pp. Paper/abstract for the 7th International AFES-PRESS Conference, Mosbach-Neckarelz, Germany, 24–27 October 1991.

Rev.1—'The Australia Group: a description and assessment', December 1991, 21 pp for publication in: Brauch, et al.

Published as—Published in *Pugwash Newsletter*, January 1992, Vol 29, no. 3. 7pp.

Published as—'The Australia Group: a description and assessment', in: Hans Günter Brauch, H. J. van der Graaf, John Grin and Wim Smit (eds), *Controlling Military Research & Development and Exports of Dual Use Technologies as a Problem of Disarmament and Arms Control Policy in the 1990s*, Amsterdam: VU University Press, and New York: St Martin's Press, 1992, pp 157–76.

271—'Chemical weapons proliferation in the Middle East'. 7 November 1991, 15 pp. Paper for King's College, London, Department of War Studies conference *Proliferation in the Middle East*, 21–22 November 1991.

Published as—Printed in Non-Conventional-Weapons Proliferation in the Middle East, Tackling the Spread of Nuclear, Chemical, and Biological Capabilities, Edited by Efraim Karsh, Martin S. Navias and Philip Sabin. Published by Clarendon Press Oxford 1993. Chapter 4, pages 69–98.

272—'The chemical weapons of Desert Storm forces, and the wider implications of tear gas and other incapacitants', 26 November 1991, 11 pp. For VUB 3rd Annual Conference on Chemical Warfare, *The 2nd Gulf War and the CBW Threat: The Impact on Threat Perception, Proliferation, Disarmament and European Security*, Brussels, 29–30 November 1991.

Outline—'The chemical weapons of Desert Storm forces, and the wider implications of tear gas and other incapacitants', 18 November 1991, 9 pp.

Rev.1—'The chemical weapons of Desert Storm forces, and the wider implications of tear gas and other incapacitants', 28 November 1991, 12 pp.

Published as—'The chemical weapons of Desert Storm forces, and the wider implications of tear gas and other incapacitants', in Jean Pascal Zanders (ed.), *The 2nd Gulf War and the CBW Threat: Proceedings of the 3rd Annual Conference on Chemical Warfare*, Vredesonderzoek Van De Vrije Universiteit Brussel, Special Issue, November 1995.

273—'The negotiations on the Chemical Weapons Convention: an historical overview', 20 November 1991, 6 pp. For the 2nd Workshop of the Rome/Turku/Frankfurt CW Project, Rome, 22–24 November 1991.

Rev.1—'The negotiations on the Chemical Weapons Convention: an historical overview', 30 September 1992, 10 pp. For the 3rd Workshop of the Rome/Turku/Frankfurt CW Project, Turku, 1–4 October 1992.

Rev.2—'The 1993 Chemical Weapons Convention', January 1993, 23pp. Revised for HSP London CBW Seminar, 10 January 1993.

Rev.3—'The Negotiations on the Chemical Weapons Convention: An Historical Overview', 7 January 1993, 17 pp.

Published as—'The negotiations on the Chemical Weapons Convention: a historical overview' in Michael Bothe, Natalino Ronzitti and Allan Rosas (eds.), *The New Chemical Weapons Convention – Implementation and Prospects*, Kluwer Law International, 1998, pp 17–36.

Rev.4—[with Thomas Stock and Ronald G Sutherland] 'Historical overview of the CWC negotiation', 8 January 1993, 12pp.

Published as—'The Chemical Weapons Convention: the success of chemical disarmament negotiations', *SIPRI Yearbook 1993: World Armaments and Disarmament*, Oxford University Press, 1993, pp 705–56.

274—[with Matthew Meselson] 'Incapacitants and the Chemical Weapons Convention', 14 December 1991, 2 pp. Working paper for 19th Workshop of the Pugwash CBW Study Group, Geneva, 11–12 January 1992.

Published as—'Editorial: no gas', *Chemical Weapons Convention Bulletin* no 14, December 1991, p 1.

275—'The Australia Group and the Chemical Weapons Convention'. 7 January 1992, 23 pp. For 19th Workshop of the Pugwash CBW Study Group, Geneva, 11–12 January 1992. [Conference circulated version]

Published as—'The Australia Group and the Chemical Weapons Convention', *Pugwash Newsletter*, January 1992, vol 29, no 3, pp 167–68.

276—[with Ralf Trapp] 'Report [on the 19th Workshop of the Pugwash CBW Study Group, Geneva, 11–12 January 1992]', 14 January 1992, 3pp.

Published as—'Report', *Pugwash Newsletter*, January 1992, vol 29, no 3, pp 156–62.

277—[with Matthew Meselson] 'Report [on the HSP/PCWSG conference *Antichemical Protection, Its Potential and Its Relation to the Spread of Chemical Weapons and Their Elimination*, El Escorial, Spain, 5–7 August 1991]', 12 August 1991, 3 pp.

Published in abridged form as—'Antichemical protection and the CWC', *Chemical Weapons Convention Bulletin* no 13, September 1991, p 18.

Published as—'Antichemical protection', *The ASA Newsletter* 91-5, 9 October 1991, pp 1, 18–19.

Published as—'18th Workshop of the Pugwash Study Group on Chemical Warfare', *Pugwash Newsletter*, vol 29 no 2, October 1991, pp 43–44 (accompanied by participant list and abstracts).

Contributions to the conference were published as

— [with Priyamwada Deshingkar and Matthew Meselson (eds.)] 'Antichemical Protection and the Chemical Weapons Convention', Harvard Sussex Program Occasional Paper no 2, 1993, 71 pp.

278—'The Chemical Weapons Convention régime: a cost-benefit balance sheet', 4 August 1991, 1 p. For the HSP/PCWSG conference *Antichemical Protection, Its Potential and Its Relation to the Spread of Chemical Weapons and Their Elimination*, El Escorial, Spain, 5–7 August 1991.

Published as—Excerpted in Pugwash Newsletter vol 29 no 2, October 1991, p 53.

279—'The Negotiations on the Chemical Weapons Convention: Outline draft of an Historical Overview', 24 January 1992, 13 pp. For the *Substance and Shadow* meeting in Washington, 3–4 February 1992.

Rev.1—'The Negotiations on the Chemical Weapons Convention: Outline draft of an Historical Overview', 9 June 1992, 15 pp.

Published as—'Origins of the Chemical Weapons Convention', in: Benoit Morel and Kyle Olson (eds), *Shadows and Substance: The Chemical Weapons Convention*, Ridgway Series in International Security Studies, Westview Press, 1992, pp 37–54.

280—'The Chemical Industry and Disarmament: Issues and Perspectives', 23 January 1992, 12 pp. For the Symposium on the Chemical Industry and Disarmament, Basel, 24–27 January 1992.

Rev.1—'The Chemical Industry and Disarmament: Issues and Perspectives', 25 January 1992, 13 pp.

281—[with Matthew Meselson] 'Questions regarding the US position seeking to exempt certain toxic chemicals from the Chemical Weapons Convention', 5 pp [private distribution].

282—[with Matthew Meselson] 'Iraqi CBW armament and the UN Special Commission', 30 September 1991.

Published as—'Iraqi CBW armament and the UN Special Commission', *Chemical Weapons Convention Bulletin* no 13, (September 1991) pp 21–22.

283—'An Update on the United Nations Inspection of Iraqi Chemical and Biological Weapons: speaking notes and vugraphs', 13 March 1992, 12 pp. For KSMA *Halabja Day* Scientific Seminar, Kurdistan Hall, London, 14 March 1992.

Abstract—'An Update on the United Nations Inspection of Iraqi Chemical and Biological Weapons', 13 March 1992, 1 p.

284—[with Matthew Meselson] 'Disabling Chemicals and the Chemical Weapons Convention', 24 March 1992, 2pp.

Published as—'What Policy for Disabling Chemicals?', *Chemical Weapons Convention Bulletin*, no 15 (March 1992) pp 4–5.

285—'Agreement on the Chemical Weapons Convention', 25 June 1992, 3pp.

Published as—'Agreement now imminent', *Chemical Weapons Convention Bulletin*, No. 16, June 1992, p l.

Published as—'Comment: Getting the Treaty Right', *Chemistry and Industry*, 1992 p 508.

Published as—'Comment: The Need for an Effective Authority', CBW News, no 9, July 1992, p 3.

286—[with Matthew Meselson] 'The 1993 Chemical Weapons Convention: A Summary', 28 November 1992.

Published as—'The 1993 Chemical Weapons Convention: A Summary', in: Elizabeth J Kirk, W Thomas Wander and Brian Smith (eds.), *1993 Science and International Security Anthology: Trends and Implications for Arms Control, Proliferation, and International Security in the Changing Global Environment*, American Association for the Advancement of Science, 1993, pp 167–74.

287—'The 1993 Chemical Weapons Convention', 26 January 1993, 3 pp.

Published as—'The 1993 Chemical Weapons Convention', *Bulletin of Arms Control*, Number 9, February 1993, pp 2–6.

288—'Science, technology and the 1993 Chemical Weapons Convention', 18 February 1993, 3 pp, for.

Published as—'Science, technology and the 1993 Chemical Weapons Convention', *SPRU's News*, March/April 1993, pp 1, 3, 14–15.

289—'Prospects for the OPCW Preparatory Commission', for presentation at the HSP London CBW Seminar, 26 February 1993. 5pp.

290—[with Peter Herby] 'The Start of International Organization for Chemical Disarmament', 23 March 1993, 5 pp.

Published as—'International Organization for Chemical Disarmament begins', *Chemical Weapons Convention Bulletin* no 19, March 1993, pp 1–4.

291—'A possible way forward for the new Pugwash CBW Study Group', 27 April 1993, 11pp. Working paper for Pugwash Meeting no 196, 1st Workshop of the Pugwash Study Group on the Implementation of the Chemical and Biological Weapons Conventions, 8–9 May 1993, Geneva, Switzerland.

292 'Seeing the future of the Chemical Weapons Conventions in its past', 21 April 1993, 18 pp. Working paper for Pugwash Meeting no 196. 1st Workshop of the Pugwash Study Group on the Implementation of the Chemical and Biological Weapons Conventions, 8–9 May 1993, Geneva, Switzerland.[2]

293—[with Peter Herby] 'Building the Organisation for the Prohibition of Chemical Weapons', 15 June 1993, 6 pp.

Published as—'Building the Organisation for the Prohibition of Chemical Weapons', Progress in the Hague Quarterly Review no 2, *Chemical Weapons Convention Bulletin* no 20, June 1993, pp 5–9.

294—'UK obligations under the Chemical Weapons Convention', 8pp. Report for HLCS no 6, Porton Down, 1 July 1993. 8pp.

---

[2] Editors' note. Julian prepared several different versions of this paper. Earlier versions appear in Benoit Morel & Kyle Olson (editors), The Chemical Weapons Convention: Shadows and Substance Chap. 3 'Origins of the Chemical Weapons Convention', Published by Westview Press, 1993. pp 37–54.; another in Micael Bothe, Natalino Ronzitti & Allan Rosas (editors), Chemical Weapons Disarmament: Strategies and Legal Problems; a third in SIPRI Yearbook 1993.

295—[with Matthew Meselson] 'Report on the 1st Workshop of the Pugwash Study Group on the Implementation of the Chemical and Biological Weapons Conventions' Pugwash Meeting no 196, 8–9 May 1993, Geneva, Switzerland. 3pp.

Published as—'1st Workshop of the Pugwash Study Group on the Implementation of the Chemical and Biological Weapons Conventions', *Pugwash Newsletter*, vol. 31 no.1 (July 1993), pp. 20–23.

296—'Negotiations Leading to the CWC', 18 November 1993, 32pp. Training Scheme for Inspectors and other OPCW Employees. UK Pilot Foundation Course [14 November–22 December 1993] Royal Military College of Science, Cranfield University.

297—'Practical National Implementation Measures for Entry into Force of the Chemical Weapons Convention', 15 pp. Regional Seminar on National Authority and National Implementation Measures for the Chemical Weapons Convention Warsaw, 7–8 December 1993.

Published as—'Practical National Implementation Measures for Entry into Force of the Chemical Weapons Convention', Summary of Proceedings, OPCW PTS Occasional Paper No.3, Annex VII, pp. 41–55.

298—For the *Seminar on National Implementation* of the Chemical Weapons Convention, organized in the Hague by the OPCW Provisional Technical Secretariat 18 December 1993. 15pp. Implementation and Pre-Implementation Measures: An Overview. Published by Provisional Technical Secretariat of the Organisation for the Prohibition of Chemical Weapons, the Hague, in Summary Proceedings Occasional Papers—No.2 as 'Implementation and Pre-Implementation Measures: An Overview', 1994. 15 pp.

299—Technology and the 1993 Chemical Weapons Convention (Draft).

300—'The CWC – One Year Later', 24 January 1994, 2 pp.

Published as—'The CWC - One Year Later', *ASA Newsletter*, 94-1, pp 1, 14.

301—'CBW weapons proliferation and control', 24 February 1994, 9 pp (corrected 28 February plus annex, 24 pp). Memorandum for the House of Commons Foreign Affairs Committee.

Published as—'Memorandum submitted by J P Perry Robinson, Science Policy Research Unit, University of Sussex: CBW weapons proliferation and control' in: House of Commons (Session 1993–94) Foreign Affairs Committee *UK Policy on Weapons Proliferation and Control in the Post-cold War Era* Minutes of Evidence, Wednesday 19 October 1994, pp 72-88, HMSO, London.

Rev.1—'What is the UK doing about chemical and biological weapons?', 26 June 1994, 25pp. Background paper for HLCS7 (CBDE Porton Down, 1 July 1994).

Rev.2—'CBW weapons proliferation and control', 22 November 1994, 16pp.

302—'Disabling chemical weapons: some technical and historical aspects', 25 May 1994, 5pp plus 53 pp appendix. Working paper for Pugwash Study Group on Implementation of the CBW Conventions Second Workshop, Den Haag/Noordwijk, Netherlands, 27–29 May 1994.

303—'Developments in 'non-lethal weapons' involving chemicals', 22 May 1994, 5pp. For the ICRC Expert Meeting on Certain Weapons Systems, Geneva, 30 May–1 June 1994.

Published as—'Developments in 'non-lethal weapons' involving chemicals', *Expert Meeting on Certain Weapon Systems and on Implementation Mechanism in International Law...Report*, (Geneva: ICRC, July 1994), pp 92–97.

304—[with Matthew Meselson] 'Report [on the 2nd Workshop of the Pugwash Study Group on the Implementation of the Chemical and Biological Weapons Conventions (27–29 May 1993, The Hague and Noordwijk, Netherlands)]', 24 July 1994. 5pp.

Published as—'Report', *Pugwash Newsletter*, vol. 32 no. 1 (July 1994), pp. 3–7.

305—'Dr. Mirzayanov and the UK National Authority', 11 July 1994, 7 pp. Write-up of oral presentation at HLCS 7, CBDE Porton Down, 1 July 1994.

306—'Chemical and Biological Weapons: Their Proliferation, and the Use of Export Controls in Response', 30 October 1994, 29 pp. Chapter for Deltac/Saferworld.

Rev.1—'Chemical and Biological Weapons: Their Proliferation, and the Use of Export Controls in Response', 16 November 1994, 29 pp.

Published as—'Chemical and Biological Weapons Proliferation and Control', in: *Proliferation and Export Controls: An analysis of sensitive technologies and countries of concern*, Deltac/Saferworld, 3 May 1995, pp 29–53.

307—'Procedures for Monitoring and Investigation', 22 November 1994, 15 pp. Paper 18.1 for Fifth Session, *Means of Ensuring Compliance with the Convention* Part 2, *Verification Mechanisms*, Hague Academy of International Law colloquium, *The Convention on the Prohibition and Elimination of Chemical Weapons: A Breakthrough in Multilateral Disarmament*, Peace Palace, 24–26 November 1994. Offprint (Printed for private circulation only), Martinus Nijhoff, Dordrecht, 1995.

Rev.1—'The verification system for the chemical weapons convention', 30 November 1994, 19 pp.

308—'Some consequences of the General Purpose Criterion having being used to define the scope of the Chemical Weapons Convention', 8 pp. Working Paper for Pugwash Meeting no. 207, 3rd Workshop of the Pugwash Study Group on Implementation of the Chemical and Biological Weapons Conventions, 19–21 May 1995, Noordwijk, The Netherlands.

Rev.1—'Some consequences of the General Purpose Criterion having being used to define the scope of the Chemical Weapons Convention', 21 June 1995, 8 pp.

309—'The Chemical Weapons Convention: Progress, and UK Ratification', 12 May 1995, 3 pp.

Published as—'The Chemical Weapons Convention: Progress, and UK Ratification', *Bulletin of Arms Control no.18*, June 1995, pp 8–13.

310—[with Matthew Meselson] 'Report' [Pugwash Meeting no 207, 3rd Workshop of the Pugwash Study Group on Implementation of the Chemical and Biological Weapons Conventions, 19–21 May 1995, Noordwijk, The Netherlands], 21 June 1995, 9 pp.

Published as—Agenda, Programme and Report (as above), *Pugwash Newsletter*, vol 33 nos 1/2, July/October 1995.

311—'CBW Events During February Through May 1995: An Excerpt from the HSP Rolling Chronology', 24 June 1995.

Published version—'News Chronology February through May 1995', *The CBW Conventions Bulletin* no 28, June 1995, pp 12–34.

312—'What can Britain do to prevent the spread of chemical and biological weapons?', 28 July 1995, 13 pp. Draft paper for the JRCT Platform for Security, *British Defence in a New World.*

Submitted version—'What can Britain do to prevent the spread of chemical and biological weapons?', 11 October 1995, 18 pp. Paper for the JRCT Platform for Security, *British Defence in a New World.*

Rev.1—'What can Britain do to prevent the spread of chemical and biological weapons?', 19 January 1996, 11 pp. Paper for the JRCT Platform for Security, *British Defence in a New World.*

Rev.2—'What can Britain do to prevent the spread of chemical and biological weapons?', 16 July 1996, 11 pp. Paper for the JRCT Platform for Security, *British Defence in a New World.*

Published as—'What can Britain do to prevent the spread of chemical and biological weapons?', in: John Gittings and Ian Davis (eds) *Britain in the 21st Century: Rethinking defence and foreign policy*, (Nottingham: Spokesman, 1997), pp 67–80.

313—'A commentary on the legislation needed to implement the Chemical Weapons Convention in the United Kingdom', 14 September 1995, 13 pp. Discussion paper for the HSP London CBW Seminar, session 8, Kings College London, 27 September 1995

314—'Implementing the Chemical Weapons Convention', 19 October 1995, 23 pp.

Published as—'Implementing the Chemical Weapons Convention' *International Affairs*, vol 72, no 1, January 1996, pp 73–89.

315—'CBW Events during May through August 1995, an excerpt from the HSP Rolling Chronology', 9 October 1995.

Published version—'News Chronology May through August 1995', *The CBW Conventions Bulletin* no 29, September 1995, pp 14–34.

316—'The UK draft CWC implementing legislation', Speaking notes for HSP London CBW Seminar, Session 8, King's College London, 27 September 1995, 4 pp.

317—[with Matthew Meselson] Draft editorial comment for CWCB29, 14 September 1995, 2 pp.

Submitted version—'Editorial comment', 22 September 1995, 2 pp.

Published as—'Defining Chemical Weapons the Way the Treaty Does', *The CBW Conventions Bulletin* no 29, September 1995, pp 1.

318—'Where is the Chemical Weapons Convention now?', 11 October 1995, 1 p.

Published as—'Where is the Chemical Weapons Convention now?', *ASA Newsletter* No. 50, 11 October 1995, pp 1+20.

319—Comment for *Chemistry in Britain*, 7 October 1995, 2pp.[3]

320—'The Structure of the OPCW', Presentation, 13 November 1995, 10pp.

321—'The Chemical Weapons Bill – Background Briefing', 20 November 1995, 16 pp. Circulated by The Royal Society of Chemistry to Members of Parliament in advance of the 2nd Reading debate scheduled for Thursday 23 November 1995.

Rev.1—'Aspects of the Chemical Weapons Bill', 28 November 1995, 21pp.

Rev.2—'Aspects of the Chemical Weapons Bill now before The Lords', 15 January 1996, 30pp.

322—[with Matthew Meselson] 'Outline for an Integrated Approach to the Problem of Biological Weapons', 1 December 1995, 4pp. Working Paper for Pugwash Meeting No 212, the 4th Workshop of the Pugwash Study Group on the Implementation of the Chemical and Biological Weapons Convention, 2-3 December 1995, Geneva, Switzerland.

323—'CBW Events during August through November 1995', an excerpt from the HSP Rolling Chronology, 18 December 1995

Published version—'News Chronology August through November 1995', *The CBW Conventions Bulletin* no 30, December 1995, pp 11–31.

324—'Aspects of the chemical weapons bill now before The Lords', 23 January 1996, 30pp.

326—'The Chemical Weapons Convention' Speaking notes for CAC/CDISS seminar, Kings College, London, 18 January 1996, 'Chemical & biological weapons: their control, capabilities, proliferation & delivery systems. What prospects for the future? 18 January 1996, 4pp.

[3] Editors' note. We cannot find any records about whether this was published.

327—[with Matthew Meselson] Report, 22 January 1996, 5pp. Pugwash Meeting No 212, 'Strengthening the Biological Weapons Convention', 4th Workshop of the Pugwash Study Group on the Implementation of the Chemical and Biological Weapons Convention, 2–3 December 1995, Geneva, Switzerland. 5 pp.

328—'Controlling Technological Change by International Treaty: The Chemical and Biological Weapons Convention', 11 February 1996, 15 pp, afternoon seminar for SPRU, 9 February 1996.

329—CBW Events During November 1995 Through February 1996, An Excerpt from the HSP Rolling Chronology, 11 March 1996.

Published version—'News Chronology November 1995–February 1996', *The CBW Conventions Bulletin* no 31, March 1996, pp 10–31.

330—'Some political aspects of the control of biological weapons' Address to the Parliamentary and Scientific Committee, London, 22 April 1996.

Published version—*Science in Parliament*, vol 53, no 3, May/June 1996 pp 6-11.

331—Implementing the Chemical Weapons Convention and the impact on the Biological Weapons Convention' for the Friedrich-Ebert-Stiftung conference, *Enchancing the BWC*, Bonn, 6 – 7 May 1996.

Rev.1—[4 May 1996] 26 June 1996.

Rev.2—[April 97] Some Lessons for the Biological Weapons Convention from Preparations to Implement the Chemical Weapons Convention, Chapter in Oliver Thränert (ed) *Enhancing the BWC*, Bonn, 1996, pp86-113.

332—Speaking notes for the Common Security Forum colloquium, *Reasoning about Disarmament: A Spectrum of Approaches*, Cambridge, 1 June 1996. For Session III: Theoretical Reflections 'Why do we have the Biological Weapons convention?' 31 May revised 26 June 1996. 3pp.

333—'CBW events during February through May 1996' An excerpt from the HSP Rolling Chronology, 23 June 1996.

Published version—'News Chronology February–May 1996', *The Chemical Weapons Convention Bulletin*, no 32, June 1996, pp 16–39.

334—Discussion paper for private distribution, *A Concept for UK implementation of the Chemical Weapons Convention in regard to 'dual use'*, 18 July 1996, 15 pp + Annex: 'Implications for the National Authority Advisory Committee', 3pp.

335—'CBW events during May through August 1996', An excerpt from the HSP Rolling Chronology, 1 October 1996

Published version—'News Chronology May through August 1996', *The Chemical Weapons Convention Bulletin*, no 33, September 1996, pp 13–35.

336—'Pugwash and the international treaties on chemical and biological warfare'. Background paper for an authors' workshop at the New York Academy of Sciences, 21–22 October 1996. 6 October 1996.

Rev.3—'Pugwash and the international treaties on biological and chemical weapons', 31 August 1997.

Rev.4—'Pugwash and the international treaties on chemical and biological warfare', 15 September 1997, 53 pp, paper for the *Scientific Cooperation and Conflict Resolution* project of the New York Academy of Sciences.

Rev.5—'Pugwash and the international treaties on chemical and biological warfare', 19 November 1997, 53 pp, paper for the *Scientific Cooperation and Conflict Resolution* project of the New York Academy of Sciences.

Rev.6—'Pugwash and the international treaties on chemical and biological warfare', 12 May 1998, 54 pp, background paper for Pugwash Meeting no.234, 9th workshop, *CWC Implementation: Balancing Transparency and Confidentiality*, Noordwijk, The Netherlands, 15–17 May 1998.

Rev.7—'Contribution of the Pugwash movement to the international regime against chemical and biological weapons', 8 November 1998, 29 pp, background paper for Pugwash Meeting no.242, Geneva Switzerland, 28–29 November 1998.

Rev.8—'The Impact of Pugwash on the Debates of Chemical and Biological Weapons', *Scientific Cooperation, State Conflict: The Roles of Scientists in Mitigating International Discord*, Vol. 866 of the Annals of the New York Academy of Sciences, 30 December 1998, pp 224–52.

Add.1—'People involved in Pugwash work on CBW disarmament, 1959-1998', May 1999, 42pp.

337—[with Treasa Dunworth and Matthew Meselson] 'Report' of Pugwash Meeting no 219, *Strengthening the Biological Weapons Convention*, 5th Workshop of the Pugwash Study Group on the Implementation of the chemical and Biological Weapons Conventions. 21 22 September 1996, Geneva, Switzerland. 8 October 1996.

Rev.1—'Report', 28 October 1996. 6pp.

338—' "Dual use", transparency, and the Chemical Weapons Convention: Approaches used in the UK implementing legislation', 8 October 1996, 16pp. Paper for the 6th Workshop of the Pugwash CBW Study Group.

339—'Review of: Leonard A Cole, *The Eleventh Plague: The Politics of Biological and Chemical Warfare*, Ridgewood, New Jersey: WH Freeman & Co, 1996 [4 Nov], 250 pp, ISBN 0-7167-2950-4-hc', for *The Economist*, 15 October 1996, 2pp.

340—[with Treasa Dunworth] 'Report', 28 October 1996, 6pp. Report of the Pugwash Meeting no 220, *The Chemical Weapons Convention in its North-South Context*, 6th Workshop of the Pugwash Study Group on the Implementation of the

Chemical and Biological Weapons Conventions, 11-13 October 1996, Noordwijk, The Netherlands.

341—'International governance of 'dual use' technology through the Chemical Weapons Convention and its national implementing legislation', 15 November 1996, 17pp. Paper for the Forum Ost-West International Pre-Workshop, Berne, 18–19 November 1996.

342—'CBW events during August through November 1996', An excerpt from the HSP Rolling Chronology, 5 December 1996.

Published version—'News Chronology August through November 1996', *The Chemical Weapons Convention Bulletin*, no 34, December 1996, pp 14–38.

343—[with Matthew Meselson] 'Draft Convention on the prevention and punishment of the crime of developing, producing, acquiring, stockpiling, retaining, transferring or using biological or chemical weapons', 15 January 1997, 9pp.

344—'Eradicating chemical and biological weapons: Environmental and political problems', 11pp. Speaking notes for a public lecture in the University of Sussex *Issues in Environmental Science* series on 28 January 1997.

345—CBW Events during November 1996 through February 1997. An excerpt from the Rolling Chronology – 16 April 1997, 96pp.

Published version—'News Chronology November 1996 through February 1997', *The Chemical Weapons Convention Bulletin*, no 35, March 1997, pp 19–43.

346—'Chemical and Biological Warfare history', 30 April 1998, 2 pp. Speaking notes for HLCS 10, Foreign and Commonwealth Office, 1 May 1997.

347—Draft contributions to a joint paper with Jordi Molas-Gallart for European Parliament STOA project, version at 21 May 1997 and version at 24 September 1997.

Published as—'Assessment of Dual-Use Technologies in the context of European Security and Defence', *Weapons proliferation and technology controls*, 21 May 1997. 11pp.

348—'Foreign Assistance for Russian Chemdemil', 5 June 1997, 7pp. Aide-memoire for the 7th Workshop of the Pugwash Study Group on Implementation of the CBW Conventions, Noordwijk, 6–8 June 97.

349—'The Status of Chemical Weapons', 31 July 1997, 12 pp. Paper commissioned for the 47th Pugwash Conference, Lillehammer, Norway, 1–7 Aug 1997.

Published in—Joseph Rotblat (ed,), *Remember your Humanity, Proceedings of the Forth-Seventh Pugwash Conference on Science and World Affairs*, World Scientific, 199, pp 397–409.

350—'CBW events during February through May 1997, An excerpt from the HSP Rolling Chronology, 17 July 1997.

Published version—'News Chronology February through May 1997', *The CBW Conventions Bulletin*, no 36, June 1997, pp 11–35.

351—'West-European Assistance for Russian Chemical Disarmament', 31 July 1997, 24pp. Paper for the Monterey-Moscow Study Group on Russian Chemical Disarmament. Revised 14 September 97. 2nd revision 27 October 97. Rev 2 also published in Russian.

352—[with Treasa Dunworth and Matthew Meselson, 'The CWC at Entry into Force', 8 August 1997, 4pp. Report of the 7th workshop of the Pugwash Study Group on Implementation of the CBW Conventions, Noordwijk, 6–8 June 1997.

Published in—Pugwash Newsletter, November 1997, pp 200–03.

353—'Controlling dual-use biotechnology: the crucial role of National Measures', 18 September 1997, 4 pp. Paper for the 8th PSCICC Workshop, Geneva, 20–21 September 1997.

Published in—Pugwash Newsletter, vol 33, no 1, April 1998.

354—'CBW events during May through August 1997. An excerpt from the HSP Rolling Chronology, 22 October 1997.

Published version—'News Chronology May through August 1997', *The CBW Conventions Bulletin*, no 37, September 1997, pp 15–34.

355—'Controlling Dual Use Technologies: The Harvard Sussex Program', 30 October 1997. Overheads and speaking notes for, talk in the series *SPRU Research: An Introduction.* 9pp.

356—'CBW events during August through November 1997, an excerpt from the HSP Rolling Chronology', 21 December 1997.

Published version—'News Chronology August through November 1997', *The CBW Conventions Bulletin*, no 38, December 1997, pp 21–39.

357—'Controlling adverse technological change: protecting biotechnology from biological weapons', Draft, 11 February 1998. Revised version 19 May 1998, 6pp.

Published as—'Protecting Biotechnology from Biological Weapons' in *IPTS Report no.26*, July 1998, pp 6–11.

358—'Iraq's germ weapons', 9 February 1998, 3 pp.

Published as—Article for *The Tablet*, published 14 February 1998.

359—'How much of the UK CWC-compliance effort are we actually seeing', 6 March 1998, 10 pp. Discussion paper for the 24 March 1998 meeting of the UK CWC National Authority Advisory Committee.

360—'CBW events during November 1997 through February 1998, an excerpt from the HSP Rolling Chronology', 21 March 1998.

Published version—'News Chronology November 1997 through February 1998', *The CBW Conventions Bulletin*, no 39, March 1998, pp 23–43.

361—'The Chemical Weapons Convention verification regime Implications for biotechnology', 22 April 1998, 15 pp. Paper for the EFB conference, *A Strengthened Biological and Toxin Weapons Convention: Potential implications for Biotechnology*, Vienna, 28–29 May 1998.

Published as—'The Chemical Weapons Convention (CWC) verification regime: Implications for biotechnology', *Proceedings of the Conference*, pp. 15–28.

362—'Transparency-confidentiality conflicts in the Chemical Weapons Convention', 13 May 1998, 3pp, working paper for Pugwash Meeting no.234, 9th workshop, *CWC Implementation: Balancing Transparency and Confidentiality*, Noordwijk, The Netherlands, 15–17 May 1998.

363—'CBW events during February through May 1998. An excerpt from the HSP Rolling Chronology', 24 June 1998.

Published version—'News Chronology February through May 1998', *The CBW Conventions Bulletin*, no 40, June 1998, pp 23–43.

364—'How the Convention on Chemical Weapons seeks to protect biotechnology from abuse and promote its peaceful application', 23 June 1998, 9pp. Paper presented at the ICGEB conference Peaceful Use of Biotechnology and the Convention on Biological Weapons, Trieste, 26–27 June 1998.

365—'Strengthening the Biological Weapons Convention', July 1998, 13pp plus annex.

Published as—'Strengthening the Biological Weapons Convention', Briefing Paper no.11, Department of Peace Studies, University of Bradford, July 1998, 16 pp.

366—'CBW Events during May through August 1998. An excerpt from HSP Rolling Chronology', 21 September 1998.

Published version—'News Chronology May through August 1998', *The CBW Conventions Bulletin*, no 41, September 1998, pp 19–43.

367—'Can university-based research promote international cooperation against the spread of chemical and biological weapons?', 12 October 1998, Speaking notes for UoS/IR research-in-progress seminar, 12 October 1998, 8pp.

368—'Is the BWC Protocol Rolling Text dealing adequately with the problem of dual use?', 19 November 1998, 7 pp, Working paper for Pugwash Meeting no.242, 10th Workshop, *The BWC Protocol Negotiation: Unresolved Issues*, Geneva, Switzerland, 28–29 November 1998.

369—[with Matthew Meselson] 'International Criminalization of Biological and Chemical Weapons', paper for the XI Amaldi Conference on Problems of Global Security, Moscow, 18–20 November 1998, 14 pp. Revised version 22 December 1998.

Published as—'A draft convention to prohibit biological and chemical weapons under international criminal law', *CBW Conventions Bulletin*, no 42, December 1998, pp 1–5.

370—'CBW Events during August through November 1998. An excerpt from HSP Rolling Chronology', 11 January 1999, 86pp.

Published version—'News Chronology August through November 1998', *The CBW Conventions Bulletin*, no 42, December 1998, pp 21–46.

371—'Chemical and biological weapons described from an historical standpoint with reference to present-day policy concerns', Reconstructed speaking notes for a UoS/CPES Thursday Colloquium on Chemical Warfare, (28 January 1999), 1 February 1999, 16pp + illustrations.

372—[with George Poste], "The Biological Weapons Convention and its projected Protocol", rough first draft of a paper for Royal Society Biological Weapons Working Group (17 March 1999), 28 February 1999, 6pp. rev.I, working paper for the Royal Society Biological Weapons Working Group, 14 April 1999, 11 pp. Revised 9 December 1999 (Rev.I), 11 pp. Published in *Measures for controlling the threat from biological weapons*, The Royal Society, July 2000, pp 9–14.

373—'CBW Events during November 1998 through February 1999. An excerpt from the HSP Rolling Chronology', 22 March 1999, 80pp.

Published version—'News Chronology November 1998 through February 1999', *The CBW Conventions Bulletin*, no 43, March 1999, pp 15–42.

374—'A warning about implementation of the Chemical Weapons Convention from the BWC Protocol negotiation', working paper for Pugwash Meeting no.246, Noordwijk, the Netherlands, (15–16 May 1999), 7 May 1999, 8pp.

375—'CBW Chronicle 44: February through May 1999. Additions and changes to the HSP Rolling Chronology and Bibliography', 9 August 1999, 119pp.

Published version—'News Chronology February through May 1999', *The CBW Conventions Bulletin*, no 44, June 1999, pp 21–46.

376—Outline note for presentation on CBW terrorism at Wilton Park Conference WP576, October 1999, 16 September 1999, 1 p.

377—[with Matthew Meselson], 'Prohibition of Chemical Weapons under International Criminal Law', *OPCW Synthesis*, November-December 1999, pp 4–5 [submitted 2 November 1999].

Published as—'A Draft Convention to Prohibit BW & CW under International Criminal Law', *Track Two*, vol 10, no.3, December 2001, pp 48–54. [A later version of this paper is registered under Item 415 below].

378—[with Emmanuelle Tuerlings] 'The Trilateral Network associated with the Chemical Weapons Convention: a case study for the Global Public Policy Networks project', 6–7 November 1999, Rev.1 17 November 1999, 64pp.

379—'CBW Chronicle 45, May through July 1999. Additions and changes to the HSP Rolling Chronology and Bibliography', 7 November 1999, 107pp.

Published version—'News Chronology May through July 1999', *The CBW Conventions Bulletin*, no 45, September 1999, pp 17–42.

380—'CBW Chronicle 46, August through October 1999. Additions and changes to the HSP Rolling Chronology and Bibliography, 14 January 2000, 88pp.

Published version—'News Chronology August through October 1999', *The CBW Conventions Bulletin*, no 46, December 1999, pp 20–42.

381—'The inquiry into Weapons of Mass Destruction', Written Memorandum for the House of Commons Foreign Affairs Committee, 17 February 2000, 7pp.

Published as—Appendix 29, Foreign Affairs Committee, *Eighth Report: Weapons of Mass Destruction*, London: Stationery Office, 25 July 2000, pp 203–206.

382—'Chemical Weapons', 3 February 2000, 7 pp. Chapter for Physicians for Social Responsibility – Finland, *War and Health.*

Published as—'Chemical Weapons', in: Ilkka Taipale (ed) *War or Health? A Reader*, New York Zed Books, 2002, pp 119–26.

383—'Effectiveness of the international treaties against chemical and biological armament, and experiences worth sharing', 5 April 2000, 6 pp. Working paper for Pugwash Meeting no. 254, Oegstgeest, The Netherlands, (8–9 April 2000).

384—'Maximizing security benefits from technical cooperation in microbiology and biotechnology: infrastructure, regulations and procedures', vugraphs for presentation at NATO Advanced Research Workshop, Piestany, Slovakia, 17–21 May 2000.

Published as—[with Graham Pearson] 'Maximizing security benefits from technical cooperation in microbiology and biotechnology: infrastructure, regulations and procedures', in: Malcolm Dando et al (eds.) *Maximizing the Security and Development Benefits from the Biological and Toxin Weapons Convention*, (The Hague: Kluwer Academic Publishers, 2002), pp 189–98.

385—'The changing nature of the threat to public health from biological and chemical weapons', draft Chap. 2 for proposed 2nd edition of *Report on Public Health Aspects of Chemical and Biological Weapons* (1970), 6pp. **385/Rev.1**. "Assessing the threat to public health from biological and chemical weapons' 24 January 2001, 7 pp.

386—'CBW Chronicle 47, November 1999 through January 2000. Additions and changes to the HSP Rolling Chronology and Bibliography', 29 May 2000, 102 pp.

Published version—'News Chronology November 1999 through January 2000', *The CBW Conventions Bulletin*, no 47, March 2000, pp 14–42.

387—'CBW Chronicle 48, February through April 2000. Additions and changes to the HSP Rolling Chronology and Bibliography', 12 September 2000, 97 pp. [*Note*:

as *The CBW Conventions Bulletin*, no 48, June 2000 was a special issue it did not include a quarterly chronology.]

388—'CBW Chronicle 49, May through July 2000. Additions and changes to the HSP Rolling Chronology and Bibliography', 23 October 2000 83pp.

Published version—'News Chronology May through July 2000', *The CBW Conventions Bulletin*, no 49, September 2000, pp 24–45.

389—'Old Wars, New Wars: chemical weapons toward disarmament', reconstruction from speaking notes of a presentation at the workshop of the *Forum per i Problemi della Pace e della Guerra* on *Perspectives toward the Chemical Weapons Convention's Fourth Year*, Firenze (4 December 2000), 7 December 2000, 6pp.

390—[with Matthew Meselson] 'The CWC and the BWC: Yesterday, Today and Tomorrow', 21 December 2000, 3pp.

Published as—'The CWC and the BWC: Yesterday, Today and Tomorrow', editorial for *The CBW Conventions Bulletin*, no 50, December 2000, pp 1–2.

391—CBW Chronicle 50, August through October 2000. Additions and changes to the HSP Rolling Chronology and Bibliography, 8 January 2001, 108pp.

Published version—'News Chronology August through October 2000', *The CBW Conventions Bulletin*, no 50, December 2000, pp 26–46.

392—[with Matthew Meselson] 'Washington and the BWC Protocol negotiation', 22 March 2001, 2pp.

Published as—'Washington and the BWC Protocol negotiation', editorial, *The CBW Conventions Bulletin*, no 51, March 2001, pp 1–2.

393—CBW Chronicle 51, November 2000 through January 2001. Additions and changes to the HSP Rolling Chronology and Bibliography, 14 May 2001, 97pp.

Published version—'News Chronology November 2000 through January 2001', *The CBW Conventions Bulletin*, no 51, March 2001, pp 24–46.

394—[with Caitríona McLeish] 'Candidate Agent TL 2636: a cautionary tale for the General Purpose Criterion', 15 June 2001, 15pp. Working paper for Pugwash Meeting no 264, Oegstgeest, The Netherlands, 23–24 June 2001.

Rev.1—'Candidate Agent TL 2636: a cautionary tale for the General Purpose Criterion', 5 October 2001, 15 pp. Working paper for the SPRU Workshop, *The General Purpose Criterion of the Chemical Weapons Convention*, University of Sussex, 12 October 2001.

395—[with Daniel Feakes] 'National implementation measures: role of the OPCW Secretariat', 18 June 2001, 3pp. Paper for Pugwash Meeting no 264, Oegstgeest, The Netherlands, 23–24 June 2001.

396—'The General Purpose Criterion and the new utility of toxicants as weapons', 17 June 2001, 8pp. Working paper for Pugwash Meeting no. 264, Oegstgeest, The Netherlands, 23–24 June 2001.

397—[with David Atwood, Malcolm Dando, Alastair Hay, Alexander Kelle, Ian Kenyon, Kathryn Nixdorff, Graham Pearson and Nicholas Sims] Written testimony to the Subcommittee on National Security, Veteran Affairs and International Relations of the House Committee on Government Reform, 10 July 2001, 9 pp.

398—'The General Purpose Criterion of the Chemical Weapons Convention', background note for the GPC Workshop at SPRU, 12 October 2001, 4pp.

399—[with Matthew Meselson] 'Strengthening the Biological Weapons Convention: what next for the Ad Hoc Group?', 26 September 2001, 4pp.

Published as—'Strengthening the Biological Weapons Convention: what next for the Ad Hoc Group?', editorial, *The CBW Conventions Bulletin*, no 53, September 2001, pp 1–2.

400—'Should we worry about chemical and biological weapons?' Speaking notes and VUGs for SPRU Friday Seminar, 19 October 2001, 19pp.

401—CBW Chronicle 52, Additions and changes to the HSP Rolling Chronology and Bibliography, 3 December 2001.

Published version—'News Chronology February through April 2001', *The CBW Conventions Bulletin*, no 52, June 2001, pp 33–58.

402—CBW Chronicle 53, May through July 2001, Additions and changes to the HSP Rolling Chronology and Bibliography 6 February 2002, 116 pp.

Published version—'News Chronology May through July 2001', *The CBW Conventions Bulletin*, no 53, September 2001, pp 25–51.

403—CBW Chronicle 54, August through October 2001, Additions and changes to the HSP Rolling Chronology and Bibliography, 1 February 2002.

Published version—'News Chronology August through September 2001', *The CBW Conventions Bulletin*, no 54, December 2001, pp 28–58.

404—[with Matthew Meselson] 'International criminal law and sanctions to reinforce the BWC', *The CBW Conventions Bulletin*, no 54, December 2001, pp 1–2.

405—'Workshop on the CWC General Purpose Criterion, SPRU – Science & Technology Policy Research, University of Sussex, 12 October 2001', Report on the proceedings, 25 February 2002, 23 pp.

406—[with Matthew Meselson] 'A Draft Convention to prohibit biological and chemical weapons under international law', presentation to the OPCW *International Symposium on Cooperation and Legal Assistance for Effective Implementation of International Agreements*, 7–9 February 2001.

Published as—'A Draft Convention to Prohibit Biological and Chemical Weapons under International Law', in: Rodrigo Yepes-Enriques and Lisa Tabassi (eds), *Treaty Enforcement and International Cooperation in Criminal Matters: With Special Reference to the Chemical Weapons Convention*, (The Hague: TMC Asser Press, 2002), pp 457–69.

407—'What Should be the Scope of the CWC?: A Workshop Report', *The CBW Conventions Bulletin* no 55, March 2002, pp 1–4.

408—'The Chemical and Biological Weapons Conventions: Will They Work?', 14 March 2002, 7pp. VUGs for talk given at Kurdish Scientific and Medical Association (KSMA) Halabja Day, Brunei Gallery Lecture Theatre, School of Oriental and African Studies, London, 15 March 2002.

409—'The Threat of Chemical and Biological Weapons': Text of presentation at the Third Swiss Student Pugwash National Conference, Geneva, 10 May 2002 [corrected and annotated 12 May 2002], 8pp.

410—'The future of the OPCW': notes for statement following José Bustani's at House of Commons briefing organised by Dr. Lynne Jones MP, Grimond Room, Portcullis House, 14 May 2002, 3pp.

411—'Should the First CWC Review address State-Party non-compliance with the National Implementation Provisions?', 13 June 2001, 11pp. Working paper for PSGICC-17, Oegstgeest, The Netherlands, 14–15 June 2002.

412—'Comments on FCO Green Paper about the BWC', 12 September 2002, 4p, in House of Commons Foreign Affairs Committee *Foreign Policy Aspects of the War Against Terrorism, Second Report of Session 2002-03*, HC 196 [incorporating HC 1196, Session 2001–02, London: The Stationery Office, 19 December 2002. Subsequently published as "Appendix 2: Memorandum from Professor J. P. Perry Robinson", in House of Commons Foreign Affairs Committee. *The Biological Weapons Green Paper: First Report of Session 2002-03*, London: The Stationery Office, 11 December 2002, pp EV38-EV39.

413—'Germs, warfare and the human impulse to keep them apart', 23 September 2002, 3pp. Outline of presentation at ICRC Montreux meeting, 23–24 September 2002, *Biotechnology, Weapons and Humanity*, in Session 2, *Ancient Norms: New Threats.*

414—'The General Purpose Criterion and the importance of its implementation', 17 September 2002, 7 pp. Slides to accompany presentation at NATO Advanced Research Workshop, Bratislava, 18–22 September 2002, *Maximizing the Security Benefits from the First Review Conference of the Chemical Weapons Convention.*

Rev.1—'The General Purpose Criterion and the importance of its implementation', 5pp. Paper presented at 19th Workshop of the Pugwash Study Group on the Implementation of the Chemical and Biological Weapons Conventions: *The First CWC Review Conference and Beyond*, Oegstgeest, The Netherlands, 26–27 April 2003.

415—"International Criminalization of Biological and Chemical Armament", abstract for participants at presentation at Wilton Park Conference, Wiston House, 27–29 September 2002, *Preventing the Proliferation of Chemical and Biological Weapons*, 26 September 2002, 3pp [= Item 377].

416—[with Malcolm Dando and Matthew Meselson] "Preventing the hostile use of biotechnology: the way forward now", *The CBW Conventions Bulletin*, no 57 (September 2002), pp 1–2.

417—[with Jordi Molas-Gallart] *Scientific and Technological Advances Relevant to Bioterrorism, and their Possible Impact on Vulnerabilities in EU Society: A Prospective Study*, Final Report, corrected and with Working Papers, for the European Science and Technology Observatory, in response to paragraph 4.2 of Commission Communication 2001/707, 29 November 2002, 439 pp.

418—Executive editor. *Public Health Response to Biological and Chemical Weapons: WHO guidance*, 2nd edition, Geneva: World Health Organization, pre-publication version posted on WHO website, November 2001, 226 pp.

419—[with Matthew Meselson] ""Law enforcement" and the CWC", *The CBW Conventions Bulletin*, no 58, December 2002, pp 1–2.

420—"Chemical and Biological Weapons", speaking notes and VUGs for Sussex Development Lecture, University of Sussex, 23 Jan 2003, 18 pp.

421—"Implications of the use of chemical incapacitants for law enforcement purposes", speaking notes for an ICRC briefing on this subject, The Hague, 28 March 2003.

Rev.2—"Law Enforcement, 'Non Lethal' Warfare and the Norm Against Biological and Chemical Armament", 10 May 2003, 23 pp.

Rev.3—Further HSP submission to the OPCW Open-Ended Working, Group on Preparations for the Second CWC Review Conference, "Non Lethal Warfare and the Chemical Weapons Convention", 24 October 2007, 37 pp.

422—[with Matthew Meselson] 'A Draft Convention to Prohibit Biological and Chemical Weapons Under International Criminal Law'. Discussion Paper for the 19th Workshop of the Pugwash Study Group on the Implementation of the Chemical and Biological Weapons Conventions: *The First CWC Review Conference and Beyond*, Oegstgeest, The Netherlands, 26-27 April 2003.

Rev.1—Published as "A Draft Convention to Prohibit Biological and Chemical Weapons under International Criminal Law", in *The Fletcher Forum of World Affairs*, vol 28, no 1 (Winter 2004) pp 57–71.

423—""Non Lethal" Warfare and the Norm against Hostile Exploitation of Disease", speaking notes for presentation at MIT in a session of the BCSIA/HSP/MIT seminar, 22 May 2003.

424—"Disabling Chemical Weapons: A Documented Chronology of Events, 1945–2003", [updating Items 302App, 302App/Add.1 and 302App/Add.2], 8 October 2003, 157 pp.

425—[with Matthew Meselson] "'Non-Lethal' Weapons, the CWC and the BWC", 1 October 2003, editorial in The CBW Conventions Bulletin no 61 (September 2003), pp 1–2. Revised and expanded as **Item 425/Rev.1**, "Towards the Sixth BWC Review Conference and Beyond: 'Non Lethal' Weapons and Implementation of the Chemical and Biological Weapons Conventions", 22 October 2003, a paper for the 20th Workshop of the Pugwash Study Group on Implementation of the CBW Conventions, Geneva, 8–9 November 2003.

426—"David Christopher Kelly, 1944–2003", 1 October 2003, an obituary published in *The CBW Conventions Bulletin*, no 61, September 2003, p 36.

427—"The Chemical Weapons Convention in a changing political context", 22 October 2003, 11 pp. Write-up of a presentation at the XV Amaldi Conference on Problems of Global Security, Helsinki, 25-27 September 2003, for which Item 427dft comprised speaking notes.

Published in—Paul Fogelberg (ed.), *Changing Threats to Global Security: Peace or Turmoil. Proceedings of the XV International Amaldi Conference, Helsinki, 25–27 September 2003*, Helsinki: Delegation of the Finnish Academies of Science and Letters, and Finnish Institute of International Affairs, 2004, pp 211-20. [Discussion of the panel of papers was also included in the book at pp 220-25.]

428—[with Daniel Feakes] CBW Chronicle 55, November 2001 through January 2002, Additions and changes to the HSP Rolling Chronology and Bibliography, 6 November 2003. [*Note*: owing to issues of availability of staff time, only a shortened version of the news chronology was published in *The CBW Conventions Bulletin*, no 55, March 2002.]

429—"Chemical terrorism", notes and vugraphs for a presentation at the British Pugwash Group public discussion meeting on *Britain and Unconventional Terrorism*, London School of Tropical Hygiene and Medicine, 10 December 2003.

Published as—Revised into speaking notes as Item 429/Rev.1 dated 10 December 2003.

430—[with Graham Pearson] "Analysis and implications", draft penultimate chapter for M. Dando, G Pearson, J. Perry Robinson, L. Rozsa, and M. Wheelis. *Deadly Cultures*, Cambridge MA: Harvard University Press. Draft sent to editors 8 January 2004.

431—'Weapons inspection regimes – biological and chemical weapons', notes for presentation in the UoS *Global Justice & Political Violence Seminar*, 28 January 2004.

432—[with Nick Dragffy, Daniel Feakes and Carolyn Sansbury] *CBW Chronicle* no 56, 3 May 2004, 125 pp.

Published version—'News Chronology February through April 2002', *The CBW Conventions Bulletin*, no 56, June 2002, pp 15-48.

433—[with Nick Dragffy, Daniel Feakes and Carolyn Sansbury] *CBW Chronicle* no 57, 4 May 2004, 137 pp.

Published version—'News Chronology May through July 2002', *The CBW Conventions Bulletin*, no 57, September 2002, pp 21–62.

434—[with Nick Dragffy, Daniel Feakes and Carolyn Sansbury] *CBW Chronicle* no 58, 5 May 2004, 95 pp.

Published version—'News Chronology August through October 2002', *The CBW Conventions Bulletin*, no 58, December 2002, pp 27–46.

435—[with Nick Dragffy, Daniel Feakes and Carolyn Sansbury] *CBW Chronicle* no 59, 6 May 2004, 93 pp.

Published version—'News Chronology November 2002 through January 2003', *The CBW Conventions Bulletin*, no 59, March 2003, pp 9-30.

436—[with Nick Dragffy, Daniel Feakes and Carolyn Sansbury] *CBW Chronicle* no 60, 7 May 2004, 122 pp.

Published version—'News Chronology February through April 2003', *The CBW Conventions Bulletin*, no 60, June 2003, pp 25–50.

437—[with Nick Dragffy, Daniel Feakes and Carolyn Sansbury] *CBW Chronicle* no 61, 8 May 2004, 106 pp.

Published version—'News Chronology May through July 2003', *The CBW Conventions Bulletin*, no 61, September 2003, pp 15–35, 37–40.

438—[with Nick Dragffy, Daniel Feakes and Carolyn Sansbury] *CBW Chronicle* no 62, 9 May 2004, 110 pp.

Published version—'News Chronology August through October 2003', *The CBW Conventions Bulletin*, no 62, December 2003, pp 29–56.

439—[with Nick Dragffy, Daniel Feakes and Carolyn Sansbury] *CBW Chronicle* no 63, 10 May 2004, 116 pp.

Published version—'News Chronology November 2003 through January 2004', *The CBW Conventions Bulletin*, no 63, March 2004, pp 16–48.

440—'WMD and US', *Falmer*, University Alumni Magazine, 16 May 2004, 4pp.

441—'Codes of conduct and individual', 31 May 2004, 3pp. A posting on ICRC extranet *Biotechnology, Weapons and Humanity*.

442—'Biological warfare', speaking notes for lecture to first year undergraduates in term 3 of Human Sciences, in the core course *Social Dimensions of Biology*, 8 June 2004, 3pp.

443—'Objectives of the workshop', 21 February 2005, a concept paper for the High Level Pugwash CBW Workshop, Oegstgeest, the Netherlands 16–17 April 2005.

444—"What's the threat?", speaking notes for presentation on 4 November 2005 in the BSMS 307 course option, Germ warfare, bioterrorism – what have they got to do with me?

445—[with Nick Dragffy, Daniel Feakes and Carolyn Sansbury] *CBW Chronicle* no 64, 28 July 2004, 168 pp.

Published version—'News Chronology February through April 2004', *The CBW Conventions Bulletin*, no 64, June 2004, pp 19–53.

446—'Chemical weapons', a presentation made during the Public Discussion Meeting, *Elimination of Weapons of Mass Destruction*, that the British Pugwash Group sponsored on 8 September 2004 during the British Association *2004 Festival of Science*, University of Exeter.

Rev.1—'Chemical Weapons and International Cooperation', 11 January 2005, 9pp + 5 pages of tables.

Rev.2—'Chemical Weapons and International Cooperation', Discussion paper given at Pugwash High Level CBW Workshop *Present Trends and Future Policy Choice*, Oegstgeest, 16–17 April 2005, 15pp.

Published as—'Chemical Weapons and International Cooperation' in: Marjorie de Reuck, (ed), *Eliminating the Threat of Weapons of Mass Destruction*, London: British Pugwash Group (2005) pp 19–30.

447—[with Matthew Meselson] "Libya and 'dual-use' ", editorial, *CBW Conventions Bulletin*, no 65, September 2004, pp 1–3.

448—"The odd neglect of toxin weapons during the 2004 Meeting of Experts", discussion paper for the 21st Workshop of the Pugwash Study Group on the Implementation of the Chemical and Biological Weapons Conventions: *The BTWC New Process and the Sixth Review Conference*, Geneva, Switzerland, 4-5 December 2004, 4pp.

449—"Martin Mark Kaplan, 1915–2004", *CBW Conventions Bulletin* no.66 (December 2004), p5.

450—"Iran's chemical and biological weapons programmes", draft for IISS dossier on Iranian WMD, 28 March 2005, 18 pp. Published as "Iran's chemical and biological programmes", in Gary Samore (ed) *Iran's Strategic Weapons Programmes: A Net Assessment* [an IISS Strategic Dossier], Abingdon, Oxon: Routledge, 6 September 2005, pp 69–85.

Rev.2—"Dual technology and perceptions of Iranian chemical and biological weapons", 26 July 2005

451—'The long view', speaking notes for ESRC Research Seminar, SPRU, 30 July 2005.

452—'Accommodating technical change: a challenge for the Second CWC Review', 11 May 2006, 3 pp, a paper presented at the 24th Workshop of the Pugwash Study Group on Implementation of the CBW Conventions, Noordwijk, the Netherlands, 13–14 May 2006.

453—[with Kai-Bastian Ilchmann, Caitríona McLeish and Tony P Phillips] 'Threat Analysis: How might RBC agents suit the purpose of terrorists?', 26 May 2006, 63 pp, a contribution by SPRU to EU Commission Framework 6 coordination action ASSRBCVUL, WP2.

454—[with Matthew Meselson] "Yellow Rain: Lessons Learned", 16 July 2006, 20 pp, typo-corrected version (Item 454 Draft) of a presentation at the King's College London conference on *Identification, Characterization and Attribution of Biological Weapons Use* (12–13 July 2006).

Rev.1—"The Yellow Rain Affair: Lessons from a discredited allegation", 13 April 2007, 45 pp.

Published in—Anne L. Clunan, Peter R. Lavoy and Susan B. Martin, (eds), *Terrorism, War, or Disease? Unraveling the Use of Biological Weapons*, Stanford University Press, 2008, pp 72–96.

455—'Vulnerability of Society: an overview of EU vulnerabilities toward deliberate release of RBC agents', 25 July 2006, 54 pp, a contribution by SPRU to EU Commission Framework 6 coordination action ASSRBCVUL, WP3.

456—'Near-term development of the governance regime for biological and chemical weapons', 14 September 2006, a rolling text circulated for private comment and then, in the versions of 4 November 2006 and subsequently, as a project paper for ESRC RES-228-25-0064.

Rev.1—[a shortened version], 'Near-term development of the governance regime for biological and chemical weapons', 14 October 2006, 34 pp.

Published as—'Chemical and biological weapons' in: Nathan E Busch and Daniel H Joyner, editors, *Combating Weapons of Mass Destruction: The Future of International Nonproliferation Policy*, University of Georgia Press (2009), pp 74–94.

Rev.2—'The governance regime for biological and chemical weapons, and the Review Conferences of 2006 and 2008', 30 October 2006, 26 pp. Paper for the 25th Workshop of the Pugwash Study Group on Implementation of the CBW Conventions, Geneva, Switzerland, 18–19 November 2006.

Rev.3—'Improving the governance regime for biological and chemical weapons', 3 January 2007. Prepared for distribution at the final session of the ESRC Research Seminar on *New Approaches to WMD Proliferation* (SPRU, 8-9 January 2007). The paper also provides a basis for Item 459.

457—'The origins of the Chemical Weapons Convention: the international perspective', 21 October 2006, 4 pp. Speaking notes for presentation at the conference *The*

History and Ethics of Porton Down: Perspectives and Work in Progress, Canterbury, 21 October 2006.

458—Speaking notes for JPPR presentation final session, ESRC Research Seminar on *New Approaches to WMD Proliferation* (SPRU, 8–9 January 2007).

459—'Categories of Challenge now facing the Chemical Weapons Convention', 12 March 2007, 21 pp, discussion paper for 52nd Pugwash CBW Workshop, Noordwijk, The Netherlands, 17–18 March 2007.

Rev.1—'Categories of Challenge now facing the Chemical Weapons Convention', 23 May 2007, 22 pp, a post-workshop revision of the paper for wider distribution.

Rev.2—'Difficulties facing the Chemical Weapons Convention', 19 December 2007.

Published as—'Difficulties facing the Chemical Weapons Convention' in *International Affairs* vol 84, no 2, March 2008, pp 223–39.

460—[Team leader and editor] 'An assessment of the risk to European societies in the possibility of terrorist resort to radiological, biological or chemical agents'. May 2007. Preliminary draft report to the EU Commission RTD Sixth Framework Programme project 502476 (ASSRBCVUL) by a team at the University of Sussex, UK.

461—[with Anthony Phillips] 'The CWC and Chemicals of Biological Origin', 8 August 2007, 9 pp, presentation at the OPCW Academic Forum, The Hague, 18–19 September 2007.

Published as—Subsequently revised for publication in the proceedings of the forum – Ralf Trapp (editor), *OPCW Academic Forum Conference Proceedings*, The Hague: Netherlands Institute of International Relations Clingendael (2008) pp 209–19.

462—'What constitutes proliferation? Who are the players? What are the modes (networks &c)?', 28 September 2007. Speaking notes for Wilton Park Conference WPC 871 (28–30 September 2007).

Rev.1—[revised 2 November 2007].

463—[with Anthony Phillips] 'Addressing the toxin problem', 27 November 2007, 7 pp, discussion paper for Pugwash Meeting no. 331, the 53rd CBW workshop.

Rev.1—'Addressing the toxin problem', 29 December 2007, a post-workshop revision of the paper for wider distribution.

Rev.2—'Addressing the toxin problem', 21 March 2008, a post-workshop revision of the paper for wider distribution.

464—'Is the CWC definition of "chemical weapons" fit for purpose?', speaking notes for IIHL Sanremo CW seminar, 15 February 2008.

Rev.1—'Is the CWC definition of "chemical weapons" fit for purpose?', 19 February 2008, 4 pp, a post-seminar revision intended for publication in the proceedings volume.

Published as—'Is the OPCW implementing the CWC definition of chemical weapons?', *The CBW Conventions Bulletin* No. 78, (February 2008), pp. 1–2.

Rev.3—'Is the OPCW implementing the CWC definition of chemical weapons?', 20 March 2008, 4 pp, submitted to Pugwash CBW Workshop, Noordwijk, 5–6 April 2008.

465—'Toxic biotechnological-process products and chemicals of biological origin', 25 March 2008, 6 pp, a discussion paper for the 28th Workshop of the Pugwash Study Group on Implementation of the CBW Conventions, *The Second CWC Review Conference and After*, Noordwijk, the Netherlands, 5–6 April 2008.

466—'Scientists and chemical weapons policies', 10 November 2008, 12 pp, a paper for the NATO Advanced Research Workshop on *The Role of Independent Scientists in Assessing the Threat of WMD*, Zagreb 13–15 November 2008. Subsequently revised for publication:

Rev.1—'Scientists and chemical weapons policies', 21 January 2009, 13 pp.

Rev.1/Rev—'Scientists and chemical weapons policies', 16 February 2009, 13 pp.

467—'From divergence to convergence', 19 November 2008, 6 pp, a discussion paper for the 29th Workshop of the Pugwash Study Group on Implementation of the CBW Conventions, *Moving towards the Seventh BWC Review Conference*, Geneva 29–30 November 2008.

Rev.1—'From divergence to convergence', 16 December 2008, 6 pp.

Published as—'Bringing the CBW Conventions closer together' in *The CBW Conventions Bulletin* no 80 (September 2008) pp 1–4.

468—'Possible OPCW futures', speaking notes for HSP Sussex Day, 15 May 2009, *The OPCW Past, Present and Future: A seminar in memory of Ian Kenyon*, as revised 18 May 2009, 5 pp.

469—'Mass destruction, chemical weapons and human security', 12 June 2009, 13 pp, a draft chapter for Narcis Serra and Mary Kaldor, editors, *From War to Human Dignity* [unformatted version, same date].

Rev.1—'Mass destruction, chemical weapons and human security' [much revised version], 30 May 2011, 13 pp.

470—(i) 'The convergence of chemistry and biology', 3 December 2009, speaking notes for PSGICC/WS-30.

470 (ii)—'The post-disarmament era of the OPCW' – seems to exist only as **Item 470 outline rev**—dated 17 June 2010 comprising 2 pp of unfinished speaking notes.

471—'Criminalization of Biological and Chemical Armament', 20 January 2011, 7 pp.

Rev.1—'Criminalization of biological and chemical armament', a background note for the 11 February 2011 HSP symposium on the subject, 31 January 2011, 7 pp

Rev.2—'Criminalization of biological and chemical armament', dated 21 February 2011, 9 pp, for the lead article in the final (special) issue of *CBW Conventions Bulletin.*

472—Speaking notes for Wilton Park Conference 1079 *Chemical Weapons Disarmament and the Future of the OPCW* (18-21 March 2011), "A contribution to the panel on 'The Challenges, 1: Science and Technology' ", 18 March 2011, 4 pp.

Rev—Speaking notes (recreating the original from which the speaking notes were cut), 23 March 2011, 6 pp.

Out—'Robinson synopsis', not dated, which seems to have been written for the WPC 1079 conference pack, 1 p.

473—[? JPR contribution to Royal Society *Brain Waves* 3?]

474—'Notes for an assessment of toxins', 25 July 2011.

474/Rev.1 'Notes for an assessment of toxins', 12 January 2012, 14 pp.

475—'The Yellow Rain: A documented chronology of events since 1964'. **Item 475Dft**—, latest version dated 21 November 2012, 168 pp, updating *Yellow Rain Chronology 1975 through 1986*, a rolling text started in 1986.

476—"Incapacitating Chemical Agents' in context: an historical overview of states' policy', 14 May 2012, 9 pp, expanded and annotated speaking notes for an ICRC expert meeting, Montreux, 24–26 April 2012.

477—'Next steps for the CWC Review Conference on convergent science and technology', 18 February 2013, 3 pp, speaking notes for Royal Society roundtable on *Scientific Convergence and the Chemical Weapons Convention*, OPCW Headquarters, The Hague, 18 February 2013. Item 477 corr, 22 February 2013.

478—'Alleged use of chemical weapons in Syria', *Harvard Sussex Program Occasional Papers new series* HSPOP 4 dated 26 June 2013, 41 pp, 20,500 words, as published on web. **Item 478 rev** is a rolling text intended as input into a follow-on publication, the newest version being dated 20 July 2013, 41pp, but subsequently recast as **Item 478 rev 2**, "Chemical warfare in Syria, part 1: Alleged use of chemical weapons prior to the Ghouta event", a rolling text of 28,000 words in the version of 20 August 2015.

479 (old)—Also intended originally as HSPOP-4 follow-on and comprising a compendium of possible input material entitled 'Ghouta chronology', this being a rolling text started in late August 2013, with annexes and an appendix, 93 pp in the version dated 18 March 2014. Subsequently fragmented into Item 478 rev 2, Item 481 rev and Item 490.

479 (new)—An alternative potential follow-on to HSPOP-4, to be developed from Item 484 rev 3 and Item 493 rev: 'Binary chemical weapons and Syrian-mode chemical warfare: Implications for the international CW control regime'.

480—'Outline of the component CBW case study [in the LSE/SPRU project *Strategic Governance of Science and Technology: Pathways to Security*]', draft of 24 March 2014, revised 22 April 2014, 7 pp.

481—'Chemical warfare in Syria, part 2: A chronological documentation of the August 2013 Ghouta event and its aftermath', cut from the 30 Mar 14 version of **Item 479 (old)** and then continued as a rolling text, 86 pp in the version of 2 May 2014. Revised into **Item 481 rev**, 'Chemical Warfare in Syria, part 2: A chronological documentation of the August 2013 Ghouta event and the subsequent international effort to eliminate the Syrian chemical-weapons programme', a rolling text of 48,000 words in its version of 22 August 2015.

482—'On the information disclosed by Syria about its chemical weapons after the Ghouta event', cut from the 30 Mar 14 version of **Item 479 (old)** where, in a draft dated 19 Feb 14, it had been Annex 1. It subsequently became a stand-alone rolling text, 10,500 words in the version of 31 August 2015, before becoming the appendix, 'On the Information Declared by Syria about its Chemical Weapons Programme', in Item 494.

483—"Omission and lies in information on the Ghouta CW event", cut from the 30 Mar 14 version of **Item 479 (old)** where it had first been Annex 1 then Annex 2, "Ghouta chronology: What next?" and is now a stand-alone rolling text, 13 pp in the version of 26 April 2014 and 10,400 words in the version, "Gaps and lies in information on the Ghouta event", of 19 February 2015. {**Item 483 rev 1** is a 10 pp version dated 30 March 2014 intended for inclusion in Item 484.}[4]

484—Drafts on Syria/CW for contribution(s) to the LSE/SPRU project outlined in Item 480, *Pathways to Security*. **Item 484 rev 2** is vignette S1, "The reappearance of binary nerve gas during the Syrian civil war", 8 pp, dated 28 July 2014. **Item 484 rev 3**, "Binary nerve gas as a stimulus to proliferation of chemical weapons, its ascendancy during the war in Syria, and the future of chemical arms control", 16 pp, dated 4 November 2015, is a draft for a possible HSPOP.

485—"The creeping legitimization of low-lethality chemical weapons for use in both war and law-enforcement", a draft for possible inclusion in the CBW case-study of the LSE/SPRU project outlined in Item 480, 8 pp, dated 20 June 2014.

486—"Hellebores and Hunyadi's arsenical smoke in the early history of CBW", vignette S1 for inclusion in the CBW case-study of the LSE/SPRU project outlined in Item 480, *Pathways to Security*, 5 pp, draft dated 6 August 2014.

---

[4] Editors' note. A version dated 5 March 2016 was distributed at the HSP workshop on Syria, 21–22 March 2016.

487—"Hamburg 1928, Bhopal 1984: the weather as component of CBW systems", vignette S5 for inclusion in the CBW case-study of the LSE/SPRU project outlined in Item 480, *Pathways to Security*,

488—"Deliberate targeting of CBW on noncombatants: Zhejiang (1942), Sardasht (1987) and Halabja (1988)", vignette S10 for inclusion in the CBW case-study of the LSE/SPRU project outlined in Item 480—, *Pathways to Security*, 5 pp, draft dated 15 January 2015.

489—"Radiological chemical weapons: polonium and the death of Alexander Litvinenko", vignette S12 for inclusion in the CBW case-study of the LSE/SPRU project outlined in Item 480—, *Pathways to Security*, 5 pp, draft dated 22 January 2015.

490—"Chemical Warfare in Syria, part 3: Alleged use of chemical weapons in Syria since culmination of the international project to eliminate the Syrian chemical-weapons programme", a rolling text of 22 pp as at 20 February 2015 and of 19,000 words as at 4 February 2016.

491—"Bioweapons Chronology: A rolling text documenting past events that could help identify drivers either of past biological armament or of the biological disarmament underpinning the 1972 Biological Weapons Convention", 56 pp as at 5 May 2015.

492—"Reported use of chemical weapons in Syria during 1982 through 2015: A documented and contextualized chronology", a rolling text formed initially from Items 478 rev 2, 481 rev and 490, 86,900 words as at 22 February 2016.[5]

493—"Chemical and biological weapons: past, present and future?", text for presentation at FCO/ACDRU 50th Anniversary seminar, London, 28 October 2015, revised 29 October 2015.

494—"The International Effort to Eliminate Chemical Weapons from Syria: A documented chronology, with an account of the CW declared by Syria", a rolling text formed initially from Items 481 rev and 490 with an appended version of Item 482, 24,800 words as at 4 March 2016.[6]

495—[with Jeanne Guillemin, Matthew Meselson and Nicholas Sims], "[contributions to] Witness Seminar:—Origins of the Biological Weapons Convention" [on 10 October 2014]. Published in F Lentzos, editor, *Biological Threats in the 21st Century*, London: Imperial College Press, 2016.

496—[Chemical warfare in Syria, part 4: Developments since 2015].

497—"CBW data capture through compilation of documented chronologies", speaking notes for the opening session of the HSP/Cranfield Syria Workshop [University of Sussex, 21–22 March 2016], 20 March 2016, 2 pp.

---

[5] Editors' note. A version dated 22 February was distributed at the HSP workshop on Syria, 21–22 March 2016.

[6] Editors' note. A version dated 4 March 2016 was distributed at the HSP workshop on Syria, 21–22 March 2016.

Rev.1—"CBW data capture through compilation of documented chronologies", 22 July 2016, 3 pp.

Unnumbered—[with Caitríona McLeish and James Revill], "Some potential implications for the chemical weapons regime resulting from the Syria case", prepared for the HSP/Cranfield Syria Workshop [University of Sussex, 21-22 March 2016], 22 pp, March 2016.

Selected non-CBW publications

XXX—'Technology, Armament and Restraint, Address to the Scientific and Technical Committee, on Friday, 28 May 1976, in Brussels', North Atlantic Assembly, June 1976.

XXX—'Trends in conventional munitions and their implications for qualitative arms limitation', Occasional Papers series, Science Policy Research Unit, University of Sussex, May 1977.

XXX—'The Neutron Bomb and Mass-Destruction Conventional Weapons'. Paper presented at the 27th Pugwash Conference on Science and World Affairs, Munich, 24–29 August 1977.

Published as—'The Neutron Bomb and Mass-Destruction Conventional Weapons', *Bulletin of Peace Proposals*, vol 8, no 4, October 1977, pp 321–27.

About the Author: Julian Perry Robinson

Julian Perry Robinson (1941–2020) made a unique contribution to protecting the world from the misuse of chemistry and biology. A trained chemist and patent lawyer, he applied his specialist knowledge to better understand the nature of chemical and biological warfare (CBW) and how best to prevent this. He pioneered open source research methods that enabled him to engage with, and inform, national and international policy making, working across the natural and social sciences, and with diverse sectors, including governments, international organizations, journalists, and non-governmental research, advocacy and trade groups. His influence can be seen in the achievement and implementation of the Biological Weapons Convention and Chemical Weapons Convention, and in his influence on CBW scholarship.

H. Wilson et al. (eds.), *Julian Perry Robinson: Pioneer in the Prohibition of Chemical and Biological Warfare*, Pioneers in Arts, Humanities, Science, Engineering, Practice 32, https://doi.org/10.1007/978-3-032-04064-0

About the Editors

Henrietta Wilson conducts policy-relevant research on arms control, disarmament and non-proliferation and teaches undergraduate and postgraduate modules on politics and international relations at UK universities. Originally trained as a physicist, her career spans three decades of collaboration with major non-governmental organizations and universities in the UK and Germany. She is a PhD student at King's College London, and Research Associate at SOAS University of London. She worked with Mary Kaldor and Julian at the Science Policy Research Unit (SPRU), University of Sussex, 1994–1996.

Mary Kaldor (CBE) is Professor Emeritus of Global Governance at The London School of Economics and Political Science. She also teaches at the Institut Barcelona d'Estudis Internacionals (IBEI). Kaldor contributed to the development of cosmopolitan democracy. She writes about globalization, international relations and humanitarian intervention, global civil society and global governance. She has pioneered the concepts of new wars and global civil society. Her elaboration of the real-world implementation of human security has directly influenced European and national governments. She met Julian in 1968; they were life partners, and co-parents to Charlie, Josh and Olly.

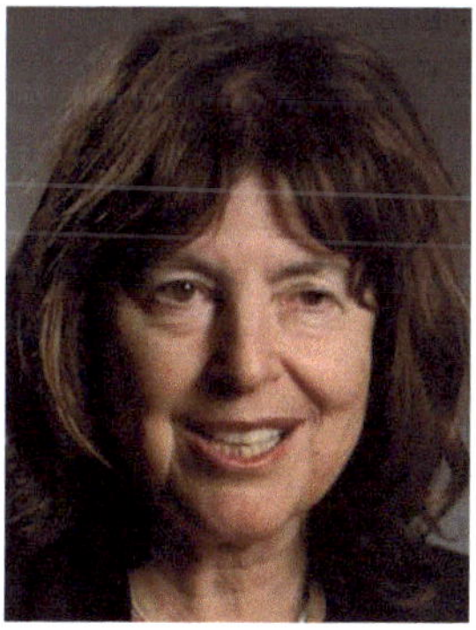

H. Wilson et al. (eds.), *Julian Perry Robinson: Pioneer in the Prohibition of Chemical and Biological Warfare*, Pioneers in Arts, Humanities, Science, Engineering, Practice 32, https://doi.org/10.1007/978-3-032-04064-0

Josh Kaldor-Robinson is one of Mary and Julian's sons. He has been a human rights activist, a pub landlord and a researcher on identities and representations of diaspora communities on the Internet.

Peter Pringle is a journalist and author. For thirty years he was a foreign correspondent for *The Sunday Times*, *The Observer* and *The Independent*, working in Europe, Africa, the Middle East, the former Soviet Union and the United States. He has also written for several US newspapers and magazines, including *The New York Times*, *The Washington Post*, *The Atlantic*, *The New Republic* and *The Nation.* He is the author and co-author of several books on science and current affairs. Peter was a close friend of Julian's and was inspired by Julian's passion for science in creating his mystery novel, *Day of the Dandelion.*

Alexander Ghionis is a Research Fellow in Chemical and Biological Security at the Harvard Sussex Program, based in the Science Policy Research Unit (SPRU), University of Sussex. His work explores emerging challenges, change and continuity in the chemical and biological weapons prohibition regimes. He is also a non-resident Fellow with UNIDIR's Weapons of Mass Destruction Program, a CBWNet Associate Researcher, and member of the Chemical Weapons Convention Coalition.

Richard Guthrie is an international security policy researcher, primarily focused on technology control and innovation issues that relate to materials and technologies that can have hostile as well as peaceful uses. He runs CBW Events, a project that aims to create a record of events to enable and encourage understanding of how policies on issues relating to chemical and biological warfare (CBW) and its prevention are developed. He worked at the Harvard Sussex Program 1990–2003.

About the HGB Foundation for Peace and Ecology in the Anthropocene

Hans Günter Brauch Foundation
for Peace and Ecology in the Anthropocene

The Hans Günter Brauch Foundation for Peace and Ecology in the Anthropocene (HGBS) is a foundation under German civil law with legal capacity which has its registered office in 74821 Mosbach in Baden-Württemberg. http://hgb-stiftung.org/index.html.

Aims and Purposes of this Foundation

The purpose of the Foundation is to promote science, research and education on international peace and ecology, and linkages of peace and ecology (peace ecology) in the Anthropocene, the new epoch in the history of the Earth. The *Hans Günter Brauch Foundation* (HGBF) *on Peace and Ecology in the Anthropocene* (PEA) was established on 8 May 2020 on the occasion of the 75th anniversary of the end of World War II and of the liberation from National Socialism to combine a joint peace research and ecology perspective on the new challenges facing humankind's survival in the Anthropocene.

H. Wilson et al. (eds.), *Julian Perry Robinson: Pioneer in the Prohibition of Chemical and Biological Warfare*, Pioneers in Arts, Humanities, Science, Engineering, Practice 32, https://doi.org/10.1007/978-3-032-04064-0

The Purpose of the Foundation Shall be Realised Through

(a) Promoting local competitions on questions of the transition to sustainability and peace at general-education grammar schools in the Neckar-Odenwald County, that is awarded annually since 2022.
(b) An international science prize for peace and ecology that will be awarded from 2023 onwards.
(c) Maintenance of the foundation home with archives, library and the foundation's office.
(d) Supporting scientific events, research projects and publications on peace research and ecology.

Members of the Governing Board

- **Dr. Hans Günter Brauch (1947), Mosbach, Germany**, Chairman of the Board, Political Scientist,
- **Prof. Dr. Úrsula Oswald Spring (1946)**, Deputy Board Chair, UNAM, CRIM, Cuernavaca, Mexico
- **Dr. Hans Friedrich Happes, M.A. (1953), Germany**, Deputy Board Chair, Former headmaster (Ret.), Nicolaus-Kistner Grammar School in Mosbach.
- **Georg Nelius (1949), Germany**, Member, former member off the State Parliament (2007–2021), Deputy Mayor of Mosbach, Member of the Town and County Council, Neckar-Odenwald District.
- **Wolfgang Roth (1964)**, Germany. Membe of the Boarrd and Member of the Town Council of Mosbach.

International Science Prize for Peace and Ecology in the Anthropocene

Starting in June 2023, an international science prize will annually be awarded by the Hans Günter Brauch Foundation (HGBF) for Peace and Ecology in the Anthropocene in Mosbach, Baden-Württemberg in South Germany. The theme of the first international Science Prize of the HGB-Foundation was: *Climate Change and Conflicts* in 2023 and the first recipient was Prof. Dr. Tobias Ide (Assoc. Prof., Murdoch University, Perth (Australia) who also teaches at the University of Hiroshima (Japan). In 2023 International Recognitions were awarded to Dr. Tom Deligiannis, Laurier University (Canada) and Dr. Richard Marcantonio, Notre Dame University (USA). The theme of the second international Science Prize in 2024 was: ***Ecocide: Impacts of Wars and/or Climate Change on Food Security Since 1945***. The second international Science prize in 2024 was awarded to Assoc. Prof. Dr. Gregory Thaler (Oxford

University, UK) for his book on: *Saving a Rainforest and Losing the World – Conservation and Displacement in the Global Tropics* (2024) and to Dr. Victor Tsilonis (lawyer) for his book on: *The Jurisdiction of the International Criminal Court* (2024). *International Recognitions* were awarded to Dr. Adong (Uganda), Dr. Quispe-Zuniga (Peru/Germany) and Dr. **Idowu Odeyemi** (Nigeria/Australia) for their Ph.D. theses. The theme of the third international Science Prize in 2025 was: ***Impacts of Wars and Global Environmental Change and Climate Change on Food Insecurity, Hunger and Migration***. The third international Science prize in 2025 was awarded to **Dr. Angela Jill Lederach** for her book on: *Feel the Grass Grow: Ecologies of Slow Peace in Colombia*, and to **Dr. Matthew G. Gillet for his book on**: *Prosecution of Environmental Harm before the International Criminal Court*.

Chairman of the Board of Trustees (*Kuratorium*) of the HGB-Foundation: Lord Mayor of the town of Mosbach, Julian Stipp. Mosbach is since 2021 a member of "Mayors for Peace".

International Recognitions were awarded in 2025 to **Dr. Bárbara Magalhães Teixeira** (Brazil), **Dr. Anselm Vogler** (Germany) and **Dr. ASM Mostafizur Rahman** (Bangladesh) from Heidelberg University.

List of Books Edited by Hans Günter Brauch

Open Access Books Published in These Five Series

List of Open Access Books in the Hexagon Series

- **HESP 1**: Truong, Thanh-Dam; Gasper, Des; Handmaker, J., Bergh, S. I. (Eds.): Migration, Gender and Social Justice - Perspectives on Human Insecurity. Hexagon Series on Human and Environmental Security and Peace, vol. 9 (Heidelberg – New York – Dordrecht – London: Springer, 2013); at: https://link.springer.com/book/10.1007/978-3-642-28012-2
- **HESP 2**: Jeroen Kool, Saeb Bamya, Samer Talozi, EcoPeace, Tel Aviv, Israel - Bethlehem, Palestine – Amman, Jordan: *Sustainable Development in the Jordan River Valley: Final Report of the Regional NGO Master Plan*. Hexagon Series on Human and Environmental Security and Peace, vol. 13 (Berlin – Heidelberg – New York: Springer, 2016); at: https://link.springer.com/book/10.1007/978-3-319-300 36-8.

List of Open Access Books in the APESS Series

- **APESS 1**: Maja Göpel: *The Great Mindshift: How a New Economic Paradigm and Sustainability Transformations go Hand in Hand* (Cham – Heidelberg – New

York – Dordrecht – London: Springer International Publishing, 2017); at: https://link.springer.com/book/10.1007/978-3-319-43766-8.

- **APESS 2**: Carl Middleton, Vanessa Lamb (Eds.) (Eds.): *Knowing the Salween River: Resource Politics of a Contested Transboundary River* (Cham: Springer International Publishing, 2019); at: https://link.springer.com/book/10.1007/978-3-319-77440-4.
- **APESS 3**: H. G. Brauch (Eds.): *Towards Rethinking Politics, Policy and Polity in the Anthropocene - Multidisciplinary Perspectives* (Cham: Springer International Publishing, 2021); at: https://link.springer.com/book/10.1007/978-3-031-71807-6.
- **APESS 4**: H. G. Brauch (Eds.): *Towards Rethinking Politics, Policy and Polity in the Anthropocene - Multidisciplinary Perspectives* (Cham: Springer International Publishing, 2021); at: https://link.springer.com/book/10.1007/978-3-031-71807-6.

List of Books in the PSP Series

- **PSP 1**: Paul J. Crutzen, Hans Günter Brauch (Eds.): *Paul J. Crutzen: A Pioneer on Atmospheric Chemistry and Climate Change in the Anthropocene.* Springer Briefs on Pioneers in Science and Practice No. 50 (Cham – Heidelberg – New York – Dordrecht – London: Springer International Publishing, 2016; at: https://link.springer.com/book/10.1007/978-3-319-27460-7.
- **PSP 2**: Nils Petter Gleditsch: *Pioneer in the Analysis of War and Peace* – Presented by Peter Wallensteen. Springer Briefs on Pioneers in Science and Practice No. 29 (Cham – Heidelberg – New York – Dordrecht – London: Springer, 2016); at: https://link.springer.com/book/10.1007/978-3-319-03820-9.
- **PSP 3**: Gleditsch, Nils Petter (Ed.): *R. J. Rummel: An Assessment of His Many Contributions.* Springer Briefs on Pioneers in Science and Practice, vol. 37 (Cham – Heidelberg – New York – Dordrecht – London: Springer-Verlag, 2017); at: https://link.springer.com/book/10.1007/978-3-319-54463-2.

List of Books in the PAHSEP Series

- **PAHSEP 1**: Nils Petter Gleditsch (Ed.): *Lewis Fry Richardson: His Intellectual Legacy and Influence in the Social Sciences.* Pioneers in Arts, Humanities, Science, Engineering, Practice (PAHSEP), No. 27 (Cham: Springer International Publishing, 2020); at: https://link.springer.com/book/10.1007/978-3-030-31589-4.

Index

H. Wilson et al. (eds.), *Julian Perry Robinson: Pioneer in the Prohibition of Chemical and Biological Warfare*, Pioneers in Arts, Humanities, Science, Engineering, Practice 32, https://doi.org/10.1007/978-3-032-04064-0

D

E

S

V

MIX
Papier aus verantwortungsvollen Quellen
Paper from responsible sources
FSC® C105338

If you have any concerns about our products,
you can contact us on
ProductSafety@springernature.com

In case Publisher is established outside the EU,
the EU authorized representative is:
Springer Nature Customer Service Center GmbH
Europaplatz 3, 69115 Heidelberg, Germany

Printed by Libri Plureos GmbH
in Hamburg, Germany